INVENTIONS

AF882422

of

JONATHAN BANNON MAHER

Featuring Clean Continuous Endless Portable Self-Powered Motors and Generators

Permanently solves global warming, reduces cost of living to alleviate poverty, provides unlimited clean energy for evaporated water purification and atmospheric carbon dioxide reduction, and eliminates the need for every other method of propulsion and energy production, including nuclear energy technology – thus reducing nuclear weapons technology proliferation.

Additional inventions disclosed include a cost and yield optimized agricultural production system, structures for automated housing construction, software for automated investing, software providing demographic targeting of advertisements to unregistered viewers, software easing computer application development, software providing audio component separation, a system for automated discovery and control of electronic devices, an optimized system for high speed transportation, a system for automated virus removal, a system for automated brain cancer surgery, and a system for an interplanetary space vehicle launch network.

Inventions USPTO filed and non-invention sections released June 20th, 2017

This book is available in print and as a free electronic download,
as well as USPTO certified and notarized versions,
and will be available in multiple languages.

http://www.bannonmaher.com/
http://www.jonathanmaher.com/

The majority of my profits from the licensing of the inventions' patents will ultimately be going toward causes that support the well being of humanity.

Your support will be providing a service to all the world.

Please read the entire book.

For those wondering why the book is published in this format,

at standard page and margin size the book was 1,135 pages,

and exceeded the physical publishing capabilities of Amazon.

TABLE OF CONTENTS

LEGAL NOTICE	11
RECIPIENTS	13
INVENTOR	23
INVENTIONS	25
PRODUCTION	49
VALUATION	237
PHILANTHROPY	241
LEGISLATION	251
LICENSING	265
ORDERING	275
GRANTS	277
INCOME OPPORTUNITIES	279
THANKS, REPORTERS, ACCOMPLISHMENTS, SUPPORTERS, HONORS, AND ATTACKERS	285
INVESTORS	327
MANUFACTURERS	337
EMPLOYMENT	339
BOARD MEMBERSHIPS	343
EVENT INVITATION	345
ADDITIONAL INVENTIONS	347
I. A system allowing content producers to provide advertisers demographic targeting of unregistered readers	349
II. Software application development system that significantly reduces the need for programmers	373
III. Automated algorithmic asset selection and trading system	393
IV. Automated surgery system utilizing small flexible mechanical arm for minimally invasive precision suction cutting and extraction of tissue including cancer – automated brain surgery	395
V. Automated system for blood analysis and removal of undesirable components including viruses and cancer	449
VI. A cost and yield optimized agricultural production system	487

VII. **An autonomous system for high speed ground based transportation** **495**

IX. **Interlocking structures that reduce building construction to assembly, thus increasing resilience, decreasing costs, and allowing for automated construction** **519**

X. **A system for the automated discovery and control of electronic devices** **535**

XII. **Song file format for storing and listening to full, acoustic, and instrumental versions** **555**

XI. **Electronic health records maintenance and collaboration system** **565**

XIII. **Sound and vibration insulation for beds** **593**

XIV. **Compressed tube distribution of temperature controlled air** **603**

XV. **A system for a reusable self-powered interplanetary space vehicle launch network** **613**

LEGAL NOTICE

These inventions have only been publicly disclosed after extremely exhaustive work to ensure their corresponding patents are impervious to endless attack with unlimited resources, the filing of each in compliance with the Patent Cooperation Treaty, which provides protection in every commercially significant country, notarization of the disclosures, consultation with a USPTO registered patent attorney and licensed professional engineer, and with a division head and engineer in person at the USPTO. I am the exclusive inventor and owner of all disclosed inventions, as well as the exclusive author and editor of this book.

I must state that the disclosed inventions are not guaranteed to function or provide benefits as described, that all information provided in this text is believed to be accurate, but its accuracy is not guaranteed, and that asserted facts, numbers, and opinions are only be relied upon to the extent they can be validated by the reader. This book may include predictions, estimates or other information that might be considered forward looking. Forward looking statements represent my current judgment on what the future holds, they are subject to risks and uncertainties that could cause actual results to differ materially. You are cautioned not to place undue reliance on these forward looking statements, which reflect my opinions only as of the date of publication. I am not obligating myself to revise or publicly release the results of any revision to these forward looking statements in light of new information or future events. I will attempt to present some important factors that may affect my predictions.

RECIPIENTS

I spent only a few minutes compiling each list, so please don't be offended by an error or omission.

The leadership of the government of each country:

Abkhazia, Afghanistan, Albania, Algeria, Andorra, Angola, Antigua and Barbuda, Argentina, Armenia, Armenia, Australia, Austria, Azerbaijan, Bahamas, Bahrain, Bangladesh, Barbados, Belarus, Belgium, Belize, Benin, Bhutan, Bolivia, Bosnia and Herzegovina, Botswana, Brazil, Brunei, Bulgaria, Burkina Faso, Burma, Burundi, Cambodia, Cameroon, Canada, Cape Verde, Central African Republic, Chad, Chile, China, Columbia, Comoros, Congo-Brazzaville, Congo-Kinshasa, Costa Rica, Cote d'Ivore, Croatia, Cuba, Cyprus, Czech Republic, Denmark, Djibouti, Dominica, Dominican Republic, East Timor, Ecuador, Egypt, El Salvador, Equatorial Guinea, England, Eritrea, Estonia, Ethiopia, Fiji, Finland, France, Gabon, Gambia, Georgia, Germany, Ghana, Greece, Grenada, Guatemala, Guinea, Guinea-Bissau, Guyana, Haiti, Honduras, Hungary, Iceland, India, Indonesia, Iran, Iraq, Ireland, Israel, Italy, Jamaica, Japan, Jordan, Kazakhstan, Kenya, Kiribati, Kosovo, Kuwait, Kyrgyzstan, Laos, Latvia, Lebanon, Lesotho, Liberia, Libya, Liechtenstein, Lithuania, Luxembourg, Macedonia, Madagascar, Malawi, Malaysia, Maldives, Mali, Malta, Marshall Islands, Mauritania, Mauritius, Mexico, Federated States of Micronesia, Moldova, Monaco, Mongolia, Montenegro, Morocco, Mozambique, Namibia, Nauru, Nepal, Netherlands, New Zealand, Nicaragua, Niger, Nigeria, North Korea, Northern Cyprus, Norway, Oman, Pakistan, Palau, Palestine, Panama, Papua New Guinea, Paraguay, Peru, Philippines, Poland, Portugal, Qatar, Romania, Russia, Saint Kitts and Nevis, Saint Lucia, Saint Vincent and the Grenadines, Samoa, San Marino, Sao Tome and Principe, Sahrawi Arab Democratic Republic, Saudi Arabia, Senegal, Serbia, South Ossetia, Seychelles, Sierra Leone, Singapore, Slovakia, Slovenia, Solomon Islands, Somalia, South Africa, South Korea, Spain, Sri Lanka, Sudan, Suriname, Swaziland, Sweden, Switzerland, Syria, Tajikistan, Tanzania, Thailand, Togo, Tonga, Trinidad and Tobago, Tunisia, Turkey, Turkmenistan, Tuvalu, Tuvalu, Uganda, Ukraine, United Arab Emirates, United Arab Emirates, United Kingdom, United States of America, Uruguay, Uzbekistan, Vanuatu, Vatican City, Venezuela, Vietnam, Wales, Yemen, Zambia, Zimbabwe.

The leadership of a global cross section of the largest media organizations:

ABC, Associated Press, Bloomberg BusinessWeek, Bild, CBS, Chosun Ilbo, Financial Times, Forbes, Huffington Post, The Intercept, NBC, News Corp, New York Times, Newsweek, ProPublica, Reuters, Reference News, Russia Today, The Sun, TechCrunch, Time Magazine, Times of India India, Wall Street Journal, Washington Post, WikiMedia, Yomiuri Shimbun.

The leadership of a global cross section of the largest public relations firms:

Adfactors, AMI, BlueFocus, Brunswick, Burson-Marsteller, Edelman, Farner, FleishmanHillard, FSB Comunicacoes, Geelmuyden.Kiese, Golin, Havas, Hill+Knowlton, ICF Mostra, iMARS Group, Ketchum, Llorente & Cuenca, Magna Carta, Media Consulta International, Mikhailov & Partners, MSLGroup, Ogilvy, Prain Global, Res Publica, Strategic Public Relations Group, TRACCS, Vector, Weber Shandwick.

The leadership of the largest investment banks:

Bank of America Merrill Lynch, Barclays Capital, BNP Paribas, Credit Suisse, Citigroup, Deutsche Bank, Goldman Sachs, HSBC, Jefferies Group, JPMorgan Chase, Lazard, Mitsubishi UFJ, Mizuho, Morgan Stanley, Nomura Holdings, RBC Capital Markets, Société Générale, Sumitomo Mitsui, UBS, Wells Fargo.

The leadership of the largest institutional investors:

Aberdeen Asset Management, Abu Dhabi Investment Authority, Affiliated Managers Group, AllianceBernstein, Allianz Global Investors Germany, Amundi, APG Netherlands, AXA Investment Managers, BlackRock, BNP Paribas Investment Partners, BNY Mellon Investment Management, Capital Group, China Investment Corporation, Columbia Threadneedle Investments, Deutsche Asset & Wealth Management, Federal Retirement Thrift, Fidelity Investments, Franklin Templeton Investments, Generali Investments, Goldman Sachs Asset Management, Government Pension Investment Fund Japan, HSBC Global Asset Management, Insight Investment, Invesco, J.P. Morgan Asset Management, Kuwait Investment Authority, Legal & General Investment Management, Legg Mason, MFS Investment Management, Morgan Stanley Investment Management, National Pension Service of Republic of Korea, Natixis Global Asset Management, New York Life Investments, Northern Trust Asset Management, Norway Government

Pension Fund, PIMCO, Pramerica Investment Management, SAFE Investment Company, SAMA Foreign Holdings, Schroder Investment Management, Social Security Trust Funds, Standard Life Investments, State Street Global Advisors, Stichting Pensioenfonds, Sumitomo Mitsui Trust Bank, T. Rowe Price, TIAA-CREF, UBS Global Asset Management, Vanguard Asset Management, Wellington Management.

The leadership of the largest energy producers:

Bharat Petroleum, British Petroleum, Centrica, Chevron, China National Aviation Fuel, China National Offshore Oil, China National Petroleum, Cosmo Oil, Ecopetrol, Engie, Eni, Enterprise Products, Exxon Mobil, Formosa Petrochemical, Gas Natural, GasTerra, Gazprom, GS Caltex, Hellenic Petroleum, Hess, Hindustan Petroleum, Idemitsu Kosan, India Oil and Natural Gas, Indian Oil, JX Holdings, Korea Gas, Lukoil, Marathon Petroleum, MOL, Motor Oil Hellas, Murphy Oil, National Iranian Oil, OMV Group, PDVSA, Pemex, Pertamina, Petrobras, Petronas, Phillips, PKN Orlen, Plains All American Pipeline, PTT, Reliance Industries, Repsol, Rosneft, Royal Dutch Shell, S-Oil, Saudi Aramco, Showa Shell Sekiyu, Sinopec, Sonatrach, Statoil, Suncor Energy, Sunoco, Surgutneftegas, Tesoro, TNK, Total, Ultrapar, Valero Energy, World Fuel Services.

The leadership of the largest air carriers:

Air Canada, Air China, Air France-KLM, All Nippon Airways, American Airlines, British Airways, Cathay Pacific Cargo, China Eastern Airlines, China Southern Airlines, Delta, Easyjet, Egyptair, Emirates, FedEx, International Airlines, Korean Air Cargo, Lufthansa, Qatar Airways, Ryanair, SkyCargo, Southwest Airlines, Turkish Airlines, United Continental, UPS.

The leadership of the largest electricity generator manufacturers:

Agder Energi, BC Hydro and Power Authority, Centrais Elétricas Brasileiras, China Yangtze Power, Duke Energy, Enercon, Gamesa, General Electric, Georgia Power, Goldwind, Honeywell, Hydro-Québec, Ming Yang, Nordex, Ontario Power Generation, RusHydro, Siemens, Stat Kraft, Sulzon Group, United Power, Vestas, Xcel.

The leadership of the largest contract manufacturers:

AsteelFlash Group, Benchmark Electronics, Beyonics Technology, Celestica, Elcoteq, Flextronics International, Hon Hai Precision Industry (Foxconn), Jabil Circuit, Kaifa Technology, Kimball Electronics Group, New Kinpo Group, Orient Semiconductor, Plexus, Shenzhen Sanmina-SCI, SIIX, Sumitronics, Universal Scientific Industrial, UMC Electronics Electronics, Venture, Zollner Elektronik.

The leadership of the largest electronics manufacturers:

Airbus, Boeing, Bosch, Caterpillar, Ericsson, General Electric, Hitachi, Honeywell, Huawei, Intel, LG, Lockheed Martin, Nokia, Panasonic, Samsung, Siemens, Sony, Toshiba, ThyssenKrupp, United Technologies.

The leadership of the largest auto manufacturers:

BMW, Changan, Daimler, Dongfeng, Ford, Fiat Chrysler, Geely, General Motors, Groupe PSA, Honda, Hyundai, Mazda, Mitsubishi, Nissan, Renault, Suzuki, SAIC, Tata, Tesla, Toyota, Volkswagen.

The leadership of the largest electronics retailers:

Aeon, Amazon, Apple, Best Buy, Bund, Casino Guichard-Perrachon, Carrefour, Centres Distributeurs E. Leclerc, Costco, Dixons Carphone, Groupe Auchan, Home Depot, IKEA, Loblaw, Metro Ag, Migros-Genossenschafts Lotte Shopping, Lowe's, Sears, Target Corporation, Tesco, Walgreen, Wal-Mart, Yonghui Superstores.

The leadership of the largest medical device manufacturers:

3M, Abbott Labs, Allergan, Baxter International, Becton Dickinson, Boston Scientific, Cardinal Health, Covidien, Essilor International, Fresenius Medical Care, General Electric, Hospira, Johnson & Johnson, Koninklijke Philips Electronics, Medtronic, Novartis, Olympus, Siemens AG, Smith & Nephew, St. Jude Medical, Stryker, Terumo, Toshiba, Zimmer Holdings.

The leadership of the largest construction equipment manufacturers:

Atlas Copce Construction Technique, Caterpillar, CNH Industrial, Deosan Infracore, Hitachi Construction Machinery, Hyundai Heavy Industries, JCB, John Deere, Kebelce Consruction Machniery, Komatsu, Liebherr, Manitowoc Crane Group, Metse, Oshkesh

Access Equipment, Sany, Terex, Volvo Construction Equipment, Wirigen Group, XCMG, Zoomilion.

The leadership of the largest charitable foundations:

Andrew W. Mellon Foundation, Azim Premji Foundation, Bill & Melinda Gates Foundation, Bloomberg Philanthropies, California Endowment, Calouste Gulbenkian Foundation, Chan Zuckerberg Initiative, Church Commissioners for England, David and Lucile Packard Foundation, Duke Endowment, Ford Foundation, Garfield Weston Foundation, Gordon and Betty Moore Foundation, J. Paul Getty Trust, John D. and Catherine T. MacArthur Foundation, Knut and Alice Wallenberg Foundation, Kresge Foundation, Leona M. and Harry B. Helmsley Charitable Trust, Li Ka Shing Foundation, Lilly Endowment, MasterCard Foundation, Mohammed bin Rashid Al Maktoum Foundation, Nemours Foundation, Pew Charitable Trusts, Realdania, Robert Bosch Foundation, Robert Wood Johnson Foundation, Rockefeller Foundation, Silicon Valley Community Foundation, Stichting INGKA Foundation, Sulaiman Abdul Aziz Al Rajhi, Tulsa Community Foundation, W.K. Kellogg Foundation, Wellcome Trust, William and Flora Hewlett Foundation.

The leadership of the largest venture capital firms:

3i, Accel, Andressen Horowitz, Austin Ventures, Bain Capital Ventures, Battery Ventures, Benchmark, Berkshire Hathaway, Bessemer Venture Partners, Draper Fisher Jurvetson, Foundation Capital, Founders Fund, Greylock, Highland Capital Partners, IDG Ventures, Insight Venture Partners, Institutional Venture Partners, Intel Capital, Intellectual Ventures, Intellectual Ventures, July Capital, Khosla Ventures, Kleiner Perkins Caufield Byers, Lightspeed Venture Partners, Menlo Ventures, Meritech Capital Partners, Morgenthaler Ventures, New Enterprise Associates, Norwest Venture Partners, Polaris Partners, Rho Ventures, Sequoia Capital, Shenzhen Capital Group, Social Capital, SV Angel, U.S. Venture Partners, Union Square Ventures, VantagePoint Venture Partners, Venrock.

The leadership of the largest venture capital firm limited partners:

Adams Street Partners, California Public Employees' Retirement System, California State Teachers' Retirement System, European Investment Fund, Greenspring Associates, HarbourVest Partners, Harvard Management Company, Hewlett-Packard Master Trust, Liberty Mutual Retirement Benefit Plan, Massachusetts Pension Reserves

Investment Management Board, Princeton University Endowment, The Regents of the University of California, Rockefeller Foundation, San Francisco Employees' Retirement System, Sherman Fairchild Foundation, Stanford Management Company, State of Wisconsin Investment Board, Credit Suisse Customized Fund Investment Group, University of Michigan Endowment, University of Texas Investment Management Company, Yale Investments Office.

The leadership of twenty award committees:

Berggruen Philosophy Prize, Blavatnik Awards, Breakthrough Prize, Crafoord Prizes, Franklin Institute Award, Fundamental Physics Prize, Ghandi Peace Award, Hilton Humanitarian Prize, Ibrahim Prize, Japan Prize, Kavli Prizes, Kyoto Prize, Nobel Committee, Queen Elizabeth Prize for Engineering, Shaw Prize, Tang Prize, Templeton Prize, Turing Prize, World Food Prize, XPrize.

Award winning physicists:

Isamu Akasaki, Hiroshi Amano, Willard S. Boyle, François Englert, Albert Fert, Roy J. Glauber, Andre Geim, David J. Gross, Peter Grünberg, John L. Hall, Theodor W. Hänsch, Serge Haroche, Peter Higgs, Takaaki Kajita, Charles K. Kao, Makoto Kobayashi, Anthony James Leggett, John C. Mather, Toshihide Maskawa, Arthur B. McDonald, Shuji Nakamura, Yoichiro Nambu, Konstantin Novoselov, Saul Perlmutter, Adam G. Riess, Brian P. Schmidt, George E. Smith, George F. Smoot, Frank Wilczek, David J. Wineland.

Award winning peace activists:

Martti Ahtisaari, Kofi Annan, Yasser Arafat, Óscar Arias, Carlos Filipe Ximenes Belo, Jimmy Carter, Kim Dae-Jung, Shirin Ebadi, Mohamed ElBaradei, Leymah Gbowee, Mikhail Sergeyevich Gorbachev, Al Gore, Tenzin Gyatso, John Hume, Tawakkul Karman, Frederik Willem de Klerk, Aung San Suu Kyi, Wangari Muta Maathai, Rigoberta Menchú, Shimon Peres, Yitzhak Rabin, José Ramos-Horta, Joseph Rotblat, Kailash Satyarthi, Ellen Johnson-Sirleaf, David Trimble, Jody Williams, Liu Xiaobo, Malala Yousafzai, Muhammad Yunus.

Award winning medical researchers:

Paul Lauterbur, Peter Mansfield, Richard Axel, Linda B. Buck, Barry J. Marshall, J. Robin Warren, Andrew Z. Fire, Craig C. Mello, Mario R. Capecchi, Martin J. Evans, Oliver Smithies, Harald zur Hausen, Françoise Barré-Sinoussi, Luc Montagnier, Elizabeth H. Blackburn, Carol W. Greider, Jack W. Szostak, Robert G. Edwards, Bruce A. Beutler, Jules A. Hoffmann, Ralph M. Steinman, John B. Gurdon, Shinya Yamanaka, James E. Rothman, Randy Schekman, Thomas C. Südhof, John O'Keefe, May-Britt Moser, Edvard I. Moser, William C. Campbell, Satoshi Ōmura, Tu Youyou, Yoshinori Ohsumi.

Award winning authors:

Octavio Paz, Nadine Gordimer, Derek Walcott, Toni Morrison, Kenzaburō Ōe, Seamus Heaney, Wisława Szymborska, Dario Fo-Cesena, José Saramago, Günter Grass, Gao Xingjian, V. S. Naipaul, Imre Kertész, J.M. Coetzee, Elfriede Jelinek, Harold Pinter, Orhan Pamuk, Doris Lessing, J. M. G. Le Clézio, Herta Müller, Mario Vargas Llosa, Tomas Tranströmer, Mo Yan, Alice Munro, Patrick Modiano, Svetlana Alexievich, Bob Dylan.

The living United States Energy Secretaries:

Spencer Abraham, Samuel Bodman, Steven Chu, Charles Duncan, Jr., John S. Herrington, Donald P. Hodel, Hazel R. O'Leary, Ernest Moniz, Federico Peña, Rick Perry, Bill Richardson.

The largest environmental groups:

350, Ceres, Environmental Defense Fund, Greenpeace, National Geographic Society, National Wildlife Federation, Natural Resources Defense Council, Rainforest Alliance, Sierra Club, Union of Concerned Scientists, World Resources Institute, World Wildlife Fund.

The leadership of the largest record companies:

Atlantic, Capitol, Columbia, Disney, EMI, Epic, Epitaph, Fueled by Ramen, Geffen, Hollywood, Interscope, Live Nation, Mercury, RCA, Sony, Universal, Virgin, Warner, Walt Disney, Warner Brothers.

The leadership of the largest publishers:

Cengage Learning Holdings, China Publishing Group, China South Publishing & Media Group, De Agostini Editore, Grupo Planeta, Hachette Livre, Harcourt, Harper Collins, Holtzbrinck, Houghton Mifflin McGraw-Hill Education, Oxford University Press, Pearson, Penguin Random House, Phoenix Publishing and Media, RELX Group, Scholastic, Simon & Schuster, Springer Science and Business Media, Thomson Reuters, Wiley, Wolters Kluwer.

The leadership of the largest movie studios:

Dreamworks, Fox (News Corporation), Lions Gate Entertainment, Marvel Studios, MGM, Netflix, Universal Studios (NBC Universal), Sony Pictures Entertainment (Columbia Pictures), Time Warner (Warner Brothers), Viacom (Paramount Pictures), Walt Disney Studios (Lucasfilm, Pixar).

Celebrities with a known interest in the environment:

Amitabh Bachchan, Ed Begley Jr., Cate Blanchett, Pierce Brosnan, Gisele Bundchen, George Clooney, Rosario Dawson, Cameron Diaz, Leonardo DiCaprio, Daryl Hannah, Julia Louis-Dreyfus, Will Ferrell, Tom Hanks, Angeline Jolie, Lady Gaga, Dave Matthews, Edward Norton, Jamie Oliver, Gwyneth Paltro, Brad Pitt, Robert Redford, Mark Rufalo.

Billionaires with a known interest in the environment:

Mukesh Ambani, John Arnold, Marc Benioff, Jeff Bezos, Alwaleed bin Talal, Michael Bloomberg, Richard Branson, Ray Dalio, Aliko Dangote, John Doerr, Larry Ellison, Bill Gates, Reid Hoffman, Chris Hohn, Vinod Khosla, Jack Ma, Dustin Moskovitz, Patrice Motsepe, Xavier Niel, Hasso Plattner, Julian Robertson, Neil Shen, Nat Simons, Carlos Slim, Masayoshi Son, George Soros, Tom Steyer, Ratan Tata, Meg Whitman, Zhang Xin, Mark Zuckerberg.

A cross section of top ranked show hosts:

Samantha Bee, Tucker Carlson, Steven Colbert, Anderson Cooper, James Corden, Jim Cramer, Lou Dobbs, Ellen Degeneres, Jimmy Fallon, Sean Hannity, Jimmy Kimmel, Matt Lauer, Rush Limbaugh, Bill Maher, Chris Matthews, Phil McGraw, David Muir, Andrew Napolitano, Trevor Noah, Conan O'Brien, John Oliver, Jeanine Pirro, Kelly

Ripa, Robin Roberts, Charlie Rose, Ryan Seacrest, Howard Stern, Michael Strahan, Brian Williams.

Celebrities with the largest social media followings:

Adele Adkins, Justin Bieber, Miley Cyrus, Ellen Degeneres, Jimmy Fallon, Rhianna Fenty, Lady Gaga, Selena Gomez, Ariana Grande, Dwayne Johnson, Kim Kardashian, Beyonce Knowles, Ashton Kutcher, Jennifer Lopez, Demi Lovato, Katy Perry, Alecia Moore, Christiano Ronaldo, Shakira Ripoll, Britney Spears, Harry Styles, Taylor Swift, Donald Trump, Justin Timberlake, Oprah Vin Diesel, Winfrey.

A cross section of celebrity media influencers:

Adele Adkins, Ben Afleck, Jennifer Aniston, Tom Brady, Sandra Bullock, Jim Carrey, Jackie Chan, Dave Chapelle, Bradley Cooper, George Clooney, Tom Cruise, Matt Damon, Johnny Depp, Leonardo di Caprio, Cameron Diaz, Vin Diesel, Robert Downey Junior, Harrison Ford, James Franco, Morgan Freeman, Tom Hanks, Kevin Hart, Paul Hewson, Hugh Jackman, Scarlet Johansson, Angelina Jolie, Mila Kunis, Jennifer Lawerence, Chris Pratt, Julia Roberts, Keanu Reeves, Madonna Ciccone, Melissa McCarthy, Matthew McConaughey, Brad Pitt, Adam Sandler, Will Smith, Ben Stiller, Taylor Swift, Arnold Schwarzenegger, Mark Walhberg.

INVENTOR

I was the first full-time employee and software engineer at a company whose software is used to protect classified U.S. military networks, and I wrote software used to scale a startup hedge fund from zero to billions of dollars of assets under management of complex credit derivatives that accurately predicted the last financial crisis two years in advance – I was also the apparently the only person in the world to predict the May 2010 flash crash in a formally copyrighted financial reform essay sent to Congress. I ran for the U.S. Senate in 2012 while waiting to be President of the United States in 2016 at the age of 35, which obviously didn't work. I am certified expert software engineer by each Microsoft (C#, T-SQL), and Oracle (Java, Pl/SQL), a prominent software entrepreneur wrote in a recommendation letter I am "extremely gifted as a computer programmer", and I have tested as a high level genius (99.997th percentile according to a 7th grade assessment). I have written and published three books, including *Building a Successful Organization*, and more recently, *The Destiny of Humanity*. *The Destiny of Humanity*, which relevant to this book, contained a chapter analyzing all known energy production systems, was implemented by world leaders, and endorsed by Kings, a Prime Minister, and Second Lady, with a Prime Minister writing that the book is "A pointing of horizons and goals to which we must be aware… the quest for harmony and a blend of attitudes that could reach the heights of the global and total dignity of human beings." I had my music licensed for shows on MTV, VH1, and Discovery Networks, and rated in the top 1% against major label releases by an independent audit of hundreds of listeners. I graduated from the University of San Diego, ranked third in the world for entrepreneurship by the Financial Times in 2015, with recommendations from the President and Dean, and completed Oxford University coursework. I have studied, over a decade and a half, Brazilian Jiu-Jitsu, Tae Kwon Do, Thai Kickboxing, Aikido, and long range target shooting. I currently own entities based on inventions later disclosed in this book, including intellectual property holding and operating companies for the energy and motor systems, PageRock, which provides the easiest system for developing sophisticated websites without technical knowledge, a soon to be released system for content producers to offer advertisers demographic targeting of unregistered readers, and a development stage electronic health records collaboration platform, and separately, an automated algorithmic trading hedge fund. I have under development, additional books, a record, a movie, and the plans for a theme park attraction. This is my third book, disclosing eighteen of my inventions.

INVENTIONS

You are about to learn the details of four separate motor and energy production systems I invented, and for which I have submitted patent filings compliant with international patent law, that satisfy the criteria of the holy grail of such systems: (1) clean (2) continuous (3) portable (4) free output after purchase (5) endless output until system failure (6) self-contained. As you will see, rather than using a wizard's wand, a sorcerer's spell, or the ceremonial sacrifice of a virgin mountain goat, the devices I've invented are implemented using principles of physics and engineering. My inventions make irrelevant all other energy production and storage systems, and all other motor systems, while permanently resolving urgent global problems.

I have included step-by-step construction instructions later in this book, and refined the inventions to utilize principally readily available commodity components, allowing each disclosed invention to be produced within a day by a competent technical manufacturer having the specified tools and components. However, in order to fully appreciate the inventions, it is important to first review their implications against predecessors. I note that I haven't been able to take the time I would have liked to refine this lengthy book, and am aware it could benefit from inconsequential optimizations.

I have prefixed the names of these energy production and motor systems with *Bannon Maher*, in honor of my parents Marianne Bannon and Richard Maher, who are no longer of optimal health, and without whose support – including, relevant to these inventions, my mother taking me to the public library every day after school, beginning in elementary school, to read books on topics including physics, engineering, and programming, and my father providing financing — I may have been less likely to invent and present these systems. I have yet to provide them an adequate return.

Background

The following matrix is my assessment all energy systems compared to the Bannon Maher systems.

	Wind	Water	Solar	Hydrogen	Geo thermal	Oil, Gas, and Coal	Nuclear	**Bannon Maher**
Continuous	No	Depends	No	Depends	Yes	Yes	Yes	Yes
Free from Toxic Byproducts	Yes	Yes	No	Yes	No	No	No	Yes
Cyber Attack Proof	Depends	Depends	Depends	Depends	Depends	Depends	No	Yes
Expensive Power Lines Required	Depends	Depends	Depends	Depends	Yes	Depends	Yes	No
Degrading Batteries Required	Yes	Depends	Yes	No	No	No	No	No
Free Output After Initial Purchase	No	Depends	No	No	No	No	No	Yes
Completely Portable	No	No	No	No	No	No	No	Yes
Self Powered	No	No	No	No	No	No	No	Yes

Self-powered. Other energy production and motor systems utilize an external fuel source, such as oil, gas, wind, sun, water currents, geothermal heat, hydrogen, or uranium. Bannon Maher systems are the first energy and motor systems in the history of the known universe to not require an external fuel source. If manufactured optimally, these systems have the potential to produce energy forever and provide endless transportation range.

Clean. Nuclear energy production, including current and proposed systems of fission, fusion, and cold (LENR), result in toxic waste and or materials, geothermal often circulates contaminants from the ground, fracking creates toxic drinking water, hydroelectric dams decompose organic matter producing the potent global warming gas methane, popular solar panel manufacturing techniques release some greenhouse gases with much longer atmospheric lifespans and that are thousands of times more potent than carbon, hydrogen takes more energy to create than it produces and is often produced using energy from less environmentally friendly clean energy sources, and other clean energy systems often use slowly degrading flammable toxic batteries to store energy for when the sun is not shining, wind is not blowing, or water is not adequately flowing. The Bannon Maher systems are completely clean.

Continuous. Wind, solar, and traditional water energy systems provide variable output, where water varies with consistency of water volume flow subject to rainfall and droughts, and wind and solar average output found to be around 15%-24% of rated output, meaning a 1 kilowatt system typically produces only around 1 kilowatt for 4 out of 24 hours of the day. Bannon Maher systems provide power output that is always completely continuous and stable. Therefore, 5 units of variable power capacity from other clean energy systems may be equivalent to 1 unit of continuous power from my systems. Additionally, wind and solar can on occasional days produce no output at all, making them an impossibly unreliable source in many areas.

Free endless output after purchase. Traditional energy systems utilize a fuel source requiring replenishment, such as carbon, hydrogen, or uranium, and other clean energy systems require batteries that degrade over time to store energy and thus require periodic replacement. The Bannon Maher systems require no external fuel source or storage battery, and so have the potential to produce energy until gravity driven orbital drift causes the Earth to be consumed by the Sun in a few billion years – assuming the units and humans are still on Earth.

Cyber-attack proof. These energy systems are self-contained, require no hackable computer to operate, and therefore are immune to computer viruses that will always be able to take down the electrical grid and nuclear and clean energy power plants – even if such plants aren't connected to the Internet, as demonstrated by the Stuxnet virus.

Blackout proof. These energy systems are designed to be kept indoors and on-site, and are thus ideal for critical facilities such as hospitals and data centers that can't afford a blackout from failed power lines or plants.

Completely portable. Existing energy systems can only function in specific environments, solar panels in the sun, wind turbines in wind, water turbines in water flow, nuclear in a stable highly controlled environment, and carbon with the aid of an emissions pipe. Unlike all other methods of energy production, Bannon Maher systems are completely portable, able to function equally well in a basement closet, as a in car, or a space ship.

Inexpensive. Aside from my systems being dramatically superior to all others, they may provide a calculated cost per kilowatt hour of 0.0084 USD (0.0068 EUR), inclusive of components, royalties, labor, and manufacturer profit, which is more than ten times less expensive per unit of output delivered to end users than every single other form of energy, current or proposed, where the cost advantage will remain even if other systems substantially decline in price, because my systems don't require installation and maintenance of power lines, an allocation of land, and degrading batteries. A part of the pricing advantage is a result of many core system components, including electrical generator heads, hydraulics, and magnets, being commodity components that have been mass produced and price optimized over nearly a century. When comparing Bannon Maher systems to other clean energy systems, a person has to multiply price of other systems by at least 10 times, 5 times to account for enough electricity generation to be stored for equivalent continuous output, 5 times for the battery storage, and if implemented as a farm, the additional cost of land, and extremely expensive power lines, for another 5-10 times. Additionally, because solar produces direct current, when 99% of the Earth's population uses alternating current, solar requires additional expense of a converter. The most expensive energy source is hydrogen, because like oil, it simply stores energy, requires machines to produce, transport, and consume, a process which consumes more energy than it produces, despite decades of development.

Weatherproof. Bannon Maher systems are designed for indoor use and unlike wind turbines, hydroelectric dams, and solar panels, my systems can't be made ineffective by environmental conditions such as freezing temperatures, snow, or rain.

Eliminates energy storage. Storage of energy is no longer relevant, even for peak output needs, since additional units of my inventions can be turned on to meet peak needs while being more economical than storage. Additionally, existing solar roof panels can be removed to allow for roof tiles designed to maintain a cool house in summer.

Eliminates power lines and fossil fuel pipes. Power line and fossil fuel pipe cost of installation per mile is reportedly anywhere from around $250,000 USD to $4,000,000 USD for say an average assuming cost minimization of around 1 million USD (814,000 EUR) which is around 15 times the average annual income in the United States, and must be replaced every 20 to 80 years, so with about 300 thousand miles of power lines, and about 200 thousand miles of pipe lines, an expenditure around 25 times the national debt will be required, passed on to energy consumers, thus dragging down the economy, with every other developed country in a similar situation. Some projections show an up to 40% increase in energy consumption over the next 20 years, which would substantially raise those costs. Additionally, power lines are not only are astonishingly expensive to install and maintain, but also a poor method of providing energy because they require environmental destruction during installation, leave visible blight, transmit power from central sites that are forever vulnerable to cyber attacks and inherently more prone to producing large scale blackouts than a decentralized system, use high voltage power lines have health consequences for those living nearby, and lose power during transmission. Power lines are required for all forms of nuclear energy, including fission, fusion and cold, for farms of solar, wind, and water, and for fossil fuels plants. Additionally, no thinking person will ever want, nor could a functional government allow, any type of nuclear reactor – fission, fusion, cold (LENR) – in their car or home, and they can't fit in consumer devices, so those sources are now wholly obsolete. Natural gas often passes through such pipelines, which is partially leaked during extraction and consumption, and is principally methane, a greenhouse gas about 20 times more potent than carbon dioxide. My inventions can permanently eliminate such costs in every country by being installed in every building for maybe 20% of the cost of just power lines and fossil fuel pipes, while allowing land currently holding power lines, as well as solar, wind, and hydroelectric installations, to be reclaimed, which is increasingly important as the population grows, all while reducing visual pollution. Eliminating power lines also increases the lifespan of consumer electronics, by eliminating surges from sources including power line lightning strikes.

Profitable. In a standard home use scenario, these may be the only energy systems that can provide a unit owner a profit, before tax credits, from selling energy back to the utility.

Prevents deaths. Plants, rigs, and pipes, for current and proposed forms of nuclear, hydrogen, gas, and oil energy can explode, coal mines can collapse, and wildlife is killed by wind turbines, hydroelectric dams, ocean and wave turbines, solar condensers, toxic battery production byproducts, and nuclear waste.

Powers atmosphere cleaning. Because an existing specialized laser can disassociate atmospheric carbon dioxide molecules into carbon molecules and oxygen molecules, and because my energy inventions provide unlimited clean energy, a power source is now available to potentially eliminate excess carbon dioxide from the air, and potentially other greenhouse gasses, if devices development is spurred by energy consumption no longer being a hindrance.

Powers water purification and pumping. 1 in 9 people lack access to clean water, with millions reportedly dying this year alone as a result of crop and animal deaths from climate change droughts. Ground water has been or is being depleted in many areas such Mexico City which has sunk 30 feet (48 meters). Evaporated water purification removes nearly every contaminant with a higher boiling point than water, and potentially all others can be removed with a standard carbon filter and ultraviolet light. Because my systems make energy nearly free when amortized over time, the energy intensive nature of evaporated water purification and pumping is no longer a hindrance, thus the inventions resolve global clean water needs. Inexpensive water purifiers can be made using an electric cooking plate and two metal kettles connected by a flexible copper pipe, where the cooking plate boils water in one kettle that condenses into the second.

Powers reduced water consumption. Electricity powered showers are available that require only a cold water pipe, and recirculate, filter, and heat water, to provide exact continuous temperature and pressure control, which may provide cleaner water when filtered than without recirculation. Additionally, there are low voltage electric showerheads that mix air with water to provide the effect of the same output using dramatically less water.

Alleviates poverty. Because energy is free after unit purchase, and the purchase price is less per unit of output than any other energy system, I've reduced the cost of living for every person on Earth, thus reducing poverty.

Reduces nuclear and hydrogen weapon proliferation. Because these energy production systems eliminate any need for any type of nuclear energy, including fission, fusion, and cold (LENR), each of which is weaponizable and one step from nuclear weapons technology, and hydrogen, which a terrorist can obtain at a hydrogen fuel station to create powerful explosives verifiable by watching an online video of balloons filled with uncompressed hydrogen being lit on fire, I have provided us all a fundamentally safer world. I must note that fusion is particularly disturbing, and research should be legally banned, as a fusion weapon could be created with the power of an exploding star, whose explosion could potentially result in an extinction level event for humans, a scenario even more disturbing when considering increasingly autonomous – and therefore inevitably hackable by individuals — weapons control.

Dramatically reduces transportation costs. Vehicles that implement my inventions no longer require fuel, therefore car companies should offer engine upgrades for all existing models, and tax credits should be provided to support transition. The first car manufacturer to license and install my engines could develop a price and quality based global market monopoly, while the first airline switching to my engines, given a quarter of airline costs are from fuel, will automatically have a price based routes monopoly. The engines also allow for travel by supersonic jets, which have been financially unsustainable as a result of fuel costs. I don't support use in the foreseeable future in flying cars, because of inherent dangers, as well as visual and noise pollution resulting from broad ownership. These units reduce the need for traditional large rockets, since they can replace a rocket's first stage, and potentially second stage as well, before requiring rocket propulsion, as detailed in a later disclosed space vehicle launcher.

Therefore, a few applications of the generators and motors include powering:

1. Every existing electric grid globally.
2. All transportation vehicles including automobiles, trains, vertical take off and landing aircraft, cargo ships, cruise ships, boats, submarines, rockets, spaceships, hover boards, and jet packs.
3. Home appliances such as televisions, washing machines, dishwashers, showers, hydroponics, and water pumps.
4. Homes, offices, factories, hospitals, and data centers that would like to disconnect from external power sources to end their recurring bill, be permanently immune from blackouts, use clean energy, and save money.
5. Portable consumer electronics, such as phones and laptops, eliminating the need to recharge, through an internally installed miniaturized unit.
6. Personal Rapid Transport, which I described in my 2011 book, *The Destiny of Humanity*, as a high speed atmosphere evacuated tunnel based transportation system over existing government controlled land using auto piloted personal pods, able to travel at thousands of miles per hour to every continent over land, using the Bearing Straight to bridge the Americas with Asia. This proposal, despite already having been confirmed as received by about 40 heads of state, was re-released in the media by someone, shortly after I saw he was an early investor in an electric car company and asked him to add his endorsement, but who instead cited predecessor ideas as his source while simultaneously publicizing other ideas from my book, as I expect he'll do again upon release of this book, and though I appreciate all of my followers, he and his team of actual rocket scientists added

multiple errors, including switching magnetic levitation and propulsion to already inefficient fan blades and using them in a partially atmosphere evacuated environment where they'd have little to push against, characterizing a transportation network as a loop, making it financially unsustainable by reducing speed, routes, privacy, and personalization, and instead of square tunnels, using round tubes, which don't properly and efficiently fit cargo containers or people. While my proposal was optimized relative to known alternatives, I have revised my design using my new engines, disclosed in the patent later in this book, which is able to provide ground based travel at thousands of miles per hour without using fuel, or air evacuated tunnels, or degrading tracks.

7. Powering home hydroponic agricultural production systems, including light, temperature control, water circulation, and a device that turns food scraps into fertilizer.
8. Powering electric cars to provide free clean unlimited range travel, while eliminating the need for refueling and batteries. Vehicle manufacturers must stop making gasoline and electric vehicles entirely, and make only Bannon Maher vehicles, or their sales will fall off a cliff. I have included as the last several pages of the licensing chapter, an explanation that auto executives will need to follow to be part of the future.
9. Powering space tourism, space resorts, space colonies, and inter galactic travel, utilizing a reusable space vehicle launch network system I disclosed later in this book. Powering live video planetary sampling probes, journeying an unlimited number of years into the universe. As I explained in my last book, each large country should fund its own domestic space organizations for commercial orbital missions, with the United States already having 4 major space companies, while government programs such as NASA focus on work with no immediate financial return, specifically deep space colonization, providing for the redundancy and resilience of the human species.

And now to retract the long red curtain…

Summary of the Inventions

The First Law of Thermodynamics, which governs the conservation of energy, states that in a closed system, energy cannot be created or destroyed, only transferred. Additionally, Newton's third law of equal and opposite reaction, reinforces that energy in a closed system can only be transferred in an equal and opposite manner. This is why an electrical motor cannot simply spin an electrical generator, since a closed system interacting only with itself produces no net energy gain. These laws remain true in the example of firewood burning to power a steam engine, where there are two energy systems interacting, with energy transferred from the wood as it burns, through the steam engine, to produce motion – so while there appears to be a loss and a gain of energy, by respectively the wood and the steam engine, the whole system simply transfers energy to produce a useful outcome.

As previously mentioned, I had analyzed all known energy systems in my last book, and in somewhat recently reviewing the section on energy, given I've always had insights no one else does when looking at the same information, and had further accumulated knowledge, I again asked myself a question which I had been fascinated with during my home electrical engineering experiments in elementary school: "how could motion be generated for a wind or water turbine to power a connected generator in such a way that generating the motion consumes less energy than captured?", or to state the question another way, "how could a self-contained net positive energy system with no external fuel source be created?". As previously stated, according to the First Law of Thermodynamics and Newton's Third Law, this can't be done by circulating wind or water over a turbine with an electric motor, as the system wouldn't be net energy positive. So how can this be achieved? There must be some way. How? Mechanical or hydraulic leverage, gravity, buoyancy, or magnetic repulsion. Using the previous example as an analogy, leverage, gravity, buoyancy, and magnetic repulsion can each be used as the firewood.

Electricity generators were first created in 1821 by Michael Faraday, producing output by rotating a generator's axle – generally using fossil fuel combustion or a fan blade rotated by the flow of wind or some state of water – which rotates magnets, and their corresponding magnetic fields, to force electrons to flow along a wire, which is electricity. Hydraulics are the compartmentalized pressure system presented in 1851 by Richard Dudgeon, which transfer pressure generated over a large area with limited force, to a small area with increased force, proportionate to the change in the containment area, allowing someone to, for example, use the force of a hand to lift a car weighing several tons off the ground. Gravity is produced by mass and causes the force produced by an object to compound as it falls toward an adequately massive object. Buoyancy causes an object in a liquid to rise, if the weight of the object is less than to the weight of the volume of liquid it displaces. Finally, magnetic repulsion, with repulsion not to be confused with the type you felt after that time you drank too much and slept with some sort of beastly woodland creature, was first formally studied in 1269 by Petrus Peregrinus de Maricourt, and causes magnets opposing each other while presenting the same polarity to provide repellant force.

Calculations of output and cost follow descriptions of the four systems. The later disclosed patents provide detailed descriptions of the components in the following system illustrations, as well as detail an assortment of alternative implementations.

In the first system, the Bannon Maher Magnetic Repulsion Motor and Generator, magnetic repulsion is used to rotate an axle, which can in turn rotate a motor axle to provide propulsion, and or rotate a generator axle to generate electricity, where the magnetic repulsion is performed by affixing permanent magnets, manufactured to retain their directional force, in a circular formation along an inner magnet holder, resembling a wheel, suspended in the center by an axle, fit inside an outer magnet holder, containing magnets in a circular formation, where all magnets are oriented to provide repellent force to the inner magnet holder, and are angled, spaced, and insulated in such a way that the non-liner nature of magnetic fields is accounted for so that each provides force on only on one side of the inner magnet holder axle, causing the inner magnet holder to spin continuously, where energy is captured in the system as a result of slowly degrading repulsive magnetic fields providing rotational force to the generator and or motor axle. The force with which the rotation occurs is controlled by the percent the repulsive fields are engaged, by sliding the outer magnet holder back and forth around the inner magnet holder. Think of the outer magnet structure as wind in the form of magnetic fields and the inner magnet structure as wind turbine blades. This system operates with 100% efficiency, which slowly degrades over time.

In the second system, the Bannon Maher Gravity Generator and Motor, force generated from electronically controlled hydraulics is used to lift a large volume of water in a tank, with the water then released through the system, to spin a turbine which powers a generator, and or motor axle, with the water flowing into a paired lowered water tank, which when full, is raised while the then empty tank is lowered, with electronic controllers looping the process forever. Linear force is used to lift the water, however as water falls, gravity compounds its force, at 0.433 pounds of pressure per square inch per foot of drop (9.81 kilopascals per meter of drop), allowing energy to be captured in the system in excess of that consumed, by means including the differential between the linear input force required to lift the water and compounding output force provided by gravity as the water falls. A human hand can provide the force to operate a hydraulic cylinder and pump that raises 100,000 pounds (45,360 kilograms) and therefore lifts 11,990 gallons (100,000 pounds of lift / 8.34 pounds per gallon) (82,718 liters) of water, a process which can be automated with electric motors and circuitry. For additional system efficiency, the tanks can be levered in a see saw formation, offloading part of the weight to the see saw support. For further system efficiency, and to optionally dramatically increase the volume of water lifted, the hydraulics may be operated by hydraulics, in a configuration disclosed in the next paragraph. This system may operate, depending on the configuration, with approximately 85% to 99% efficiency, meaning the electricity required to power the components, principally the motor providing the force of a human hand to operate the hydraulics, may utilize less than 1% of the electricity produced. As it may be decided my inventions ought to be described by their own laws of physics, I submit Bannon Maher's First Law, which states that energy captured may exceed energy consumed within the operational union of energy systems.

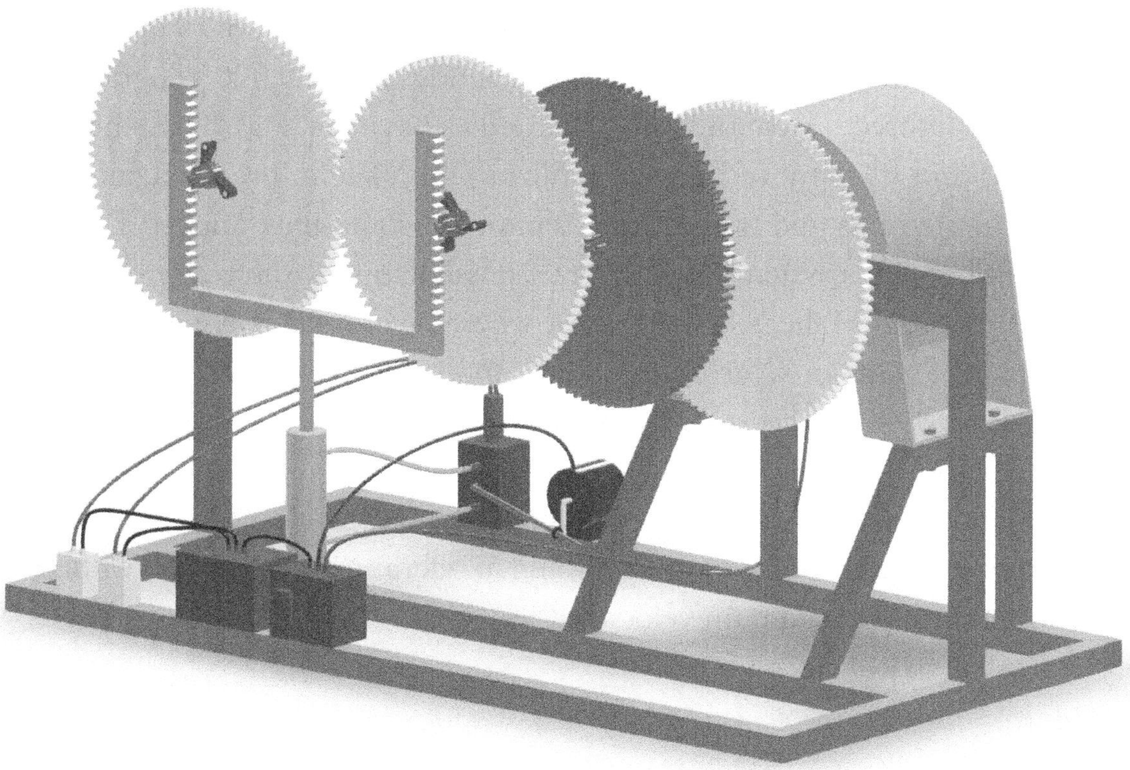

In the third system, the Bannon Maher Leveraged Generator and Motor, an axle to power a generator and or function as a motor is spun by the force generated by hydraulics. A commodity hand operable hydraulic cylinder and pump, can generate force of 100,000 pounds, with the hydraulic pump handle powered by an electric motor. Gears allow force to be increased in speed proportionate to the reduction in force. In this system, the force generated by the electronically controlled continuous push pull motion of the hydraulic piston is transferred through gears, which convert it to unidirectional motion at an increased speed and proportionately reduced force, to power a motor axle and generator, which respectively provide propulsion and electricity. Energy is captured in the system in excess of that consumed from the differential of the input force required to operate the hydraulics and the output force provided. This system in this configuration may operate with greater than 85% efficiency, and is not only compact but if optimally manufactured could potentially run forever. Efficiency of the system may be brought to greater than 99% by operating hydraulics with hydraulics, where the piston of a hydraulic cylinder is connected to operate additional hydraulic pump handles and corresponding cylinders. For example, if 10 pounds (4.5 kilograms) of input force are used to operate a hand operable hydraulic cylinder and pump providing 100,000 pounds (45,360 kilograms) of output force, that cylinder's piston may in turn operate 10,000 additional hand operable hydraulic pump handles with corresponding cylinders each providing 100,000 pounds (45,360 kilograms) of output force, thus producing total output force of 1,000,000,000 pounds (453,600,000 kilograms), utilizing 10 pounds of input force. Archimedes is recorded as having stated "Give me a lever and a place to stand and I will move Earth." Utilizing this system, braced against a celestial body of appropriate mass and trajectory, Archimedes could have moved Earth with the force of his hand. Such a system may be made possible by my discovery of an additional law of physics, Bannon Maher's Second Law, where layered leverage provides efficiency gains as a result of gains in layer output force (total output = (unit output force / unit input force) ^ layers) exceeding gains in layer cycle time (total cycle time = unit cycle time ^ layers). Therefore, this system with adequately powerful hydraulics, can provide any level of output force, using comparatively effectively no input force, to provide any level of electricity generation or propulsion, including accelerating any vehicle, without fuel, to effectively instantaneous maximum velocity.

In the final system, the Bannon Maher Buoyancy Generator and Motor, buoyancy is utilized, which causes an object in a liquid to rise if the weight of the object is less than to the weight of the volume of liquid it displaces, where the buoyant force provided by liquid displacement is used to raise a buoyant weight, which upon reaching the top of the liquid container is pushed off to fall into an open air chain connected chamber, providing force as it drops, to rotate the axle of a generator to produce electricity or function as a motor, and the cycle repeats, when the buoyant weight reenters the liquid container through a compartment at the bottom, where in order for the buoyant weight to not have to displace the weight of all the liquid in the container, the bottom compartment is sealed until the buoyant weight has entered, where the buoyant weight enters at an angle and utilizes weight and gravity to push the water contained in the bottom compartment through a pipe back to the top of the liquid container, then the compartment entry door is closed, the bottom compartment seal is opened and the buoyant weight again floats to the top to repeat the cycle, allowing energy to be captured in excess of that consumed as a result of the differential between the lack of force required to raise the buoyant weight and the force produced as it falls, where the taller the liquid container, the higher the raising of the buoyant weight, and the greater the energy captured by the system as it descends. Energy is consumed in the system fixed and expended only when assisting with the entry of the buoyant weight into the bottom compartment of the liquid container, and when pushing the weight off the top of the liquid container, thus when using an adequately tall liquid container the system operates with efficiency of over 99%.

In the non-liquid systems system, when implemented as a flying vehicle engine, I have included in the full patents an electronically controlled hydraulic piston and pump, for vertical take off and landing, to control the angle at which the unit operates, where vertical take off and landing to and from cruising altitude will overwhelmingly eliminate noise pollution and airport runways.

Each system, when implemented to produce electricity, can be built around existing generators decoupled from their wind, water, or steam turbine, where steam turbines are generally used in coal, nuclear, natural gas, and geothermal power plants. This allows, for example, existing generators coupled to a wind turbine, which typically only produce around 20% of the potential capacity, to now continuously produce 100% output. To upgrade, simply buy units without a generator, that are licensed for use with a generator and rated to provide the speed and force required by the existing generators, have a service technician decouple the generators, attach a gear to the existing generators axle that connects with the gear on my inventions output axle, and connect the two. As power lines go out of service, the units can be individually sold off over time to customers, with utilities making money from financing units sales and leases and providing service plans to customers.

Calculations of Electricity and Propulsion Output and Cost

According to the Goldman Sachs 2016 report "The Low Carbon Economy", the cost per kilowatt hour for residential solar, which may be the primary alternative to these energy systems in delivering energy directly to end users, was found to be about 120 British pounds per megawatt hour, which converts to about $0.093 per kilowatt hour. According to the United States Energy Information Administration March 2018 report "Independent Statistics and Analysis", the lowest cost energy source is geothermal, at $41 per megawatt hour, which is $0.041 (EUR) per kilowatt hour, however the cost of electricity is much higher by the time it reaches consumers given residential electricity rates reported by United States Energy Information Administration for May 2017 found the lowest average cost per kilowatt hour of electricity of charged in any state in America is $0.0969 (EUR) per kilowatt hour. In the following section, you will see how the components and other unit costs result in these systems providing an expected cost per kilowatt hour of as low as $0.0076, which is more than ten times less expensive than the alternatives when delivering energy to the end user.

I preface this section with background on referenced terms. To describe electricity, amperes are volume, volts are pressure, watts are volts multiplied by amperes, and a kilowatt is a thousand watts. To describe propulsion, horsepower is used, where horsepower is pounds of rotational force around an axis at a distance of 1 foot (.3 meters), multiplied by revolutions per minute, with the result divided by the horsepower constant 5252.

Calculation of the outputs and costs of the systems is done by utilizing specific commodity components with known energy consumptions, outputs, and costs. Before analyzing the output calculations of the systems, it's important to establish the context of useful output levels.

Energy Consumption by Homes

The world's largest generator manufacturer by revenue currently sells a 10,000 watt home gas powered backup generator, requiring rotation at 1,800 revolutions per minute at 13.3 horsepower, that it recommends for homes up to 5,000 square feet (465 square meters).

Energy Consumption by Electric Cars

The world's best selling electric car uses roughly an average of around 4,000 watts.

Energy Production by Utilities

Output requirements for utilities vary by customer base, but because any number of units can be installed, any level of output can be produced. Since units are less expensive than all other energy production and storage systems, it makes financial sense to buy the number of units needed for peak output, then turn off unneeded units at non-peak times, and I have written and included software code for automated management of large scale installations.

Horsepower of Vehicles

Car and truck engines produce a minimum of around 100 horsepower, an average of around 250 horsepower, and in the world's most powerful road vehicle, a maximum of around 1,250 horsepower, with revolutions per minute averaging around a maximum of 7,000, and the fastest car engine providing a maximum of about 10,500.

Horsepower of Jet Engines

Jet engines on the largest commercial airplanes each produce around 185,000 horsepower.

Calculations

I must note that I wrote out these simplified calculations and explanations rather quickly, they may be imperfect, and more comprehensive and accurate calculations are contained in the patents. Additionally, actual performance will vary based upon quality of unit components and manufacture. However, even if these calculations are off by many times, which they're not, my inventions are still dramatically superior to every other method of energy production and propulsion.

Bannon Maher Magnetic Repulsion Motor & Generator

In calculating the output and cost of the Bannon Maher Magnetic Repulsion Motor and Generator, which utilizes the force provided by magnetic repulsion to produce continuous rotational force, the primary system components to be considered are the magnets and generator when included. A 1 cubic inch (16 cubic centimeter) permanent neodymium magnet weighing 0.4 pounds (0.18 kilograms), sells for around 20 USD (16 EUR), and at a distance of a half of an inch from an opposing magnet of the same specifications, generates 20 pounds (9 kilograms) of repulsive force. Given the magnets will be rotating and engaged only approximately an average of half of the time, each magnet pair can be assumed to provide 10 pounds (4.5 kilograms) of repulsive force. The inner magnet holder, which resembles a wheel, rotates inside the fixed outer magnet holder, with all magnets mounted and angled to provide maximum repulsive force to a single side of the inner magnet holder's center axle. The fastest motor currently available spins at 100,000 revolutions per minute with no load, and thus given the rotational speed of the unit is limited only by air resistance when there is no load, the unit can be assumed to have a similar maximum rotational rate, however when there is a load, the average rotational rate will be reduced by the load, where the target rate for a generator is 1,800 revolutions per minute, with the decreasing speed provided by the resistance of the generator axle. Therefore with 5 magnets in the inner magnet holder and a compliment of 12 magnets in the outer magnet holder, the axle may rotate with around 50 pounds (23 kilograms) of continuous force at 10,000 revolutions per minute, per cubic inch (16 cubic centimeters) of magnet, at a magnet component price of 340 USD (17 magnets x 20 USD) (277 EUR). A 10 kilowatt generator may operate optimally at 1,800 revolutions per minute with the previously determined 38.8 pounds (17.6 kilograms) of force, so one complete set of 1 cubic inch (16 cubic centimeter) magnets may spin the generator, providing a total magnet cost of 680 USD (20 USD x 17 magnets * 2 inches to compensate for degradation over time) (553 EUR), with the engagement of the inner and outer magnet structures adjusted to provide a few watts under the maximum output of the generator to ensure it's not overworked. To determine the force required to provide a specific level of horsepower, 250 horsepower at 7,000 revolutions per minute may be targeted, which is that of the average road car, where 250 horsepower (250 = (n pounds of force * 7,000 revolutions per minute)/5252 horsepower constant), solving for n pounds of force, requires 0.75 pounds (0.34 kilograms) of force per unit of horsepower, thus requiring a total 187.5 pounds (85 kilograms) of force, and therefore the 5 inner magnet structures should then be 7.5 inches (19 centimeters) in length having a magnet cost of around 5,100 USD (7.5 inches of magnets * 2 times the length of magnets to compensate for degradation over time * 17 magnets * 20 USD per inch) (4,148 EUR), additional unit component costs of 1,000 USD (813 EUR), horsepower royalty of 4,997 USD (19.95 USD * 250) (4,064 EUR), labor of say 2,500 USD (2,033 EUR) and manufacturer profit of say 2,500 (2,033 EUR), for a total of around 16,079 USD (13,078 EUR), for a an average horsepower motor, but that provides unlimited range without ever refueling or recharging. Any level of horsepower can be provided by lengthening or increasing the number magnets, to provide a car engine faster than any on the road, or a jet engine.

Bannon Maher Gravity Motor & Generator

In calculating the output of the Bannon Maher Gravity Motor and Generator, where energy is captured in excess of that consumed by means including the differential between the linear force required to raise water and the compounding force provided by gravity as water falls, 1,000 gallons (3,790 liters) of water are used weighing approximately 8,340 pounds (3,783 kilograms), flows through the system transitioned by the previously cited hydraulic piston and pump, with the same force at the same speed provided by an equivalent electric car jack which can utilize a maximum of 180 watts (12 volts * 15 amperes) before a car cigarette lighter fuse blows. The energy provided by the water flow may be calculated by utilizing a water drop speed of 20 miles per hour (9 meters per hour), and a force of 0.433 pounds of pressure per square inch per foot of drop (9.81 kilopascals per meter of drop), with the total square inches (centimeters) of water flowing through the pipe calculated as the pipe radius squared times the constant pi of 3.14. To power a 10,000 watt generator requiring 1,800 revolutions per minute at 13 horsepower requires 38.8 pounds of force ((13.3 horsepower = n pounds of force rotating around an axle at a distance of 1 foot * 1,800 axle revolutions per minute)/5252 horsepower constant). Therefore, targeting 38.8 pounds of force, water passing through piping providing a diameter of 10 inches (0.25 meters) to rotate a turbine with a corresponding diameter of almost 10 inches (0.25 meters), where the 10 inch (0.25 meters) diameter pipe providing a drop of 6 feet (1.83 meters), may provide 2.6 pounds of pressure per square inch (6 feet * 0.433 pounds) (17.9 kilopascals), over an area of 78.5 square inches ((5 inch radius ^ 2) * 3.14 constant pi) (32,260 square millimeters), for a total of 204 pounds (93 kilograms) of force at 20 miles (32 kilometers) per hour, with rotations per minute of the turbine axle calculated where water as determined by gravity flows at about 20 miles (32 kilometers) per hour or 21,020 inches (536 meters) per minute, and an 10 inch (0.2 meters) diameter turbine has a circumference of 31.4 inches (10 inch diameter * 3.14), with a turbine therefore providing around 336 revolutions per minute ((21,120 inches of water flow per minute / 31.4 turbine diameter) * 50% efficiency), where a 10,000 watt generator may require 1,800 revolutions per minute and 13.3 horsepower, requiring the turbine rotational force be connected to the generator through a gear pair with a teeth ratio that approximately increases speed 5 times and decreases force by 5 times, therefore providing around 41 pounds (23 kilograms) of force at 1681 revolutions per minute, which is 13.1 horsepower ((41 pounds of force * 1681 revolutions per minute))/5252 horsepower constant), therefore allowing each system cycle to sustain generator output of 10,000 watts, where if looking for a self powered engine, a smaller generator may be used with the excess horsepower used as a motor. For additional validation, an alternative method for calculating the watts of output able to be provided by water flow is watts of output is equal to water drop in meters multiplied by water flow in liters per second multiplied by gravity of 9.8 meters per second per second. For example, a drop of 72 inches (1.82 meters) providing flow of 120 gallons (454 liters) per second multiplied 9.8 results in a power output calculation of approximately 8,100 watts (1.82 meter drop * 454 liters per second * 9.8 meters per second of gravity). The total cost of the system at the time of disclosure can be calculated as the cost of the generator, 5,000 USD (4,068 EUR), plus the costs of the hydraulic pump and cylinder, the repeat cycle timers to ensure continuous motion in the system, the battery to start the unit, a support structure, and a power converter, for an additional 2,000 USD (1,627 EUR), plus the cost of royalties, which are the greater of 0.99 USD or CHF per watt for a total of 9,950 USD (8,096 EUR), plus manufacturer labor expenses of 2,500 USD (2,035 EUR) and profit of 2,500 USD (2,035 EUR), for a total of around 21,950 USD (17,852 EUR), with an expected useful life of 30 years when utilizing a commodity generator and current standard manufacturing processes, providing a cost per kilowatt hour of about 0.0084 USD (21,950 USD / (30 years * 365 days * 24 hours * 10 kilowatts)) (0.0068 EUR), making it not only a source of energy dramatically superior to all others, but more than ten times less expensive, and therefore an effectively free source of energy when amortized over time.

Bannon Maher Leverage Motor & Generator

In calculating the energy output of the Bannon Maher Leverage Motor and Generator, which captures energy in excess of that consumed as a result of the differential between input force required and output force provided by certain force providing device configurations, the components to be considered are the generator output, the hydraulic pump power consumption, and corresponding hydraulic piston force. A commodity generator producing 10,000 watts may be used, where the average cost per watt of such generators is around 0.50 USD (41 EUR), and the optimal speed of generator axle rotation is 1,800 rotations per minute at 13.3 horsepower, which requires 38.8 pounds of force ((13.3 horsepower = n pounds of force rotating around an axle at a distance of 1 foot * 1,800 axle revolutions per minute) / 5252 horsepower constant). The generator is powered by a commodity electric hydraulic force providing device, such as a hydraulic cylinder and pump, which may provide force moving at approximately 1 inch per second which converts to 60 inches (152 centimeters) per minute. By utilizing gears, force can be transitioned to be unidirectional, and increased in speed proportionate to the reduction in force, therefore, when targeting 1,800 rotations per minute, when rotating a 1 inch axle with a circumference of 3.14 inches (8 centimeters) will provide revolutions per minute of 19, and therefore require an increase in speed of about 95 times, so the force will have to be passed through gears having a teeth ratio of 1:95, whereby the force is also decreased by a factor of 95. Therefore, the lift to rotate the 10,000 watt generator for maximum output will need to provide a force of around 3,686 pounds (1,843 kilograms) – 38.8 pounds (19.4 kilograms) of required force multiplied by a 95 times speed reduction – or 1.85 tons, which is less than the force provided by the functional equivalent of a hydraulic piston and pump, a commodity electric hydraulic car jack, which in providing such force uses less than 180 watts (12 volt x 15 amps) or it would blow out the cigarette lighter fuse, therefore resulting in a dramatically net positive energy production system. The total cost of the system at the time of disclosure can be calculated as the cost of the generator, 5,000 USD (4,068 EUR), plus the costs of the hydraulic pump and cylinder, the repeat cycle timers to ensure continuous motion in the system, the battery to start the unit, a support structure, and a power converter, for an additional 2,000 USD (1,627 EUR), plus the cost of royalties, which are the greater of 0.99 USD or CHF per watt for a total of 9,950 USD (8,096 EUR), plus manufacturer labor expenses of 2,500 USD (2,035 EUR) and profit of 2,500 USD (2,035 EUR), for a total of around 21,950 USD (17,852 EUR), with an expected useful life of 30 years when utilizing a commodity generator and current standard manufacturing processes, providing a cost per kilowatt hour of about 0.0084 USD (21,950 USD / (30 years * 365 days * 24 hours * 10 kilowatts)) (0.0068 EUR).

In considering the calculations of the Bannon Maher Leverage Motor, to determine the force required to provide a specific level of horsepower, I'll use the horsepower of the average car on the road, 250 horsepower at 7,000 revolutions per minute, with the 1,800 revolutions per minute of the gear powering the generator, increased by a factor of 4 utilizing a gear with the corresponding teeth ratio, bringing revolutions per minute to 7,200, and therefore 250 horsepower (1 horsepower = (n pounds of force rotating around an axle at a distance of 1 foot * 7,200 axle revolutions per minute)/5252 horsepower constant), solving for n pounds of force, requires 0.75 pounds (0.34 kilograms) of force per unit of horsepower, times the 4 times and earlier 95 times speed increase and force reduction for 285 pounds (129 kilograms) of initial force per horsepower, therefore requiring a total of 71,250 pounds (32,318 kilograms) of force, or 36 tons of force, with an additional 4 tons for the generator, for 40 tons of force required, with a horsepower royalty cost of 4,997.50 USD (19.99 USD per unit of horsepower * 250 horsepower) (4,067 EUR), plus at the time of disclosure an estimated 8,000 USD for the hydraulics and generator, 2,500 USD (2,035 EUR) labor expenses, and 2,500 USD (2,035 EUR) manufacturer profit, for a total cost of around 17,997 USD (14,646 EUR) for a self-powered motor that's of standard horsepower, that provides endless range without refueling or recharging, and therefore provides transport that is nearly free over time. The engine can of course easily be made dramatically more powerful than the most powerful car engine ever made, or manufactured as a jet engine, by very simply using more powerful components at approximately proportionately increased costs.

Bannon Maher Buoyancy Motor & Generator

In calculating the output and cost of the Bannon Maher Buoyancy Motor and Generator, the principal components to consider are the generator, the construction of the liquid container, buoyant weight, and support structure, the repeat cycle timers, and the motors. The total cost of the system at the time of disclosure can be calculated as the cost of the generator, 5,000 USD (4,068 EUR), plus the costs of the hydraulic pump and piston, the repeat cycle timers to ensure continuous motion in the system, the battery to start the unit, a support structure, and a power converter, for an additional 2,000 USD (1,627 EUR), plus the cost of royalties, which are the greater of 0.99 USD or CHF per watt for a total of 9,950 USD (8,096 EUR), plus manufacturer labor expenses of 2,500 USD (2,035 EUR) and profit of 2,500 USD (2,035 EUR), for a total of around 21,950 USD (17,852 EUR), with an expected useful life of 30 years when utilizing a commodity generator and current standard manufacturing processes, providing a cost per kilowatt hour of about 0.0084 USD (21,950 USD / (30 years * 365 days * 24 hours * 10 kilowatts)) (0.0068 EUR).

PRODUCTION

The following are the production instructions for the systems, which reference the numbers in the accompanying illustrations to provide additional clarity, and allow for a competent technical manufacturer with a relevant background and the specified tools and components to produce each invention within a day. You may find modest redundancy in the instructions if you have read everything to this point, because to reduce preparation time, these instructions have not been reformatted from patent format. I note these are not my final submitted patents, though formatting and structure has been optimized since initial filing based on feedback from patent experts. The actual filings don't contain the figures providing computer code, because in order to provide for easier international translation, a significant number of words aren't allowed in illustrations under Patent Cooperation Treaty rule 11.11, though I have included such code for educational purposes and ease of production, allowing this book to be used as an educational textbook. I wrote all code in my favorite language, Python, which offers extensive libraries, and provides for concise syntax, dynamically typed variables, and automated memory management, where processing inefficiencies from such features are rarely relevant to modern computing – I think of it as being for people who value efficiency, aesthetics, and not having to clean up. Additionally, I am releasing these production instructions in a comprehensive step-by-step format for the following reasons:

1. These are the production instructions to be used by the manufacturers who produce the units, and I want production to be as easy as possible.
2. Patents require, in exchange for a twenty year exclusive right to produce an invention, the full disclosure of an invention's construction instructions – and patent filings become public after eighteen months anyway.
3. I don't want any thinking person to have any question about the validity of my inventions.
4. It will be beneficial for others to see in detail how physics, engineering, programming, medicine, and law, can be usefully integrated. Additionally, too much of what's taught in programs of top ranked schools, related to these topics, is outdated and useless, while being time consuming to learn, which prevents development of skills, and work on projects, that optimally contribute to the well being of humanity. There are countless other beneficial inventions, across countless fields, that can be derived from the software and hardware construction

details I have provided, with the automated robotic cancer surgery system perhaps being the most educational, and allowing for others to derive other robots that perform countless other useful functions, such as a robot that goes through landfills and uses image recognition to collect recyclable metal and plastic. The disclosure of my work will yield global waves of innovation.

5. I want energy systems to be easily produced from the instructions in verifiable extreme poverty situations without payment to me, as described in the chapter on philanthropy.

SELF-POWERED MOTOR AND GENERATOR

INVENTOR JONATHAN BANNON MAHER

TECHNICAL FIELD

[0001] Embodiments of the invention relate to the fields of motors, generators, physics, engineering, and programming.

ABSTRACT

[0002] Systems, methods, apparatuses, and in some embodiments computer programs encoded on a computer storage medium, provide for clean continuous portable self-powered propulsion and electricity generation, utilizing repulsive magnetic fields, consistent with the laws of physics, where in some embodiments, including one complete embodiment, a first structure holding magnetic fields is able to be engaged inside or around a complimentary second structure holding magnetic fields, with magnetic fields angled, spaced, and shielded to allow one or both structures to spin continuously, where speed and force may be determined by percent the two structures are engaged, with the structures rotation in turn providing electricity generator axle rotation and or providing the propulsion of a traditional motor. The invention permanently solves global warming, provides reduced cost of living and cost of goods to alleviate poverty, provides unlimited clean energy for evaporated water purification and atmospheric carbon dioxide splitting, and eliminates the need for every other method of energy production, including nuclear technology – thus reducing nuclear weapons technology proliferation.

BACKGROUND

[0003] The majority of the proceeds from the licensing of this patent will be going to causes that support the well being of humanity, and your support in ensuring the patent is forever in every way as strong as possible, will be providing a service to all the world.

[0004] This section is intended to introduce the reader to various aspects of the art that may be related to various aspects of the present techniques, which are described and or claimed. This discussion is believed to be helpful in providing the reader with background information to facilitate a better understanding of the various aspects of the

present disclosure. Accordingly, it is understood that these statements are to be read in this light, and not a citation of any prior art.

[0005] Patent filings on structurally differentiated and fundamentally deficient disclosures may exist that may attempt to claim any invention that is self-powered, based on previously publicly known failed attempts to build such devices, however any such disclosures do not enable the purported inventions, with overly broad claims not supported by the disclosure that also fail to distinctly claim the invention, and any such patent filings are inherently invalidated by prior public disclosure, the enablement requirement, and the claims support requirement.

[0006] Known and proposed energy production, transmission, and storage systems have some or all of the following deficiencies:

[0007] 1. External fuel source required: an external fuel source is utilized such as oil, gas, coal, wind, sun, water currents, geothermal heat, hydrogen, or uranium, where the cost of providing fuel in the form of electricity or gasoline to a vehicle over its useful life potentially exceeds the cost of the vehicle, and about a quarter of airline costs are from fuel.

[0008] 2. Environmentally unfriendly: nuclear energy production, including fission, fusion, and cold (LENR), results in toxic waste and or materials, geothermal often circulates contaminants from the ground, fracking creates toxic water, hydroelectric dams decompose organic matter producing the potent global warming gas methane, solar panel manufacturing often releases toxic byproducts including greenhouse gases far more potent and long lived than those from fossil fuels, hydrogen takes substantially more energy to produce and transport than it provides, while clean energy sources may utilize slowly degrading flammable toxic batteries to store energy for when the sun is not shining, wind is not blowing, or water is not adequately flowing.

[0009] 3. Intermittent: wind, solar, and traditional water energy systems provide variable output, with average output often found to be around 20% of rated output, as a result of environmental conditions, meaning a 5 kilowatt system typically produces average output of only 1 kilowatt.

[0010] 4. Not portable: solar panels can only function in the sun, wind turbines in wind, water turbines in water currents, nuclear in a stable highly controlled environment, and carbon with the aid of an emissions pipe.

[0011] 5. Extremely expensive transmission costs: power lines are required for all forms of nuclear energy, including fission, fusion and cold, for farms of solar, wind, and water, and for fossil fuels plants, while pipelines are generally required for oil and gas, including natural gas, which is principally methane, a greenhouse gas more than twenty times more potent than carbon dioxide that may be leaked during extraction, transport, and consumption. Power line and fossil fuel pipe line cost of installation per 1 mile (1.6 kilometers) has been found to be up to around 20 times the average annual income in the United States, and lines must be replaced every 30 to 50 years, so in the United States alone, with 300 thousand miles (480 thousand kilometers) of power lines, and 200 thousand miles (320 thousand kilometers) of pipelines, replacement would require an expenditure around 25 times the national debt, passed on to consumers, and dragging down the economy, with every other developed nation in a similar situation. Power lines require environmental destruction during installation, leave visible blight, are forever vulnerable to cyber attacks, transmit power from central sites that are inherently more prone to failure and blackout than a decentralized system, dissipate power during transmission, with high voltage power lines having health consequences for those living nearby.

[0012] 6. High initial costs: in addition to previously cited expense of power lines, fossil fuel pipes, and ongoing fuel costs, clean energy farms require an allocation of land, and degrading batteries requiring periodic replacement, which is why tax credits are often required for clean energy systems to be affordable. In addition to those factors, comparing a 1 kilowatt rated clean energy system to a 1 kilowatt rated traditional system, may require multiplying the cost of the clean energy system by approximately 10 times, 5 times to account for enough electricity generation to be stored for the equivalent continuous output, and 5 times for the battery storage.

[0013] 7. Vulnerable to weather: wind turbines, hydroelectric dams, and solar panels, can be made ineffective by environmental conditions such as freezing temperatures, snow, and rain, and similar to power lines, may be taken down by extreme weather and lightning strikes.

[0014] 8. Vulnerable to black outs: power transmitted over power lines creates vulnerability for critical facilities such as hospitals and data centers.

[0015] 9. Vulnerable to cyber attack: utility scale energy systems often require a hackable computer to operate, and are therefore forever vulnerable to computer viruses able to take down and or destroy nuclear power plants and the electrical grid, even if such systems aren't connected to the Internet, as demonstrated by the Stuxnet virus.

[0016] 10. Causes deaths: plants, rigs, and pipes, for current and proposed forms of nuclear, hydrogen, gas, and oil energy can explode, and coal mines can collapse, while wildlife is killed by wind, ocean, wave, and river turbines, hydroelectric dams, solar condensers, solar panel and battery manufacturing byproducts, and nuclear waste.

[0017] 11. Encourage nuclear weapon proliferation: fission, fusion, and cold (LENR) nuclear energy are or can be one step from weaponizable, while hydrogen can be obtained by a terrorist at a hydrogen fuel station to create a powerful compressed hydrogen explosion, as verified by reviewing a video of a balloon filled with hydrogen being lit on fire. Fusion is particularly disturbing, as a fusion weapon could be created with the power of an exploding star, able to take out the planet, and resulting in an extinction level event for humans, a scenario even more likely when considering increasingly autonomous – and therefore inevitably hackable by individuals – weapons control.

BRIEF DESCRIPTION OF THE ILLUSTRATIONS

[0018] Illustrations are presented by way of example, and not by way of limitation, where some embodiments may not contain all components, may contain additional components, and may contain functionally similar components.

[0019] FIG. 1 is an illustration of an example of an embodiment, with a rotatable magnet structure inside a rotationally fixed outer magnet structure, with all magnets angled, spaced, and insulated, to provide magnetic repulsion to rotate one of the magnet structures to spin an electricity generator and or motor axle.

[0020] FIG. 1A is an illustration of a magnified view of part of the magnetic fields of a pair of magnets in FIG. 1.

[0021] FIG. 2 is an illustration of an example of an embodiment of a pattern which may be used to form insulation material that directs a magnetic field.

DETAILED DESCRIPTION

[0022] The disclosure is related to the field of clean continuous portable self-powered energy and propulsion. It is understood that any reference to a person skilled in the art, recognizes that at the time of filing, there is no one else skilled in the art of this particular field, or a closely related field. Given the extraordinary nature of the disclosure, regardless of how full, clear, concise and exact the disclosure in enabling the production and use of embodiments of the disclosure, what could be construed to be undue experimentation during production and use, is simply the ordinary effort required in the assembly and use of an embodiment of such a disclosure.

[0023] It is understood that, as in any engineering or design project, the development of any actual implementation will include numerous implementation specific decisions made to achieve the developers' specific goals, such as compliance with business related and system related constraints, which may vary from one implementation to another. It is understood that such a development effort might be complex and time consuming, but is nevertheless a routine undertaking of design, fabrication, and manufacture for those skilled in the art having the benefit of this disclosure. The disclosed steps may be read as prefaced by "In some embodiments, including one complete embodiment, ", may be executed or performed in other orders or sequences, and are not limited to the order and sequence shown and described, which are provided to enable ease in constructing an embodiment, and along with each components of each step, may be removed, modified, combined, or rearranged, and other steps and or step components may be added, without departing from the scope of this disclosure and or invention. Although embodiments of the invention have been described and illustrated in the disclosed implementations, it is understood that the present disclosed subject matter, including apparatuses, methods, specification, and illustrations, has been made only by way of example, not by way of limitation, and the methods and apparatuses may be used in other systems, and that numerous changes and optimizations in the details of implementation of the invention and or embodiment are made without such modifications departing from the spirit and scope of this disclosure and or embodiments of the invention. Although the disclosure has been shown and described with respect to one or more embodiments, features of the disclosed embodiments can be combined and rearranged in various ways, and changes including equivalent alterations, substitutions, modifications, and additional efficiencies will of course occur to someone of ordinary skill in the art without departing from the spirit and scope of this disclosure and or invention. In particular regard to the various functions performed by the described components, the terms used to describe such components are intended to correspond, unless otherwise indicated, to any component which performs the specified function of the described component, or is functionally equivalent to the described component, even though not structurally equivalent to the disclosed structure which performs the function in the implementations described in this disclosure. In addition, while a particular feature of the disclosure may have been provided with respect to only one of several embodiments, such feature may be combined with one or more other features of other embodiments as may be desired and advantageous for any given or particular application. In some instances, well-known circuits, structures and techniques have not been shown in detail in order not to obscure the understanding of this disclosure. Articles in this disclosure such as "a" "an" and "the" may allow for both singular and plural forms. Verbs in this disclosure such as "is" may be read as "may be". Conjunctions in this disclosure such as "or" as used herein may be interpreted as inclusive or meaning any one or any combination, where "A, B or C" means "any of the following: A; B; C; A and B; A and C; B and C; A, B and C". Relational terms in this disclosure, for example first and second, top and bottom, left and right, are to distinguish one entity or action from another, and may not necessarily require or imply a relationship, or order between, such entities or actions. The disclosure includes the best mode contemplated by the inventor, a completely described specific embodiment, along with optional components and alternative embodiments to best suit the implementer, measurements in imperial and metric units to support universal understanding, and dramatically exceeds claims support requirements and enablement requirements by allowing for selection and or construction of the required components to be carried out easily, quickly, and routinely by persons of ordinary skill in the art, who are provided the additional benefit of utilizing readily available commodity components whenever possible. The present disclosure includes material protected by copyrights, and the owner of the copyrights hereby reserves all rights, but with authorization for publication as required by government patent offices. Various embodiments of the present invention may provide all, some or none of the disclosed technical advantages.

[0024] The computer code descriptions disclosed, in order to provide comprehensive enabling disclosure, rather than utilizing flow charts, which according to Patent Cooperation Treaty 11.11a are prohibited from containing "text matter, except a single word or words, when absolutely indispensable, such as... a few short catchwords indispensable for understanding", are provided in a text only format where the number of arrows preceding a line indicate logical block level, semicolons indicate a new segment of a logical block, and periods indicate the closure of one or more logical blocks. It is understood that any computer code representations in this disclosure are merely illustrative, rather than restrictive. While code may be written in nearly any computer language, including Java and C++, the illustrative computer code descriptions were derived from code written the Python language, which may be run through the Python interpreter, with appropriate supportive libraries, which at the time of disclosure, may run on nearly any computer, for example one with an Intel or AMD processor, running a current version of Linux, Windows, or Mac OS. All code components may read as if prefaced by "In some embodiments, including one complete embodiment, ". In some embodiments, functionality may be modified, rearranged, excluded, and added. To provide more fundamental computer system details, in some embodiments, the functionality associated with the disclosed computer code descriptions may be referred to as a script, module, software, software application, or code, and can be written in any form of language, including compiled, interpreted, declarative, or procedural, able to be deployed in any form suitable for use in a computing environment, including as an independent or integrated program, module, component, or subroutine, for execution by the computer system, implemented on one or more independent or integrated computers, utilizing a central processing unit in the form of one or more general or special purpose microprocessors, in conjunction with digital electronic circuitry, which may include special purpose logic circuitry such as a field programmable gate array or application specific integrated circuit, with the computer controlled by and operatively coupled to tangibly embodied software and or firmware, which may include code that creates an environment for code execution, including individual or combined use of processor firmware, a protocol stack, a database management system, and an operating system, where such software and or firmware may exist in one or more parts in memory on one or more computers, and is encoded on one or more tangible non transitory software carriers, such as individual or combined use of a random or serial access device or substrate, a semiconductor memory device, transient or persistent random access memory, a magnetic, magnetic optical, or optical disk, or encoded on an artificially generated transmitted signal, for example, optical, electrical, or electromagnetic, transmitted using a sending and a receiving apparatus, where the interaction between the user and the software may be implemented by operatively coupling, to the local implementing computer, or a local computer connected to one or more remote computers through a local or wide area network, a display device which may implement liquid crystals or light emitting diodes, a keyboard, and a pointing device.

[0025] The inventor retains absolutely no liability for any implementation of this invention, and the invention is implemented exclusively at the risk and liability of the implementer.

[0026] In some embodiments, for quality control purposes, all components may be manufactured from scratch.

[0027] Calculations, formulas, and specific units are not in any way restrictive, are not be relied upon, are not required as presented to produce an embodiment of the invention, are provided exclusively as a courtesy to enhance enablement for those resizing components and or constructing alternative embodiments, and may contain inaccurate assumptions easily modified during practice, with all calculations utilized to select and estimate components and unit output being rough estimates that may vary greatly based on factors that include the type and quality of purchased and or manufactured components and embodiment construction, where embodiments may be constructed utilizing an effectively endless range of output and component configurations and selection processes.

[0028] Embodiments of the invention provide some or all of the following benefits over previously discussed predecessors:

[0029] 1. Self-powered for free output: embodiments provide the first energy and motor system in the history of the known universe to not require an external fuel source. Embodiments can produce endless energy, and provide endless transportation range, until there is a system failure, and therefore may potentially produce energy and propulsion until gravity driven orbital drift causes the Earth to be consumed by the Sun in a few billion years – assuming the units and humans are still on Earth. Embodiments reduce transportation costs by allowing implementing vehicles to operate without fuel, allowing for effectively free and endless until system failure transportation range after purchase, and additionally allow for travel by supersonic jets and flying cars, which have been impractical principally as a result of fuel costs.

[0030] 2. Clean: the manufacture and use of embodiments produces no notable harmful environmental byproducts, nor the potential for deaths associated with predecessors.

[0031] 3. Continuous: embodiments produce electricity and propulsion that is continuous and stable.

[0032] 4. Portable: embodiments are able to function as well in a basement closet as in a car.

[0033] 5. Cyber attack proof: embodiments are self-contained thus require no hackable computer to operate and are therefore immune to computer viruses.

[0034] 6. Blackout proof: embodiments are designed to be kept indoors and on-site, and are thus ideal for critical facilities such as hospitals and data centers that can't afford a blackout from failed power lines or plants.

[0035] 7. Inexpensive: embodiments can be manufactured and operated at the lowest cost total cost possible, because they don't require fuel, installation and maintenance of power lines, an allocation of land, or degrading batteries.

[0036] 8. Weatherproof: embodiments are self-contained for indoor use and therefore aren't vulnerable to environmental factors such as freezing temperatures, snow, rain, lightning strikes, or extreme weather events.

[0037] 9. Eliminates energy output storage: because additional embodiments of this system can be utilized at peak times with limited cost, storage of energy is no longer relevant, for either utilities or homes, even for peak output needs.

[0038] 10. Eliminates expensive power lines and fossil fuel pipe lines: because embodiments are designed to be kept on site, and any number of units can be utilized to meet peak power needs, power and fossil fuel transmission lines and their associated costs are now rendered irrelevant, thus substantially unburdening all economies globally of associated costs. Land currently holding power lines, as well as arrays of solar panels, wind turbines, and hydroelectric turbines, can be reclaimed to reduce visual pollution and make space for a growing population. Roofs of solar panels can be removed to allow for roof tiles that reflect heat to maintain a cool house in summer. No thinking person will ever want, nor could a functional government allow, any type of nuclear reactor – fission, fusion, cold (LENR) – in a car or home, leaving only now irrelevant power lines for transmission, thereby making those sources wholly irrelevant.

[0039] 11. Potentially profitable: in a standard home use scenario, embodiments may be the only way for a unit owner to make a profit from selling energy back to the utility at wholesale rates.

[0040] 12. Alleviates poverty: because energy is effectively free after embodiment purchase, and the purchase price is less per unit of output than other energy systems, embodiments reduce the cost of living and the cost of goods for every person on Earth, thus reducing poverty.

[0041] 13. Powers water purification and pumping: 1 in 10 people live without access to clean water, while climate change driven droughts fuel conflicts. Because embodiments make energy nearly free over their useful lives, the energy intensive nature of evaporated water purification — which removes nearly every contaminant with a higher boiling point than water, and potentially all others can be removed with a standard carbon filter and ultraviolet light – is no longer a barrier, nor is pumping, thus embodiments provide for the global resolution of clean water needs for individual consumption and agriculture.

[0042] 14. Powers reduced water consumption: effectively eliminates the energy cost of operating electricity powered showers that require only a cold water pipe, and recirculate, filter to potentially cleaner than direct from pipe, and heat water, to provide exact continuous temperature and pressure control, as well as powering low voltage electric showerheads that mix air with water to provide the effect of the same output using dramatically less water.

[0043] 15. Powers atmosphere cleaning: because an existing specialized laser can disassociate atmospheric carbon dioxide molecules into carbon molecules and oxygen molecules, and because embodiments can provide effectively unlimited and continuous completely clean energy, a power source is now available to reduce carbon dioxide in the air, and potentially other greenhouse gasses, if corresponding devices are developed.

[0044] 16. Reduces nuclear and hydrogen weapon proliferation: because embodiments eliminate any need for any type of nuclear energy, including fission, fusion, and cold (LENR), each of which is or may be weaponizable and one step from nuclear weapons technology, and also eliminate the need for hydrogen, which a terrorist can obtain at a hydrogen fuel station to create a powerful compressed hydrogen explosion verifiable by watching a video of a balloon filled with hydrogen being lit on fire, I have provided us all a fundamentally safer world.

[0045] 17. Therefore, a few applications of embodiments include powering: all transportation vehicles including automobiles, trains, jets, spaceships, cargo ships, cruise ships, tugboats, boats, submarines, hover boards, jet packs, including in vertical take off and landing configurations; powering electrical grids as well as homes, offices, factories, hospitals, and data centers that would like to disconnect from external power sources to end their recurring bill, be permanently immune from blackouts, use clean energy, and save money; televisions, washing machines, dishwashers, showers, and water pumps; portable consumer electronics, such as phones and laptops, through an internally installed miniaturized embodiment eliminating the need to recharge; personal rapid transport; home hydroponic production systems, including light, temperature control, and nutrient water circulation; spaceship electro magnetic ion drives using a fraction of the fuel of traditional rockets; high intensity laser powered solar sails, with the laser powered by a large number of these energy units on the surface of any space based body with a limited atmosphere; space tourism; video streaming planetary sampling probes journeying an unlimited number of years into the universe; space colonies; and inter galactic travel.

[0046] **In some embodiments, including one complete embodiment, given the desired output of the motor and or generator, select and obtain corresponding magnets, and generator if used.** In some embodiments, including one complete embodiment, the magnet structures are designed to support the magnets, and the magnets are selected to correspond to desired unit output, where sizes and types of magnets may vary upon factors including availability from magnet manufacturers and cost. In some embodiments, when implementing the motor and the generator, because the generator may optimally require a fixed rotational speed, it may be best to create each as a separate unit.

[0047] In some embodiments, including one complete embodiment, in reference to FIG. 1, because magnets may slowly have their field strength degraded when facing an opposing magnetic field for an extended of period of time, magnets are used to the extent practically possible that were manufactured to have the highest possible magnetic coercivity, meaning that the mix of materials used, and the manufacturing process, have made it as difficult as possible for a magnet to lose its directional force, by means which may include using specific materials and aligning material directional polarities prior to baking. In some embodiments, including one complete embodiment, Neodymium permanent magnets may be used. In some embodiments, Samarium-Cobalt magnets may be used, which may be four times as resistant to magnetic coercivity as Neodymium, but are significantly more expensive. In some embodiments, iron nitride permanent magnets may be used, produced utilizing abundant inexpensive iron, rather than rare earth metals, but such magnets are not broadly available at the time of disclosure. In some embodiments, substantially less expensive ferrite magnets may be used, which lose their directional force faster than standard permanent magnets. In some embodiments, other compositions, types, numbers, and sizes of magnets may be used. In some embodiments, which may include one complete embodiment, given magnets may slowly lose directional strength over time, additional magnet force output may be targeted to maximize longevity, and magnets may be longer than necessary, so that when magnetization degrades, the magnet structures can simply be further engaged to produce the same output.

[0048] In some embodiments, including one complete embodiment, permanent Neodymium magnets are used, which per square inch (645 square millimeters), may weigh around 0.3 pounds (0.13 kilograms), and may be of high grade type N52 which may produce around 20 pounds (9 kilograms) of repulsive force when at a half an inch (13 millimeters) distance opposing a magnet with the same specifications, however given that the magnets may be on average engaged based on distance at about half strength as they rotate, an average of 10 pounds (4.5 kilograms) of repulsive force may be assumed. The amount of force during engagement may be better approximated at the time magnets are obtained by securing one magnet on a scale, and placing an opposing magnet in a position representative of where it will be held in the magnet structure, and reading the weight, less the weight of the magnet alone on the scale.

[0049] In some embodiments, to determine the number of magnets to use, it may be the case that the greater the number of magnets used in the rotatable magnet structure, per unit of area, with a corresponding number in the rotationally fixed magnet structure, the smoother the magnetic repulsion, while also recognizing the greater the number the greater the design and assembly complexity, and that commodity magnets may only be readily available in certain sizes. In some embodiments, to approximate the number of magnets that may be in the rotationally fixed magnet structure, the width of the face of the magnet, as it sits within the circumference of the magnet structure, plus the width of the magnetic shielding, may be divided by the circumference of the rotationally fixed magnet structure interior. In some embodiments, including one complete embodiment, the design as it's built in CAD (Computer Aided Design) software may determine the exact number of magnets, ensuring angling and insulation so that the magnets do not providing repellent force on approach that prevents the inner magnet structure from effectively immediately spinning upon engagement with the outer magnet structure, or the rotating magnet structure may need to be provided initial rotational force. In some embodiments, including one complete embodiment, 5 magnets are held in the rotatable magnet structure, and a compliment of 12 magnets are held in the rotationally fixed magnet structure, where the inner magnet count multiplied by the repellent force of an inner structure magnet against an outer structure magnet, may be used for calculating the output.

[0050] To implement a generator, in some embodiments, including one complete embodiment, in reference to FIG. 1, 10,000 watt generator 1300 is obtained that may produce optimal output at 1800 revolutions per minute at 13.3 horsepower, which requires force of 38.8 pounds (18 kilograms) ((13.3 horsepower = n pounds of force * 1,800 axle revolutions per minute)/5252 horsepower constant). In some embodiments, which may include one complete embodiment, a generator is selected that outputs the volts, amperes, and hertz of the desired final output for the relevant application, with a converter utilized to power internal embodiment components. In some embodiments, to approximately calculate the force required by the magnets to power the generator, or where an existing generator is to be powered by the embodiment, for example, in the place of a wind or water turbine, magnets may be selected that correspond to the force required to power such generator, which may at the time of disclosure be around 1.33 horsepower per 1,000 watts. In some embodiments, including one complete embodiment, 5 magnets in the rotatable magnet holder powering the output, with a complimentary 12 magnets in the rotationally fixed magnet holder, where each magnet segment of dimensions 1 inch (25 millimeters) width by 1 inch (25 millimeters) height by 1 inch (25 millimeters) length may provide around 10 pounds (4.5 kilograms) of continuous rotational force, for around 50 pounds (23 kilograms) of rotational force, with the exact engagement of the inner and outer magnet structures to be determined at the time of unit use, to produce stable optimal output by the generator. In some embodiments, the length of the magnets used in the embodiment may be extended to any length necessary to provide the required and or desired force. In some embodiments, including one complete embodiment, when implementing only a generator, the magnets may be of 1 inch (25 millimeters) height by 1 inch (25 millimeters) width by 5 inches (127 millimeters) in length, to produce 250 pounds (113 kilograms) of rotational force to power the generator, with the engagement of the magnets limited to that required to power the generator at optimal speed, which as the magnets degrade in force over time, structured may be further engaged to maintain ideal force and revolutions per minute. In some embodiments, if smaller magnets are inserted in sequence, if there is any repulsive force between the magnets, insulation as later described in this disclosure may be placed between them. In some embodiments, which may include one complete embodiment, to convert pounds of force to pound-feet of force, for the calculations of horsepower and corresponding components, identical gears may be placed on the axle of each the rotatable magnet structure and the generator, with the gears interconnecting, and with each having a radius of 1 foot. In some embodiments, to accurately measure the output of the embodiment, a torque gauge may be used to measure pound feet of force, and a tachometer may be used to determine revolutions per minute, or alternatively a dynamometer may be used which measures pound feet of force as well as revolutions per minute, and then horsepower may be calculated as torque multiplied by revolutions per minute with the result divided by the horsepower constant 5252.

[0051] In some embodiments, including one complete embodiment, when implementing a motor, 5 magnets in the rotatable magnet holder and 12 magnets in the rotationally fixed magnet holder may be used, where the rotatable magnet holder operates inside the rotationally fixed magnet holder, with the 5 rotatable magnet holder magnets at 1 cubic inch (16 cubic centimeters) providing 50 pounds (23 kilograms) of rotational force, the 5 rotatable magnet holder magnets of width 1 inch (25 millimeters), height of 1 inch (25 millimeters), and a total length of 24 inches (610 millimeters), where length may be achieved by a single continuous magnet, or sequential magnets appropriately insulated between each other using insulation in a later described manner if they repel each other at their sides, or in embodiments where additional lengths are used, length may be split between separate embodiments on each the front axle and rear axle. In some embodiments, the rotationally fixed magnet structure may instead also rotate, in turn rotating a generator axle or providing propulsion, where for example to allow for a generator or wheel to be mounted in place, rotational force may be first provided to a rotating an intermediary to allow for the sliding engagement of the magnet structures.

[0052] In some embodiments, to approximately calculate the pounds of force to use when implementing the embodiment as a motor, a horsepower may be targeted. The fastest motor currently available spins at around 100,000 revolutions per minute, and thus given the rotational speed of the unit is limited only by air resistance, when there is no load, an embodiment may be assumed to have a similar rotational rate, however accounting for a load and air resistance, an embodiment may be assumed to provide up to 10,000 revolutions per minute. To target a horsepower, that of a standard road car may be used, which is 250 horsepower at a maximum of 7,000 revolutions per minute, solving for 250 horsepower requires approximate pounds of force of 333 ((250 = n pounds of force * 7,000 revolutions per minute)/5252 horsepower constant), or 151 kilograms. Therefore, given the unit uses five magnets on the rotatable structure, providing a total of 50 pounds (23 kilograms) of repulsive force per inch (25 millimeters), and to provide more force than required to compensate for gradual degradation, magnets 13 inches (330 millimeters) in length may be used for 650 pounds (295 kilograms) of force. This will provide a car with unlimited range and no external fuel requirements. In embodiments where a tremendous amount of force is required, such as in jet engines, in addition to or in place of extending magnet lengths, a greater number of magnets may be used on the rotatable magnet holder with a corresponding number in the rotationally fixed magnet holder, to achieve effectively any level of horsepower.

[0053] In some embodiments, countless variations of the number, size, positioning, spacing, angling, insulation, and type of magnets may be used, which may, along with the quality of the magnets, and the quality of construction of the unit, together determine the method and accuracy of the calculations and the output of the unit.

[0054] In some embodiments, the magnet structures and magnets may be manufactured as a single continuous component, with magnetic fields instilled throughout the structures to produce a functionally equivalent effect. In some embodiments, the magnetic fields may be instilled throughout the structures by means which may include using specific materials suitable to create magnets, for example as neodymium, iron, and boron, and using a manufacturing process which may include powder metallurgy, sintered magnet process, rapid solidification, or bonded magnet process. In some embodiments, for example, sintered magnetic fields may be instilled in the structures by melting raw materials in a furnace, cooling them into ingots, pulverizing them into powder, sintering them into blocks within the structures, heat treating, and appropriately aligning and directing magnetic fields. In some embodiments, for example, bonded magnet fields may be instilled in the structures by preparing raw materials using melt spinning, cooling the melted raw materials, then pulverizing them into a powder, mixing them with a polymer, compression or injection molding them into the structure, and appropriately aligning and directing magnetic fields. A limitation of fully integrating the magnetic fields within structures, rather than using individual magnets, is that when the magnetic fields eventually lose adequate directional force, they can't be replaced without replacing the structure.

[0055] **In some embodiments, including one complete embodiment, design and produce the rotatable magnet holder and the rotationally fixed magnet holder along with side panels.** In some embodiments, including one complete embodiment, in reference to FIG. 1, a rotatable magnet holder 1100 and a rotationally fixed magnet holder 1101 are produced, along with magnet securing side panels 1102 1103, where the structures are designed to secure and support the magnets that were previously acquired. In some embodiments, including one complete embodiment, the rotatable magnet structure is fixed in place except for the ability to rotate by a rotatable axle, while the rotationally fixed magnet structure is mounted by an attached ring to a sliding bar to allow for changing its engagement with the rotatable structure. In some embodiments, where the outer magnet structure is the rotatable structure, the inner magnet structure may be fixed in place by a static axle, and the outer magnet structure may be connected to a rotatable axle, which may be connected through an axle hole in its reinforced side panel. In some embodiments, the rotatable magnet structure is on a sliding axle that interconnects with laterally fixed rotatable axles at either end. In some embodiments, instead of one of the magnet structures sliding, each magnet and any corresponding insulation within one of the magnet structures has its angle facing the opposing rotatable magnet structure, by means which may include hinging the magnet to the structure, and may include connecting all of the hinged magnet structures by a series of adjustable rods so they are all operable as one, with the degree of the rotations controlling the engagement of the rotating magnet structure, where the rotation degree may be controlled manually, or may by be individually controlled by means including linear actuators or hydraulic pistons, or may be collectively controlled by a linear actuator or hydraulic piston, and while such an embodiment has the benefit of reduced space consumption, it has the cost of adding a one or more point(s) of potential mechanical failure. In some embodiments, when implemented as a jet engine, the rotatable magnet structure may have a traditional jet turbine structure on its inside, or the outer magnet structure may rotate and have a traditional jet turbine structure on its outside. In some embodiments, magnet structures may be produced that can hold more magnets than are required, with empty space filled by non-magnetic blocks to achieve desired force. In some embodiments, including one complete embodiment, the rotationally fixed magnet holder extends its insulating encasement around the rotatable magnet structure when it's disengaged, to reduce magnetic forces from acting on anything outside the unit.

[0056] In some embodiments, including one complete embodiment, in reference to FIG. 1, the rotatable magnet holder 1100 holds magnets 1200 1201 1202 1203 1204, and the rotationally fixed magnet holder 1101 holds magnets 1205 1206 1207 1208 1209 1210 1211 1212 1213 1214 1215 1216, with the magnets arranged so that the rotatable and rotationally fixed magnet holders' magnets render repelling forces. In some embodiments, the magnet holders may be designed to hold, and do hold, smaller magnets in between the larger magnets, which benefit from the same manner of insulation and angling as the larger magnets. In some embodiments, which may include one complete embodiment, the magnets in the outer magnet holder may be of greater depth than those in the inner magnet holder, to provide greater repulsive force to the rotatable magnet holder without increasing the rotatable magnet holder's weight.

[0057] In some embodiments, including one complete embodiment, in reference to FIG. 1, to determine the angling of each magnet within the magnet structures, each aligns with the center axle in such a way that each provides the majority of its magnetic field force to one side of the center axle, where one side of the magnet may be aligned with the perimeter of the axle, such that the magnetic force of the magnet, when visualized as a line from magnet edge to axle, just misses the center axle, and furthermore making sure the magnets are not wasting force by pushing directly on the axle, or causing resistance on approach, and are pushing on one side of the axle in a consistent direction, additionally recognizing that magnetic fields are not perfectly linear and curve outward, and adding a compensating adjustment past the previous axle alignment, and any further adjustment past the previous adjustments to limit or eliminate resistance on approach to optimize rotation, where the angling to optimize continuous repulsive force may be later refined in conjunction with insulating materials that further direct the magnetic fields. In some embodiments, other means of angling magnets to provide continuous rotational force may be used.

[0058] In some embodiments, including one complete embodiment, in reference to FIG. 1, the diameter of the inner magnet structure may be roughly two times the magnet diameter along with insulation, plus the width of the axle, plus two times the minimum thickness required at any point, plus the magnet inset, and where the internal diameter of the outer magnet holder may be sized to allow for the opposing angled magnets to be insulated, spaced, and distanced to provide optimized rotational force.

[0059] In some embodiments, including one complete embodiment, in reference to FIG. 1, the rotatable magnet holder 1100 has a hole through the exact center that snugly fits axle 1006, with one end of axle 1006 passing through bearing 1007 mounted in axle support beam 1008, and the other end of axle 1006 passing through bearing 1009 mounted in support beam 1010.

[0060] In some embodiments, including one complete embodiment, in reference to FIG. 1, the thickness of the magnet holders is such that when all magnets are fitted and angled, there is not less than adequate thickness at any point in the support structure to support the magnetic forces without deforming.

[0061] In some embodiments, including one complete embodiment, in reference to FIG. 1, the spacing, angling, and insulation of magnets in the magnet structures, is based on ensuring engagement between the two magnet structures produces rotational force that is immediate, consistent, and continuous, including adequate insulation on the approach side of the magnets that eliminates repulsive force on approach so that the magnet structures will spin immediately when engaged. In some embodiments, which may include one complete embodiment, to control the non-linear nature of magnetic fields, the magnetic shielding on the approach side of each magnet may cover up to and including around the middle of the magnetic field. In some embodiments, including one complete embodiment, a pair of opposing magnetic fields from the embodiment in FIG. 1 is rotated for clarity and magnified in FIG. 1A to illustrate a rough approximation of directed magnetic field trajectories on the exposed side of the magnets. In some embodiments, including one complete embodiment, the magnets in the opposing magnet structures are insulated, spaced, and angled to optimize rotational consistency. In some embodiments, which may include one complete embodiment, all magnets, no magnets, or one structures magnets, may be spaced equally. In some embodiments, which may include one complete embodiment, each magnet in the inner magnet structure or the outer magnet structure is spaced to be engaged at an evenly spaced percent of maximum force. For example, if there are 5 magnets in the inner structure, they might be spaced with an approximate target of outer magnet structure engagement pattern as a percent of maximum force of 100% 80% 60% 40% 20% or 100% 75% 50% 25% 0%. In some embodiments, other means may be used to determine the spacing, angling, and insulation of the magnets to provide optimal rotation of the axle. In some embodiments, the shaped opposing magnetic fields provided for in this disclosure may instead be secured to the underside of a vehicle, and along a track, in corresponding magnetic field holders, with the magnetic fields angled, insulated, and directed in repulsion, to support vehicle propulsion and or lift.

[0062] In some embodiments, including one complete embodiment, in reference to FIG. 1, the magnets are inset within the magnet structures, where the inset is angled inward to secure each magnet in place, and additionally to adequately limit any potential repulsive forces inhibiting approach, the shielding and inset are of an amount that allows one or both of the opposing magnet structures when engaged to rotate continuously, optimally, and immediately. In some embodiments, including one complete embodiment, the minimum angle of the inset may be around 30 degrees and the length of distance of the inset covering the magnet surface on the non-approach side may be around a tenth of the magnet surface, and on the approach side may cover up to around half of the magnet surface. In some embodiments, which may include one complete embodiment, the magnet structures' inset size and angle on the non-approach side may be no more than is required to physically hold the magnet in place. In some embodiments, which may include one complete embodiment, the outsides of the magnet structures engagement approach sides have an angled inset to limit repelling forces as the structures move laterally to become initially engaged. In some embodiments, alternative means of securing the magnets may be used, including magnets with concave sides, with matching magnet holder shapes, where the midpoint of the curve may be of adequate depth to hold the magnets in place, or the magnets having slits matching slit holders on the structures. In some embodiments, other methods of insulating the magnets may be used. In some embodiments, other means may be used that adequately secures of the magnets within the magnet structures. In some embodiments, which may include one complete embodiment, to overcome any initial repulsive forces, and optionally to provide for smoother sustained rotation, 1400 is a motor connected to battery switch 1801 which provides power from battery 1800 to initiate rotation of the rotatable magnet structure.

[0063] In some embodiments, a weighted momentum wheel is attached to the rotatable magnet structure, or the rotatable magnet structure holds adequate weight to sustain momentum, and once the desired speed and force of axle rotation has been achieved, the engagement of the magnet structures may be reduced, optimizing magnetic field preservation, while sustaining force and speed.

[0064] In some embodiments, including one complete embodiment, in reference to FIG. 1, to secure the magnets against lateral slippage, magnet holders are designed to be structurally enclosed except on the magnetic field exposure side and the magnet insertion side, with covers attached to the insertion side of each magnet holder, matching the size and shape of the sides of the magnet holders. In some embodiments, including one complete embodiment, the magnet holder covers are 0.5 inches (13 millimeters) thick and made of the same material as the magnet holders, where magnet holder cover 1102 is attached to rotatable magnet holder 1100 through inner magnet holder with screw holes drilled through both, and secured with screws 1104 1105, and magnet holder cover 1103 is attached to rotationally fixed magnet holder 1101 with screw holes drilled through both and attached with screws 1106 1107. In some embodiments, including one complete embodiment, the magnet holding structures or magnet holder covers, on the engagement approach side of the magnet structures, have the same inset insulation of the magnet holders, to limit repulsive forces on initial engagement.

[0065] In some embodiments, including one complete embodiment, in reference to FIG. 1, the magnet holders and magnet holder covers may be made out of a strong non-magnetic material that may also provide magnetic field shielding, and or redirection, and or absorption. In some embodiments, including one complete embodiment, a non-magnetic metal may be used such as Aluminum, which may be in the form of high strength Aluminum 7075, which has a yield strength of around 20,000 pounds (9,070 kilograms) of pressure per square inch (645 square millimeters). In some embodiments, which may include one complete embodiment, non-metal 3D printed filament may be used, for example polythermide, where such filaments, at the time of disclosure, may have rated strengths between 4,700 pounds (2130 kilograms) and 16,500 pounds (7,480 kilograms) of pressure per square inch (645 square millimeters), but may diminish in strength under extreme environmental conditions. In some embodiments, which may include one complete embodiment, MuMetal is to be used, which is defined by United States Military Specification 1441C, as primarily Nickel with a yield rating of up 44,000 pounds (19,960 kilograms) of pressure per square inch (645 square millimeters). In some embodiments, titanium may be used, if it is adequately non-magnetic, since it is stronger than high grade steel but may currently be around eight times more expensive. In some embodiments, which may include one complete embodiment, to reduce the weight of the magnet structures, hollow areas may be left to the extent they do not detract from structural resilience. In some embodiments, including one complete embodiment, one or more layers of magnetic field shielding material is added around the magnets either directly or by being placed in magnet holder compartments, to shape the magnetic field in the previously explained manner, by limiting or preventing magnetic fields from emanating in undesired directions. If no additional magnetic insulation is used, the magnet holders may be of a material and thickness at all points that adequately shapes magnetic fields. In some embodiments, because magnetic fields may saturate materials and those materials diminish in insulation effectiveness, multiple materials may be layered within the magnet holding compartments to limit saturation. In some embodiments, including one complete embodiment, in reference to FIG. 1A, the magnets are enclosed by sheets 0.04 inches (0.1 millimeters) thick from the magnet outward of each iron + iron + tin + stainless steel + tin + aluminum, a layering combination that has been found to reduce external emission of permanent magnet fields by more than 90%, with additional layers added as needed until such shielding is adequate. In some embodiments, including one complete embodiment, the pattern to bend the insulation materials resembles that in FIG. 2, where the insulation materials may be formed using standard sheet metal fabrication tools, where insulation is provided by magnet bottom insulation 2000, magnet approach side insulation 2001, magnet approach side angled insulation 2002, magnet non-approach side insulation 2003, magnet non-approach side angled insulation 2004, magnet structure non-engagement side insulation 2005, magnet structure engagement side insulation 2006, magnet structure engagement side angled insulation 2007, and where each layer beyond the first will need to be appropriately extended to fit around the thickness of the previous layers. In some embodiments, any other type, thickness, pattern, and arrangement of insulation materials may be used that direct the magnetic fields in such a way that supports continuous rotation of either or both magnet holders. In some embodiments, which may include one complete embodiment, the insulation materials may only be on the approach sides of the magnets or magnetic fields. In some embodiments, any material or materials of any thickness may be used that allow the magnets or magnetic fields to provide optimal rotational force. In some embodiments, where there is residual magnetic resistance, the magnets may be inserted into the holders with power tools or inserted in smaller segments.

[0066] In some embodiments, which may include one complete embodiment, after having accounted for these details, specifications, and obtained components, molds consisting of an appropriate material may be made to cast the required components. The molds may be manually etched, or may be setup in CAD (Computer Aided Design) files, to then be processed by a CNC (Computer Numerical Control) machine, to etch each mold. While there are many versions of CNC machines, for many, the design process may allow for each part to be designed in a CAD program, which may export the design to the vendor neutral Initial Graphics Exchange Specification (IGES) format, developed by the United States Air Force for parts manufacturing, and may have the CNC machine's software convert the designs to G-Code instructions, which may control the movement of the CNC machine tools. The selected metal may then be melted and poured into the graphite molds. In some embodiments, including one complete embodiment, the components including the magnet holders may be 3D printed.

[0067] In some embodiments, which may include one complete embodiment, in reference to FIG. 1, because there are so many possible matches of commodity materials, components, and manufacturers, the best way to determine optimal output in practice may be for the primary design to be adjusted with varying spacing, angling, and insulation, with magnet holders created for each variation, and several different materials used, to determine which design provides optimal output force and rotational consistency with the purchased magnets and their corresponding magnetic fields. In some embodiments, which may include one complete embodiment, a gauss meter and a magnetic field visualizer are used to ensure the magnetic field is being insulated and shaped in an optimized manner, with corresponding adjustments made to the design and manufacture of corresponding components including the magnet holders and insulating materials.

[0068] In some embodiments, designs may be miniaturized, and produced using molds or 3D printing, to produce portable power units for embedding in portable consumer electronics such as phone and laptops, or to fit inside of the casing of standard batteries such as A, AA, AAA, C, D, or a watch batteries. If such unit sizes don't allow for full power needs to be met, embodiments can at least allow for self-recharging batteries.

[0069] **In some embodiments, which may include one complete embodiment, optionally add an electric field around the magnets.** In some embodiments, it may be desirable to wind uninsulated electrical wire around each magnet, and provide it with an electrical charge, to increase, and or maintain, and or restore magnetic directional force, where additional space may be created in the magnet holder to accommodate the wire, and an additional coating of a non-conductive material may need to be applied to the magnet holders' insulation or compartments to prevent the wires from sending electricity through the magnet holders, and the wiring is provided electricity from an attached generator or a separate power source. This may be beneficial in situations where the magnets must maintain directional force forever, or where extraordinarily high output is required, using minimal space and weight, such as in engines for jets or space vehicles while in atmosphere.

[0070] **In some embodiments, including one complete embodiment, insert magnets in rotatable and rotationally fixed magnet holders and secure with side magnet holder covers.** In some embodiments, including one complete embodiment, in reference to FIG. 1, the previously purchased and prepared magnets are each placed into each magnet holder's magnet compartments. In some embodiments, including one complete embodiment, in reference to FIG. 1, rotatable magnet holder 1100 receives 5 magnets 1200 1201 1202 1203 1204, and rotationally fixed magnet holder 1101 receives 12 magnets 1205 1206 1207 1208 1209 1210 1211 1212 1213 1214, with the inner and outer magnet holder magnets inserted and aligned to be magnetically repulsive, with the inner and outer magnet holder covers screwed on, where magnet holder cover 1103 is attached to rotatable magnet holder 1100 with screws 1104 1105 through threaded holes drilled through each, and magnet holder cover 1103 is attached to rotationally fixed magnet holder 1101 with screws 1106 1107 through threaded holes drilled through each. In some embodiments, the magnet holder covers may be secured by alternative means, including being bolted or welded in place.

[0071] **In some embodiments, including one complete embodiment, build the support structure of the unit.** In some embodiments, including one complete embodiment, in reference to FIG. 1, a support structure holds the axles that connect to the rotatable magnet structure, supports the sliding rotationally fixed magnet holder, and the remaining components, including a generator if implemented, where the support structure components are connected by means which may include welding. In some embodiments, which may include one complete embodiment, the support structure may be 3D printed or cast as a single component, which may additionally be an extension of the outer magnet holder 1101. In some embodiments, including one complete embodiment, in reference to FIG. 1, beams 1000 1001 1002 1003 are secured in a rectangle to form the base, with beam 1004 running through the center of the structure resting on beams 1000 and 1002 that has connected to it axle holders beams 1005 1006 with those beams having holes drilled in them to support bearings 1007 1009, through which the axle rotates. In some embodiments, which may include one complete embodiment, where a generator is used, in reference to FIG. 1, beam 1011 is attached to support beam 1002.

[0072] **In some embodiments, including one complete embodiment, connect the rotatable magnet holder to its axle.** In some embodiments, including one complete embodiment, in reference to FIG. 1, axle 1006 is inserted through rotatable magnet holder 1100 and welded into place, and in order to transfer force from the rotatable magnet holder as it rotates, rotatable magnet holder axle 1006 is slipped through its support beams 1005 1006 on either side which contain appropriately sized holes that may include a bearing, and rotatable magnet holder axle 1006 is coupled by means which may include welding to the axle of generator 1300, and or wheel or jet turbine 1400, and where to control the engagement of the rotatable and rotationally fixed magnet structures, rotationally fixed magnet holder 1100 has at its base rectangular ring 1012 around support beam 1004, which allows it to slide back and forth along the beam, utilizing handle 1013, to engage and disengage rotatable magnet holder 1101. In some embodiments, which may include one complete embodiment, axle 1006 is cast or 3D printed as an extension of magnet holder 1100, and axle support structures are cast or 3d printed as a continuous part of the support structure. The force and speed of the rotation may be primarily determined by the extent to which the magnet holders surround each other and the resistance of connected components. In some embodiments, the rotatable magnet holder axle may or may not be connected to additional axles or gears, as required by the application. In some embodiments, the rotatable magnet structure slides in and out of the rotationally fixed magnet structure, where the sliding axle of the rotatable magnet structure interconnects with fixed axles connected to the generator or the motor output axle.

[0073] In some embodiments, an engagement lever of the same material as the magnet structures moves the rotatable magnet structure 1100 in and out of a rotationally fixed magnet structure 1101 by connecting to and moving rotatable magnet structure axle 1006, where engagement lever 1014 holds onto a space on the axle 1006 between two extended axle perimeter extension rings, to move the axle without changing its speed, and is able to move back and forth.

[0074] **In some embodiments, including one complete embodiment, connect rotatable magnet structure axle to the generator and or allow it to function as a motor axle.** In some embodiments, including one complete embodiment when a generator is implemented, in reference to FIG. 1, generator 1300 is attached to the support structure beam 1011, by means which may include bolting or welding, with the axle of generator 1300 coupled to rotatable magnet structure axle 1006, by means which may include welding, to be able to receive rotational force from the axle connected to the rotating rotatable magnet structure, with the generator being of sufficient distance from the magnet structures to prevent magnetic interference, or may have placed before it additional magnetic shielding of the same type used for the magnet structures. In some embodiments, including one complete embodiment when implemented as a motor, a rotatable magnet holder axle 1006 functions as a motor axle by connecting, for example, on either side to tires or to a jet turbine blade, to provide propulsion.

[0075] **In some embodiments, which may include one complete embodiment, attach electronically controlled engagement controller.** The percent output of the embodiment is determined primarily by the rotatable magnet structure's percent engagement with the rotationally fixed magnet structure, and the resistance of connected components, with for example, full engagement producing maximum output, and no engagement in effect turning the unit off.

[0076] In some embodiments, which may include one complete embodiment, the engagement handle is pushed and pulled by linear actuator 1600, or a hydraulic piston and pump, which are wired to an external energy source, or to a generator and or a battery, with control provided by a computer controlled relay board and software, as described in a later step.

[0077] In embodiments where the unit is used as a vehicle motor, there may be a mechanical connection between the gas petal and brake pedal and the engagement handle, thus eliminating the requirement that a vehicle have any electrical power source to control propulsion. In some embodiments, the engagement handle may be controlled by any number of alternative means.

[0078] **In some embodiments, which may include one complete embodiment, add electric hydraulic piston with controller.** In some embodiments, an electric hydraulic piston 1500 powered by hydraulic pump 1501, and controlled by hydraulic pump switch 1502, allows the embodiment to be angled, for applications including vertical take off and landing of a vehicle including a car, airplane, or spaceship. If an electric hydraulic piston is added, it is attached to the same structure as the embodiment, with the embodiment connected to the vehicle with hinges so that it can be angled. In embodiments where the hydraulic fluid would be exposed to freezing temperatures, insulation may be provided around the hydraulics and or an anti-freeze agent may be added to the hydraulic fluid. The hydraulic pump may be attached to an external power source, or an internal battery and or generator.

[0079] **In some embodiments, which may include one complete embodiment, create computer control code.** In some embodiments, which may include one compete embodiment, the engagement handle, and or the vertical take off and landing hydraulic piston, may be computer controlled, with the engagement handle adjusted through a directly or indirectly motorized device connected the Ethernet controlled relay board controlled by and connected to through a Ethernet crossover cable, a computer included in the unit, that runs custom software on startup, where when operating the engagement handle, the axle rotation is continuously monitored, and the engagement handle adjusted to achieve the desired speed, and for automatically setting and maintaining unit rotation speed when used as a generator and or motor, where the speed of rotation of the axle of the unit may be monitored by means which may include a hall effect sensor is aligned with a magnet on the rotating axle, which counts the time elapsed between magnet detections to calculate revolutions per minute. In some embodiments, which may include one complete embodiment, in the instance of energy generation, to maintain optimal rotation speed over time, because the magnets may lose force over time, the engagement controller may be adjusted to maintain optimal speed, based on readings of rotations per minute provided to the generator. In some embodiments, the continuous reading of a digital power consumption meter may be done by computer code to adjust the engagement controller to provide output to match demand, and thus optimize preservation of the units' magnets. In some embodiments, the computer may be a Raspberry Pi set on startup to run software developed in the Python language from the disclosed description, where the software is installed by connecting to the computer through telnet or secure shell, then at the command prompt typing "nano run.py" and adding and saving the software code, then at the command prompt typing "nano /etc/rc.local" and adding and saving the line "python /$location/run.py", where $location is the path to the directory containing the previously created software file.

[0080] The software code to operate the computer connected relay board, in some embodiments, including one complete embodiment, provides functionality comprising:
> import a library to connect to the relay;

> import a library for accessing system resources;
> import a library for hardware access to read the revolutions per minute of the axle;
> import a library to pause program execution;
> import a library to connect to the power consumption meter;
> create a value holding whether or not the unit is to function as a generator;
> create a variable holding the maximum watt output capacity;
> create a variable holding the generator maximum revolutions per minute;
> create a variable holding the axle circumference;
> initialize a variable holding axle revolutions per minute;
> create a variable holding the time the unit was started;
> create variables holding the relay board connection, IP address, username, and password;
> create variables holding the relay board on off state values;
> create variables holding the state of each relay including the forward and backward engagement controller states, and the unit forward and backward states;
> create a variable indicating whether or not a power consumption meter is to be used;
> create variables holding the power consumption reader IP address, username, and password.
> create a function to update the relay board, first establishing a connection to the relay board, if it has not been initialized, or has been dropped, then creating a string sequence of relay states, and sending the states to the relay board.
> create a function to calculate the revolutions per minute at which the generator is being rotated;
>> calculate the time elapsed as the current time minus the time started;
>> set the time started to the current time;
>> calculate the distance the axle has rotated;
>> calculate the revolutions per minute.
> create a function to set the revolutions per minute of the unit by controlling the magnet structures engagement;
>> if the axle revolutions per minute is less than the desired revolutions per minute, set the engagement forward state to on for a moment;
>> if the axle revolutions per minute is greater than the desired revolutions per minute, set the engagement backward state to on for a moment.
> create a function to listen for user input;
>> ready in any user key press;
>> if the key pressed is the forward arrow, set the engagement backward state to off, and the engagement forward state to on;

>> if the key pressed is the backward arrow, set the engagement forward state to off, and engagement backward state to on;
>> if the key pressed is the up arrow, set the unit backward state to off, and unit forward state to on;
>> if the key pressed is the down arrow, set the unit forward state to off, and the unit backward state to on;
>> if the key pressed is the space bar, set all relay states to off;
>> if a key was pressed, call the function to send the relay states to the relay board.
> create the function called upon program start;
>> read in any command line parameters;
>> initialize a connection to the computer hardware, and register the rotational speed sensor.
>> create a function binding the speed calculation function to the rotational speed sensor.
>> continuously loop;
>>> proceed if functioning as a generator;
>>>> if the power consumption meter is used, retrieve the current power consumption from the power consumption meter, and call the function to increase or decrease engagement and thus rotational speed based on current consumption;
>>>> if the power consumption meter is not in use, run the unit at maximum output;
>>> listen for user input to control motor output and if received call the function to control engagement and thus rotational speed.

[0081] In some embodiments, where the engagement controller is computer controlled, Ethernet connected relay board 1700 is connected through an Ethernet crossover cable to computer 1709, while linear actuator 1600, which may utilize a reversible brushed direct current motor, in order to allow for its back and forth control, is wired to relays on the relay control board 1700 by wiring motor positive terminal to relay 1701 and relay 1703, and wiring motor negative terminal to relay 1702 and 1704, and hall effect sensor 1710 is aligned with magnet 1711 on the rotating axle 1006, with the hall effect sensor connected to computer 1709.

[0082] In some embodiments, where hydraulic piston 1500 and hydraulic pump 1051 control vertical take off and landing, motor 1504, which may be a reversible brushed direct current motor, is attached to hydraulic pump switch 1502, after having rod 1503 welded to its axle to push the buttons of the hydraulic pump switch 1502, and held in place by metal clamp 1504, with the wires from motor 1504 wired to positive terminal to relay 1701 and relay 1703, and wiring motor negative terminal to relay 1702 and 1704, or cut off the hydraulic pump switch and directly connect its wires to the relays in the pattern required to control operation.

[0083] **In some embodiments, including one complete embodiment, optionally create and attach a protective enclosure.** In some embodiments, including one complete embodiment, a protective enclosure 1900 houses all of the components, to protect the operator from the force of the embodiment and protect the electronic components from external elements, where the enclosure is made of sheets of a strong lightweight metal, such as aluminum, or a magnetically insulating material such as previously discussed layers of insulating material, or MuMetal, formed by means which may include fabrication or welding in the shape and dimensions of the support structure, and where the enclosure allows when appropriate, power connector 1800 to be externally accessible to the operator. In some embodiments, which may include one complete embodiment, the base of the enclosure may be manufactured as a continuous part of the support structure.

[0084] **In some embodiments, including one complete embodiment, enjoy clean continuous portable self-powered energy and or propulsion.** In some embodiments, including one complete embodiment, in reference to FIG. 1, there is no power switch because the unit doesn't require a battery to start, and so as soon as the unit is fully constructed, and the engagement handle appropriately engages the magnet structures, it provides clean free continuous self-powered energy and or propulsion. In some embodiments, when used to produce electrical power, once the unit is verified as correctly constructed, an experienced licensed electrician may attach the unit to a central power grid and engage the unit's engagement controller to produce optimal output. In some embodiments, when used to provide propulsion, mount the embodiment in the engine compartment, first removing any existing engine, and appropriately couple the embodiment to the vehicle and its controls to provide propulsion. In some embodiments, including one complete embodiment, should the magnets directional force ever degrade unacceptably, they may be switched for ones of acceptable directional force.

CLAIMS

What is claimed is:

1. Structures designed to be capable of acting as a motor to provide propulsion and or power an electricity generator, with the invention comprising:
   a first structure designed to hold magnetic fields, and designed to be able to rotate in opposition to a second structure designed to hold magnetic fields;
   said structures capable of directing magnetic fields in such a way as to allow said first magnetic field holding structure to rotate in opposition to said second magnetic field holding structure, with or without the assistance of insulation of said magnetic fields.
2. Claim 1 further comprising structures capable of directing magnetic fields utilizing one or more layers of material able to direct magnetic fields.
3. Claim 1 further comprising the ability to control the engagement of said structures.
4. Claim 1 further comprising a weighted structure attached to an axle to be able to retain momentum.
5. A method performed by an apparatus comprising:
   producing motion from repelling magnetic fields.
6. A method for constructing an apparatus comprising:
   arranging magnetic fields such that when a first set of magnetic fields is in opposition to a second set of magnetic fields motion is produced.
7. A structure designed to be able to direct magnetic fields, with the invention comprising:
   one or more layers of material able to direct magnetic fields.
8. A non-transitory computer-readable recording medium holding stored instructions, which when executed by one or more processing devices, cause the one or more processing devices to implement a method comprising:
   detecting the current rotational speed of a rotating magnetic field structure, then to achieve a desired speed, increasing or decreasing engagement of the magnetic field structures by utilizing an electronic controller.
9. A non-transitory computer-readable recording medium holding stored instructions, which when executed by one or more processing devices, cause the one or more processing devices to implement a method comprising:
   reading a power consumption meter, and or a utilizing a desired rotational speed, then adjusting the magnet structures engagement to support the desired power output or rotational speed.

FIG. 1

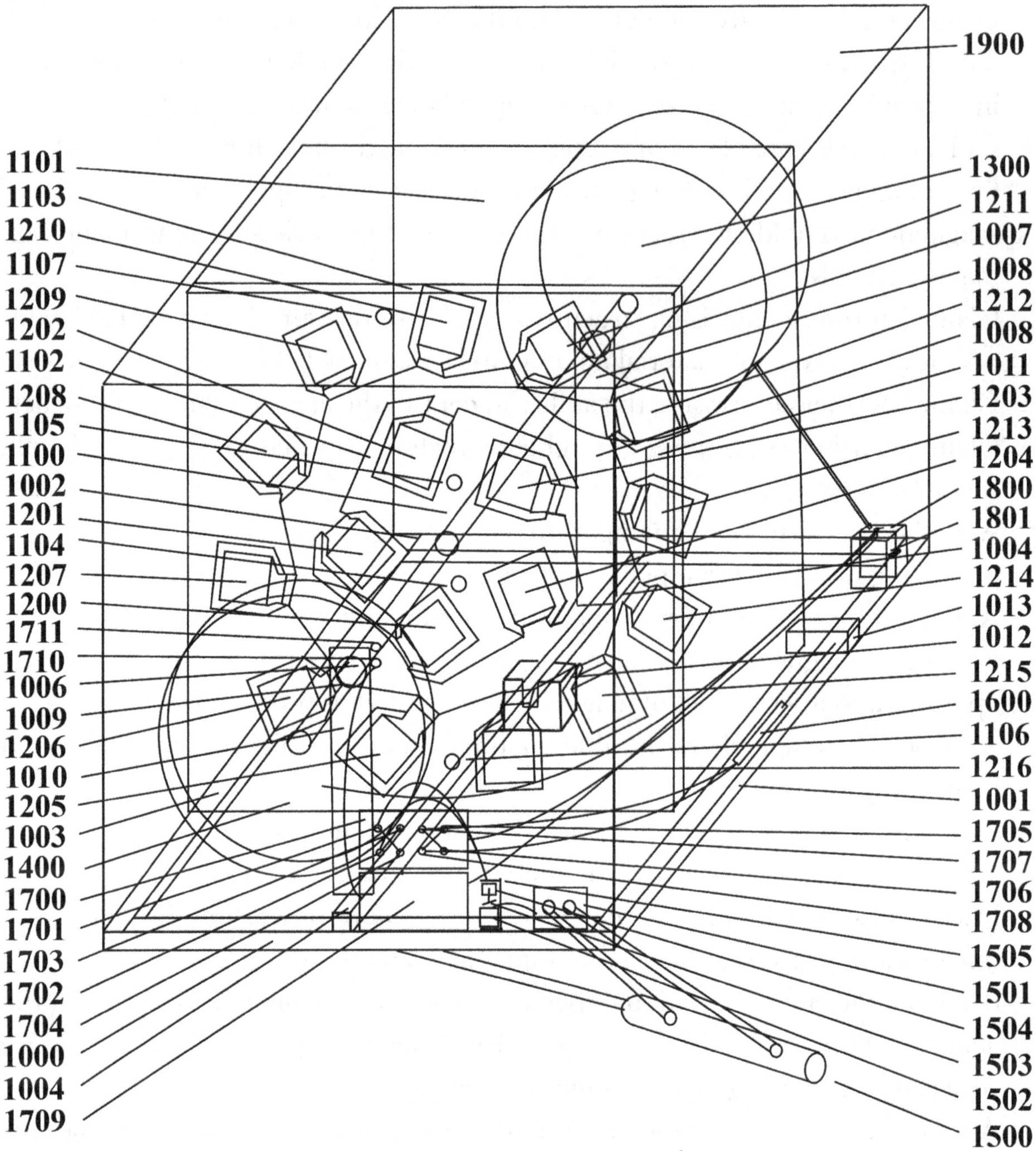

Jonathan Bannon Maher

FIG. 1A

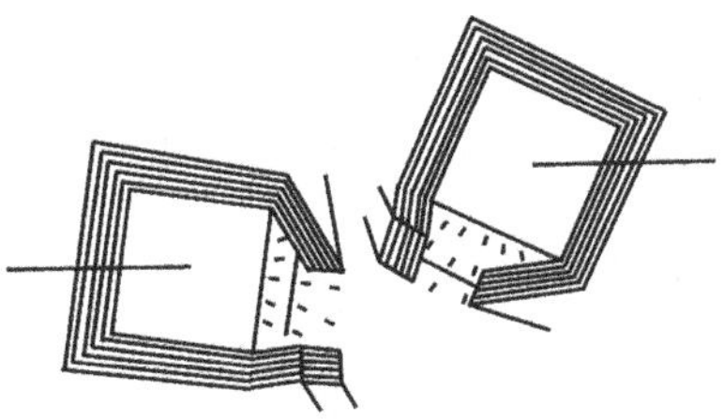

Jonathan Bannon Maher

FIG. 2

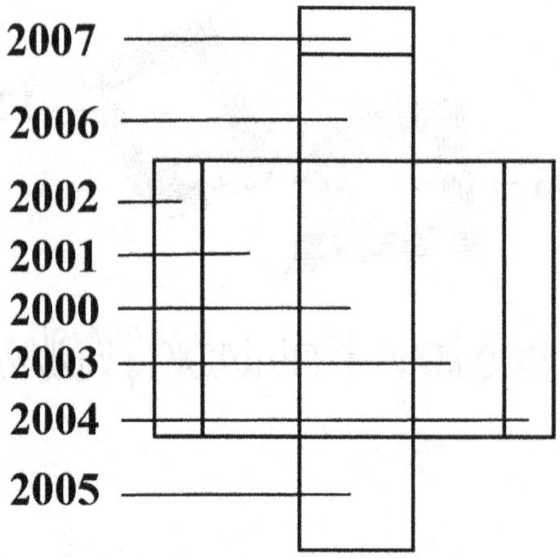

Jonathan Bannon Maher

FIG. 3

3000 – import a library to connect to the relay

import telnetlib

3001 – import a library for accessing system resources

import sys

3002 – import a library for hardware access to read the revolutions per minute of the axle

from RPi import GPIO

3003 – import a library to pause program execution

import time

3004 – import a library to connect to the power consumption meter

import urllib

3005 – create a value holding whether or not the unit is to function as a generator

generator = True

3006 – create a variable holding the maximum watt output capacity

generator_maximum_watts = 10000

3007 – create a variable holding the generator maximum revolutions per minute

```
generator_revolutions_per_minute = 1800
```

3008 – create a variable holding the axle circumference

```
axle_circumference_centimeters = 6
```

3009 – initialize a variable holding axle revolutions per minute

```
axle_revolutions_per_minute = 0
```

3010 – create a variable holding the time the unit was started

```
unit_time_started = time.time()
```

3011 – create variables holding the relay board connection, IP address, username, and password

```
relay = None
relay_ip = "169.254.1.1"
relay_username = "admin"
relay_password = "admin"
```

3012 – create variables holding the relay board on off state values

```
off = 0
on = 1
```

3013 – create variables holding the state of each relay including the forward and backward engagement controller states, and the unit forward and backward states

relay_1_engagement_forward_state = off
relay_2_engagement_backward_state = off
relay_3_unit_forward_state = off
relay_4_unit_backward_state = off

3014 – create a variable indicating whether or not a power consumption meter is to be used

power_meter = False

3015 – create variables holding the power consumption meter IP address, username, and password

power_meter_ip = "156.254.1.2"
power_meter_username = "admin"
power_meter_password = "admin"

3016 – create a function to update the relay board, first establishing a connection to the relay board, if it has not been initialized, or has been dropped, then creating a string sequence of relay states, and sending the states to the relay board

```
def update_relay():

    if not relay:
        relay = telnetlib.Telnet(relay_ip, 23)
        relay.read_until(b"User Name: ")
        relay.write(
            relay_username.encode('ascii') + b"\n")
        relay.read_until(b"Password: ")
        relay.write(relay_password.encode('ascii') + b"\n")

  command =
    str(relay_1_engagement_forward_state)
  command+=
```

```
    str(relay_2_engagement_backward_state)
  command+=
    str(relay_3_unit_forward_state)
  command+=
    str(relay_4_unit_backward_state)
  relay.write(
    command.encode('ascii') + b"\n")
```

3017 – create a function to calculate the revolutions per minute at which the generator is being rotated

```
def calculate_speed(channel):
```

3018 – calculate the time elapsed as the current time minus the time started

```
  time_elapsed = time.time() – unit_time_started
```

3019 – set the time started to the current time

```
  unit_time_started = time.time()
```

3020 – calculate the distance the axle has rotated

```
  distance_kilometers =
    axle_circumference_centimeters/100000
```

3021 – calculate the revolutions per minute

```
  axle_revolutions_per_minute =
    (distance_kilometeres/time_elapsed) * 3600
```

3022 – create a function to set the revolutions per minute of the unit

```
def set_speed(revolutions_per_minute):
```

3023 — if the axle revolutions per minute is less than the desired revolutions per minute, set the engagement forward state to on for a moment

```
    if axle_revolutions_per_minute * 1.01 <
        revolutions_per_minute:
        relay_1_engagement_forward_state = on
        update_relay()
        time.sleep(.1)
        relay_1_engagement_forward_state = off
        update_relay()
```

3024 — if the axle revolutions per minute is greater than the desired revolutions per minute, set the engagement backward state to on for a moment

```
    if axle_kilometers_per_hour * 0.99 >
        kilometers_per_hour:
        relay_2_engagement_backward_state = on
        update_relay()
        time.sleep(.1)
        relay_2_engagement_backward_state = off
        update_relay()
```

3025 — create a function to listen for user input

```
def listen():
```

3026 — ready in any user key press

```
    key_pressed = ord(sys.stdin.read(1))
```

3027 — if the key pressed is the forward arrow, set the engagement backward state to off, and the engagement forward state to on

```
if key_pressed == 37:
    relay_2_engagement_backward_state = off
    relay_1_engagement_forward_state = on
```

3028 — if the key pressed is the backward arrow, set the engagement forward state to off, and engagement backward state to on

```
if key_pressed == 37:
    relay_1_engagement_forward_state = off
    relay_2_engagement_backward_state = on
```

3029 — if the key pressed is the up arrow, set the unit backward state to off, and unit forward state to on

```
if key_pressed == 38:
    relay_4_unit_backward_state == off
    relay_3_unit_forward_state == on
```

3030 — if the key pressed is the down arrow, set the unit forward state to off, and the unit backward state to on

```
if key_pressed == 40:
    relay_3_unit_forward_state == off
    relay_4_unit_backward_state == on
```

3031 — if the key pressed is the space bar, set all relay states to off

```
if key_pressed == 32:
    unit_forward_state == off
    unit_backward_state == off
    relay_1_engagement_forward_state = off
    relay_2_engagement_backward_state = off
```

3032 – if a key was pressed, call the function to send the relay states to the relay board

```
if key_pressed:
    update_relay()
```

3033 – create the function called upon program start

```
def __main__():
```

3034 – read in any command line parameters

```
  parameter = sys.argv
```

3035 – initialize a connection to the computer hardware, and register the rotational speed sensor

```
  GPIO.setwarnings(False)
  GPIO.setmode(GPIO.BCM)
  GPIO.setup(sensor, GPIO.IN,
  pull_up_down=GPIO.PUD_UP)
```

3036 – create a function binding the speed calculation function to the rotational speed sensor

```
  GPIO.add_event_detect(sensor, GPIO.FALLING,
    bouncetime = 15, callback = calculate_speed)
```

3037 – if functioning as a generator, continuously loop

```
  if generator:
    while True:
```

3038 – if the power consumption meter is used, retrieve the current power consumption from the power consumption meter, and call the function to increase or decrease engagement based on current consumption

```
if power_meter:
    power_meter_address =
        "http://" + power_meter_ip +
        "/?username=" +
        power_meter_username +
        "&password=" +
        power_meter_password +
        "&command=consumption"
        power_meter_consumption =
        urllib.open(
        power_meter_address).read()
        consumption_kilometers_per_hour =
         (power_meter_consumption/
        generator_maximum_watts) *
        generator_maximum_kilometers_per_hour
        set_speed(consumption_kilometers_per_hour)
```

3039 – if the power consumption meter is not in use, run the unit at maximum output

```
if not power_meter:
    set_speed(generator_revolutions_per_minute)
```

3040 – if the generator is not in use, listen for user input to control motor output

```
if not generator:
    while True:
        listen()
```

SELF-POWERED MOTOR AND GENERATOR

INVENTOR JONATHAN BANNON MAHER

TECHNICAL FIELD

[0001] Embodiments of the invention relate to the fields of motors, generators, physics, engineering, and programming.

ABSTRACT

[0002] Systems, methods, apparatuses, and in some embodiments computer programs encoded on a computer storage medium, provide for clean continuous portable self-powered energy generation and propulsion, consistent with the laws of physics, by in some embodiments, including one complete embodiment, transferring force generated through hydraulics and or pneumatics and or mechanical leverage and or motorized mechanical leverage, to cause the flow of a medium, to spin a turbine, to power an electricity generator, and or rotate an axle to provide the propulsion of a traditional motor, where energy may be captured in excess of that consumed as a result of the differential between input force required and output force provided by certain force providing device configurations, as well as a result of the differential between the linear force used to elevate the medium and the compounding force provided by gravity during medium drop. The invention permanently solves global warming, provides reduced cost of living to alleviate poverty, provides unlimited clean energy for evaporated water purification and atmospheric carbon dioxide splitting, and eliminates the need for every other method of energy production, including nuclear technology – thus reducing nuclear weapons technology proliferation.

BACKGROUND

[0003] The majority of the proceeds from the licensing of this patent will be going to causes that support the well being of humanity, and your support in ensuring the patent is forever in every way as strong as possible, will be providing a service to all the world.

[0004] This section is intended to introduce the reader to various aspects of the art that may be related to various aspects of the present techniques, which are described and or claimed. This discussion is believed to be helpful in providing the reader with background information to facilitate a better understanding of the various aspects of the present disclosure. Accordingly, it is understood that these statements are to be read in this light, and not a citation of any prior art.

[0005] Patent filings on structurally differentiated and fundamentally deficient disclosures may exist that may attempt to claim any invention that is self-powered, based on previously publicly known failed attempts to build such devices, however any such disclosures do not enable the purported inventions, with overly broad claims not supported by the disclosure that also fail to distinctly claim the invention, and any such patent filings are inherently invalidated by prior public disclosure, the enablement requirement, and the claims support requirement.

[0006] Known and proposed energy production, transmission, and storage systems have some or all of the following deficiencies:

[0007] 1. External fuel source required: an external fuel source is utilized such as oil, gas, coal, wind, sun, water currents, geothermal heat, hydrogen, or uranium, where the cost of providing fuel in the form of electricity or gasoline to a vehicle over its useful life potentially exceeds the cost of the vehicle, and about a quarter of airline costs are from fuel.

[0008] 2. Environmentally unfriendly: nuclear energy production, including fission, fusion, and cold (LENR), results in toxic waste and or materials, geothermal often circulates contaminants from the ground, fracking creates toxic water, hydroelectric dams decompose organic matter producing the potent global warming gas methane, solar panel manufacturing often releases toxic byproducts including greenhouse gases far more potent and long lived than those from fossil fuels, hydrogen takes substantially more energy to produce and transport than it provides, while clean energy sources may utilize slowly degrading flammable toxic batteries to store energy for when the sun is not shining, wind is not blowing, or water is not adequately flowing.

[0009] 3. Intermittent: wind, solar, and traditional water energy systems provide variable output, with average output often found to be around 20% of rated output, as a result of environmental conditions, meaning a 5 kilowatt system typically produces average output of only 1 kilowatt.

[0010] 4. Not portable: solar panels can only function in the sun, wind turbines in wind, water turbines in water currents, nuclear in a stable highly controlled environment, and carbon with the aid of an emissions pipe.

[0011] 5. Extremely expensive transmission costs: power lines are required for all forms of nuclear energy, including fission, fusion and cold, for farms of solar, wind, and water, and for fossil fuels plants, while pipelines are generally required for oil and gas, including natural gas, which is principally methane, a greenhouse gas more than twenty times more potent than carbon dioxide that may be leaked during extraction, transport, and consumption. Power line and fossil fuel pipe line cost of installation per 1 mile (1.6 kilometers) has been found to be up to around 20 times the average annual income in the United States, and lines must be replaced every 30 to 50 years, so in the United States alone, with 300 thousand miles (480 thousand kilometers) of power lines, and 200 thousand miles (320 thousand kilometers) of pipelines, replacement would require an expenditure around 25 times the national debt, passed on to consumers, and dragging down the economy, with every other developed nation in a similar situation. Power lines require environmental destruction during installation, leave visible blight, are forever vulnerable to cyber attacks, transmit power from central sites that are inherently more prone to failure and blackout than a decentralized system, dissipate power during transmission, with high voltage power lines having health consequences for those living nearby.

[0012] 6. High initial costs: in addition to previously cited expense of power lines, fossil fuel pipes, and ongoing fuel costs, clean energy farms require an allocation of land, and degrading batteries requiring periodic replacement, which is why tax credits are often required for clean energy systems to be affordable. In addition to those factors, comparing a 1 kilowatt rated clean energy system to a 1 kilowatt rated traditional system, may require multiplying the cost of the clean energy system by approximately 10 times, 5 times to account for enough electricity generation to be stored for the equivalent continuous output, and 5 times for the battery storage.

[0013] 7. Vulnerable to weather: wind turbines, hydroelectric dams, and solar panels, can be made ineffective by environmental conditions such as freezing temperatures, snow, and rain, and similar to power lines, may be taken down by extreme weather and lightning strikes.

[0014] 8. Vulnerable to black outs: power transmitted over power lines creates vulnerability for critical facilities such as hospitals and data centers.

[0015] 9. Vulnerable to cyber attack: utility scale energy systems often require a hackable computer to operate, and are therefore forever vulnerable to computer viruses able to take down and or destroy nuclear power plants and the electrical grid, even if such systems aren't connected to the Internet, as demonstrated by the Stuxnet virus.

[0016] 10. Causes deaths: plants, rigs, and pipes, for current and proposed forms of nuclear, hydrogen, gas, and oil energy can explode, and coal mines can collapse, while wildlife is killed by wind, ocean, wave, and river turbines, hydroelectric dams, solar condensers, solar panel and battery manufacturing byproducts, and nuclear waste.

[0017] 11. Encourage nuclear weapon proliferation: fission, fusion, and cold (LENR) nuclear energy are or can be one step from weaponizable, while hydrogen can be obtained by a terrorist at a hydrogen fuel station to create a powerful compressed hydrogen explosion, as verified by reviewing a video of a balloon filled with hydrogen being lit on fire. Fusion is particularly disturbing, as a fusion weapon could be created with the power of an exploding star, able to take out the planet, and resulting in an extinction level event for humans, a scenario even more likely when considering increasingly autonomous – and therefore inevitably hackable by individuals – weapons control.

BRIEF DESCRIPTION OF THE DRAWINGS

[0018] Illustrations are presented by way of example, and not by way of limitation, where some embodiments may not contain all components, may contain additional components, and may contain functionally similar components.

[0019] FIG. 1 is an illustration of an example of embodiment of the invention, which utilizes hydraulics and mechanical leverage operating to circulate a medium through piping between two containers on a levered support structure, to rotate a turbine to provide rotational force to an electricity generator axle to generate electricity and or to rotate or function as a motor axle.

[0020] FIG. 2, containing some components in FIG. 1, is an illustration of an example of an embodiment of an automated hydraulic pump and cylinder whose piston operates another hydraulic pump handle to operate its corresponding cylinder piston to improve input output efficiency.

[0021] FIG. 3 is an illustration of an example of an embodiment of a traditional electric hydraulic pump and cylinder which in some embodiments may be used in place of the converted manual to automatic hydraulic pump and cylinder in FIG. 1.

DETAILED DESCRIPTION

[0022] The disclosure is related to the field of clean continuous portable self-powered energy and propulsion. It is understood that any reference to a person skilled in the art, recognizes that at the time of filing, there is no one else skilled in the art of this particular field, or a closely related field. Given the extraordinary nature of the disclosure, regardless of how full, clear, concise and exact the disclosure in enabling the production and use of embodiments of the disclosure, what could be construed to be undue experimentation during production and use, is simply the ordinary effort required in the assembly and use of an embodiment of such a disclosure.

[0023] It is understood that, as in any engineering or design project, the development of any actual implementation will include numerous implementation specific decisions made to achieve the developers' specific goals, such as compliance with business related and system related constraints, which may vary from one implementation to another. It is understood that such a development effort might be complex and time consuming, but is nevertheless a routine undertaking of design, fabrication, and manufacture for those skilled in the art having the benefit of this disclosure. The disclosed steps may be read as prefaced by "In some embodiments, including one complete embodiment, ", may be executed or performed in other orders or sequences, and are not limited to the order and sequence shown and described, which are provided to enable ease in constructing an embodiment, and along with each components of each step, may be removed, modified, combined, or rearranged, and other steps and or step components may be added, without departing from the scope of this disclosure and or invention. Although embodiments of the invention have been described and illustrated in the disclosed implementations, it is understood that the present disclosed subject matter, including apparatuses, methods, specification, and illustrations, has been made only by way of example, not by way of limitation, and the methods and apparatuses may be used in other systems, and that numerous changes and optimizations in the details of implementation of the invention and or embodiment are made without such modifications departing from the spirit and scope of this disclosure and or embodiments of the invention. Although the disclosure has been shown and described with respect to one or more embodiments, features of the disclosed embodiments can be combined and rearranged in various ways, and changes including equivalent alterations, substitutions, modifications, and additional efficiencies will of course occur to someone of ordinary skill in the art without departing from the spirit and scope of this disclosure and or invention. In particular regard to the various functions performed by the described components, the terms used to describe such components are intended to correspond, unless otherwise indicated, to any component which performs the specified function of the described component, or is functionally equivalent to the described component, even though not structurally equivalent to the disclosed structure which performs the function in the implementations described in this disclosure. In addition, while a particular feature of the disclosure may have been provided with respect to only one of several embodiments, such feature may be combined with one or more other features of other embodiments as may be desired and advantageous for any given or particular application. In some instances, well-known circuits, structures and techniques have not been shown in detail in order not to obscure the understanding of this disclosure. Articles in this disclosure such as "a" "an" and "the" may allow for both singular and plural forms. Verbs in this disclosure such as "is" may be read as "may be". Conjunctions in this disclosure such as "or" as used herein may be interpreted as inclusive or meaning any one or any combination, where "A, B or C" means "any of the following: A; B; C; A and B; A and C; B and C; A, B and C". Relational terms in this disclosure, for example first and second, top and bottom, left and right, are to distinguish one entity or action from another, and may not necessarily require or imply a relationship, or order between, such entities or actions. The disclosure includes the best mode contemplated by the inventor, a completely described specific embodiment, along with optional components and alternative embodiments to best suit the implementer, measurements in imperial and metric units to support universal understanding, and dramatically exceeds claims support requirements and enablement requirements by allowing for selection and or construction of the required components to be carried out easily, quickly, and routinely by persons of ordinary skill in the art, who are provided the additional benefit of utilizing readily available commodity components whenever possible. The present disclosure includes material protected by copyrights, and the owner of the copyrights hereby reserves all rights, but with authorization for publication as required by government patent offices. Various embodiments of the present invention may provide all, some or none of the disclosed technical advantages.

[0024] The computer code descriptions disclosed, in order to provide comprehensive enabling disclosure, rather than utilizing flow charts, which according to Patent Cooperation Treaty 11.11a are prohibited from containing "text matter, except a single word or words, when absolutely indispensable, such as... a few short catchwords indispensable for understanding", are provided in a text only format where the number of arrows preceding a line indicate logical block level, semicolons indicate a new segment of a logical block, and periods indicate the closure of one or more logical blocks. It is understood that any computer code representations in this disclosure are merely illustrative, rather than restrictive. While code may be written in nearly any computer language, including Java and C++, the illustrative computer code descriptions were derived from code written the Python language, which may be run through the Python interpreter, with appropriate supportive libraries, which at the time of disclosure, may run on nearly any computer, for example one with an Intel or AMD processor, running a current version of Linux, Windows, or Mac OS. All code components may read as if prefaced by "In some embodiments, including one complete embodiment, ". In some embodiments, functionality may be modified, rearranged, excluded, and added. To provide more fundamental computer system details, in some embodiments, the functionality associated with the disclosed computer code descriptions may be referred to as a script, module, software, software application, or code, and can be written in any form of language, including compiled, interpreted, declarative, or procedural, able to be deployed in any form suitable for use in a computing environment, including as an independent or integrated program, module, component, or subroutine, for execution by the computer system, implemented on one or more independent or integrated computers, utilizing a central processing unit in the form of one or more general or special purpose microprocessors, in conjunction with digital electronic circuitry, which may include special purpose logic circuitry such as a field programmable gate array or application specific integrated circuit, with the computer controlled by and operatively coupled to tangibly embodied software and or firmware, which may include code that creates an environment for code execution, including individual or combined use of processor firmware, a protocol stack, a database management system, and an operating system, where such software and or firmware may exist in one or more parts in memory on one or more computers, and is encoded on one or more tangible non transitory software carriers, such as individual or combined use of a random or serial access device or substrate, a semiconductor memory device, transient or persistent random access memory, a magnetic, magnetic optical, or optical disk, or encoded on an artificially generated transmitted signal, for example, optical, electrical, or electromagnetic, transmitted using a sending and a receiving apparatus, where the interaction between the user and the software may be implemented by operatively coupling, to the local implementing computer, or a local computer connected to one or more remote computers through a local or wide area network, a display device which may implement liquid crystals or light emitting diodes, a keyboard, and a pointing device.

[0025] The inventor retains absolutely no liability for any implementation of this invention, and the invention is implemented exclusively at the risk and liability of the implementer.

[0026] In some embodiments, for quality control purposes, all components may be manufactured from scratch.

[0027] Calculations, formulas, and specific units are not in any way restrictive, are not be relied upon, are not required as presented to produce an embodiment of the invention, are provided exclusively as a courtesy to enhance enablement for those resizing components and or constructing alternative embodiments, and may contain inaccurate assumptions easily modified during practice, with all calculations utilized to select and estimate components and unit output being rough estimates that may vary greatly based on factors that include the type and quality of purchased and or manufactured components and embodiment construction, where embodiments may be constructed utilizing an effectively endless range of output and component configurations and selection processes.

[0028] Embodiments of the invention provide some or all of the following benefits over previously discussed predecessors:

[0029] 1. Self-powered for free output: embodiments provide the first energy and motor system in the history of the known universe to not require an external fuel source. Embodiments can produce endless energy, and provide endless transportation range, until there is a system failure, and therefore may potentially produce energy and propulsion until gravity driven orbital drift causes the Earth to be consumed by the Sun in a few billion years – assuming the units and humans are still on Earth. Embodiments reduce transportation costs by allowing implementing vehicles to operate without fuel, allowing for effectively free endless transportation range after purchase, and additionally allow for travel by supersonic jets and flying cars, which have been impractical principally as a result of fuel costs.

[0030] 2. Clean: the manufacture and use of embodiments produces no notable harmful environmental byproducts, nor the potential for deaths associated with predecessors.

[0031] 3. Continuous: embodiments produce electricity and propulsion that is continuous and stable.

[0032] 4. Portable: embodiments are able to function as well in a basement closet as in a car.

[0033] 5. Cyber attack proof: embodiments are self-contained thus require no hackable computer to operate and are therefore immune to computer viruses.

[0034] 6. Blackout proof: embodiments are designed to be kept indoors and on-site, and are thus ideal for critical facilities such as hospitals and data centers that can't afford a blackout from failed power lines or plants.

[0035] 7. Inexpensive: embodiments can be manufactured and operated at the lowest cost total cost possible, because they don't require fuel, installation and maintenance of power lines, an allocation of land, or degrading batteries.

[0036] 8. Weatherproof: embodiments are self-contained for indoor use and therefore aren't vulnerable to environmental factors such as freezing temperatures, snow, rain, lightning strikes, or extreme weather events.

[0037] 9. Eliminates energy output storage: because additional embodiments of this system can be utilized at peak times with limited cost, storage of energy is no longer relevant, for either utilities or homes, even for peak output needs.

[0038] 10. Eliminates expensive power lines and fossil fuel pipe lines: because embodiments are designed to be kept on site, and any number of units can be utilized to meet peak power needs, power and fossil fuel transmission lines and their associated costs are now rendered irrelevant, thus substantially unburdening all economies globally of associated costs. Land currently holding power lines, as well as arrays of solar panels, wind turbines, and hydroelectric turbines, can be reclaimed to reduce visual pollution and make space for a growing population. Roofs of solar panels can be removed to allow for roof tiles that reflect heat to maintain a cool house in summer. No thinking person will ever want, nor could a functional government allow, any type of nuclear reactor – fission, fusion, cold (LENR) – in a car or home, leaving only now irrelevant power lines for transmission, thereby making those sources wholly irrelevant.

[0039] 11. Potentially profitable: in a standard home use scenario, embodiments may be the only way for a unit owner to make a profit from selling energy back to the utility at wholesale rates.

[0040] 12. Alleviates poverty: because energy is effectively free after embodiment purchase, and the purchase price is less per unit of output than other energy systems, embodiments reduce the cost of living and the cost of goods for every person on Earth, thus reducing poverty.

[0041] 13. Powers water purification and pumping: 1 in 10 people live without access to clean water, while climate change driven droughts fuel conflicts. Because embodiments make energy nearly free over their useful lives, the energy intensive nature of evaporated water purification — which removes nearly every contaminant with a higher boiling point than water, and potentially all others can be removed with a standard carbon filter and ultraviolet light – is no longer a barrier, nor is pumping, thus embodiments provide for the global resolution of clean water needs for individual consumption and agriculture.

[0042] 14. Powers reduced water consumption: effectively eliminates the energy cost of operating electricity powered showers that require only a cold water pipe, and recirculate, filter to potentially cleaner than direct from pipe, and heat water, to provide exact continuous temperature and pressure control, as well as powering low voltage electric showerheads that mix air with water to provide the effect of the same output using dramatically less water.

[0043] 15. Powers atmosphere cleaning: because an existing specialized laser can disassociate atmospheric carbon dioxide molecules into carbon molecules and oxygen molecules, and because embodiments can provide effectively unlimited and continuous clean energy, a power source is now available to reduce carbon dioxide in the air, and potentially other greenhouse gasses, if corresponding devices are developed.

[0044] 16. Reduces nuclear and hydrogen weapon proliferation: because embodiments eliminate any need for any type of nuclear energy, including fission, fusion, and cold (LENR), each of which is or may be weaponizable and one step from nuclear weapons technology, and also eliminate the need for hydrogen, which a terrorist can obtain at a hydrogen fuel station to create a powerful compressed hydrogen explosion verifiable by watching a video of a balloon filled with hydrogen being lit on fire, I have provided us all a fundamentally safer world.

[0045] 17. Therefore, a few applications of embodiments include powering: all transportation vehicles including automobiles, trains, jets, spaceships, cargo ships, cruise ships, tugboats, boats, submarines, hover boards, jet packs, including in vertical take off and landing configurations; powering electrical grids as well as homes, offices, factories, hospitals, and data centers that would like to disconnect from external power sources to end their recurring bill, be permanently immune from blackouts, use clean energy, and save money; televisions, washing machines, dishwashers, showers, and water pumps; portable consumer electronics, such as phones and laptops, through an internally installed miniaturized embodiment eliminating the need to recharge; personal rapid transport; home hydroponic production systems, including light, temperature control, and nutrient water circulation; spaceship electro magnetic ion drives using a fraction of the fuel of traditional rockets; high intensity laser powered solar sails, with the laser powered by a large number of these energy units on the surface of any space based body with a limited atmosphere; space tourism; video streaming planetary sampling probes journeying an unlimited number of years into the universe, recharging by utilizing an embodiment when resting on a mass producing gravity or in conjunction with a device creating artificial gravity; space colonies; and inter galactic travel.

[0046] **In some embodiments, including one complete embodiment, obtain medium containers.** In some embodiments, including one complete embodiment, the principal limiting factor in providing output is medium flow to rotate the turbine to rotate a generator axle or function as a motor, and therefore the principal limiting factor in providing output is ultimately the physical size of the medium containers, and so they may be selected first, and then supporting components and output determined by working backwards from the size of the medium containers. In some embodiments, including one complete embodiment, the medium to be used is water, which allows the medium containers to be easily filled on site, thus reducing shipping costs while allowing for the medium containers to be emptied at any time without causing environmental harm. In some embodiments, the medium to be used is a liquid metal, for example zinc which is over six times as heavy as water, to allow for faster unit cycles while utilizing smaller containers. In some embodiments, the medium may be any other liquid, though when accounting for safety and durability, a non-corrosive non-toxic medium may be used. In some embodiments, if the unit is to be run outside, in temperatures below freezing, a liquid with a lower freezing point than water may be used, or an anti-freeze agent may be added. In some embodiments, the medium may be a non-liquid, such as metal balls. In some embodiments, any medium that allows for rotational force to be generated may be used. In some embodiments, components including the medium containers are selected to provide optimal flow to the selected turbine.

[0047] In some embodiments, including one complete embodiment, two low profile medium containers are used to hold the transferrable medium, secured to the ends of a support platform, and are raised and lowered on a cycle that allows each to become approximately either full or empty each cycle. In some embodiments, the medium containers may be constructed from scratch, where a material such as sheets of polyethylene or stainless steel may be welded together, with each wall of adequate thickness so as not to deform over time. In some embodiments, where the medium is water, to approximately calculate the dimensions of the containers based on medium volume, utilize calculations accounting for water having a volume of 1 cubic foot per 7.48 gallons (1 cubic meter per 1000 liters), where the total volume of water in cubic feet (meters) may be spread across any combination of tank length, width, and height. In some embodiments, which may include one complete embodiment, the height of the tanks may be reduced while extending the widths and maintaining or reducing the lengths to reduce the raising and lowering distance and cycle time and thus increase output efficiency. In some embodiments, an alternative container or conduit for circulating a medium in the system may be used, including a water wheel.

[0048] In some embodiments, including one complete embodiment, the containers are two 1,000 gallon (3,790 liter) low profile polyethylene tanks, each having a volume of 135 cubic feet (1,000 gallons / 7.48 cubic feet per gallon) (3.83 cubic meters), spread across dimensions of 1.5 feet (0.5 meters) high by 9 feet (2.7 meters) wide by 10 feet (3 meters) long, with each weighing approximately 400 pounds (181 kilograms). In some embodiments, which may include one complete embodiment, to reduce the raising and lowering distance and cycle time, alternative sizes may be used, for example, 1 foot (0.3 meters) high by 1 foot (0.3 meters) long by 134 feet (41 meters) wide. In some embodiments, medium containers may be comprised of countless other dimensions, volumes, and materials.

[0049] In some embodiments, which may include one complete embodiment, in order to maintain continuous output, medium flow interruption is prevented during medium container transition by emptying the elevated outer medium container into a center top medium container before transitioning, therefore providing continuous medium flow to the drop pipe and thus the turbine during outer medium container transition, with a center bottom medium container receiving the medium flow, after it passes the turbine, before it flows into the lowered outer medium container, and where the flow between the center and outer medium containers may be at a higher rate than between the top center medium containers and the turbine, to allow for the outer tanks to quickly exchange the medium with the center medium containers. However, the addition of center medium containers may make self-construction of embodiments in extreme poverty situations prohibitively complex, as well as making the embodiment of greater height, thus consuming more space which may not be readily available indoors, and therefore attaching an uninterruptable power supply to the unit may be the simplest option for continuous output, though utilities may not require continuous output because multiple embodiments operating simultaneously in a staggered manner can provide continuous output.

[0050] In some embodiments, to provide a specific horsepower at a specific number of revolutions per minute, to provide rotational force as required by a generator axle and or to function as a motor, in order to determine the pounds of force required, horsepower is equal to pounds of force multiplied by revolutions per minute with the result divided by the horsepower constant of 5252. In some embodiments, which may include one complete embodiment, to convert pounds of force from the previous calculation to pound feet of force, as used in the standard calculation of horsepower, a gear may be placed on the generator and or motor axle, which is coupled to the force providing gear immediately before it, where each of those gears have a radius of 1 foot (0.3 meters), while maintaining all gear ratios in the system. In some embodiments, to accurately measure the rotational force and speed provided by the turbine, a torque gauge may be used to measure pound feet of force, and a tachometer may be used to determine revolutions per minute, or alternatively a dynamometer may be used which measures pound feet of force as well as revolutions per minute, and then horsepower may be calculated as pound feet of force multiplied by revolutions per minute with the result divided by the horsepower constant 5252.

[0051] The set of factors required to validate the embodiment as providing a self-powered generator, are the speed and power consumption of the medium container raising and lowering force providing devices, the speed and force provided by the medium flow volume and drop to rotate the turbine, the speed and force required by the generator, and that gears increase speed in proportion to reduction of force. In some embodiments, to calculate the approximate embodiment output, water may be used, where water has a weight of 8.34 pounds per gallon (1 kilogram per liter), 1 cubic foot of water is 7.48 gallons (1 cubic meter of water is 1000 liters), water drops as a result of gravity at approximately 20 miles per hour (32 kilometers per hour) or 29.3 feet (8.9 meters) per second, with a force of 0.433 pounds of pressure per square inch per foot of drop (9.81 kilopascals per meter of drop), and the total square inches (centimeters) of water flowing through the pipe calculated as the pipe radius squared times the constant Pi of 3.14, where the rotational speed provided by the turbine may be increased utilizing gears with gear ratios that increase speed in proportion to a reduction in force. In some embodiments, to calculate the approximate cycle time of the specified gallons, where the water flow per second is calculated as the pipe radius in cubic feet (meters) multiplied by the 29.3 feet ((20 miles per hour * 5,280 feet per mile) / (60 minutes * 60 seconds)) per second that water flows multiplied by the gallons of water in the system. In some embodiments, to approximately calculate total output of the system, the generator output is multiplied by the cycle time of the medium flow in seconds, and then subtracted from the total medium container transition height which includes the heights of the tanks and the pipes multiplied by the speed of the medium container raising and lowering force providing devices in inches per second. In some embodiments, energy may be captured in excess of that consumed, as a result of linear force being utilized to elevate the medium, while gravity creates compounding force on the medium as it falls.

[0052] In some embodiments, including one complete embodiment, to approximately calculate the cycle time of the embodiment, 1000 gallons (3,790 liters) of water are used weighing approximately 8,340 pounds (3,783 kilograms), with each tank weighing 400 pounds (181 kilograms), for a total of 9,140 pounds (4,145 kilograms), which is raised then held then lowered on a continuous cycle, utilizing later described force providing devices, which for example may be a hydraulic piston and pump, with the water passing through piping providing a diameter of 10 inches (0.25 meters) to rotate a turbine with a corresponding diameter of almost 10 inches (0.25 meters), where the 10 inch (0.25 meters) diameter pipe providing a drop of 6 feet (1.83 meters), may provide 2.6 pounds of pressure per square inch (6 feet * 0.433 pounds) (17.9 kilopascals), over an area of 78.5 square inches ((5 inch radius ^ 2) * 3.14 constant pi) (32,260 square millimeters), for a total of 204 pounds (93 kilograms) of force at 20 miles (32 kilometers) per hour, with rotations per minute of the turbine axle calculated where water as determined by gravity flows at about 20 miles (32 kilometers) per hour or 21,020 inches (536 meters) per minute, and an 10 inch (0.2 meters) diameter turbine has a circumference of 31.4 inches (10 inch diameter * 3.14), with a turbine therefore providing around 336 revolutions per minute ((21,120 inches of water flow per minute / 31.4 turbine diameter) * 50% efficiency), where a 10,000 watt generator may require 1,800 revolutions per minute and 13.3 horsepower, requiring the turbine rotational force be connected to the generator through a gear pair with a teeth ratio that approximately increases speed 5 times and decreases force by 5 times, therefore providing around 41 pounds (23 kilograms) of force at 1681 revolutions per minute, which is 13.1 horsepower ((41 pounds of force * 1681 revolutions per minute))/5252 horsepower constant), therefore allowing each system cycle to sustain generator output of 10,000 watts, where the cycle time of the specified gallons (liters) can be approximately calculated, utilizing a pipe radius of 10 inches (0.25 meters) providing the previously calculated area of 78.5 square inches or 0.55 square feet (0.05 square meters), results in water flow per second of 16 cubic feet (29.3 feet per second * 0.55 square feet of pipe) (0.45 cubic meters), which is 120 gallons (16 cubic feet

* 74.8 gallons per cubic foot) (454 liters) per second, therefore 1,000 gallons (3,785 liters) will flow through system over approximately 8.3 seconds (1,000 gallons / 120 gallons per second) (3,785 liters / 291 liters per second). In some embodiments, including one complete embodiment, to approximately calculate total output of the system, a total transition height may be used of 120 inches (18 inch tank heights x 2 + 72 inch drop pipe + 12 inches of intermediary piping) (3 meters), where the force providing devices may move at 1 inch (2.5 centimeters) per 5 seconds over 120 inches (3 meters) while the motor to operate the force providing device pump handle consumes 55 watts, resulting in a net positive output per cycle of 50,000 watts ((10,000 watts provided * 8.3 seconds) – (55 watts consumed * 5 seconds per inch * 120 inches)). For validation through comparative reference, a wind turbine typically provides optimal rotational force at around 20 revolutions per minute which is then passed through gears to increase speed while reducing force to power a generator, where this is providing approximately 204 pounds (93 kilograms) of rotational force at the same speed. For additional validation, an alternative method for calculating the watts of output able to be provided by water flow is watts of output is equal to water drop in meters multiplied by water flow in liters per second multiplied by gravity of 9.8 meters per second. For example, a drop of 72 inches (1.82 meters) providing flow of 120 gallons (454 liters) per second multiplied 9.8 results in a power output calculation of approximately 8,100 watts (1.82 meter drop * 454 liters per second * 9.8 meters per second of gravity), which is close to the output previously calculated, and would therefore provide net positive output per cycle of approximately 34,200 watts ((8,100 watts provided * 8.3 seconds) – (55 watts consumed * 5 seconds per inch * 120 inches)).

[0053] In some embodiments, to extend power generation output and or cycle time, a greater volume of medium is circulated each time, where the extra force and power consumption required of the motor to operate the hand operable hydraulics is minimal relative to the increase in embodiment output. In some embodiments, which may include one complete embodiment, to reduce unit cycle time, more powerful force providing devices may be used, which may be additionally moved inward to reduce transition time, but may require greater force as they don't receive the full benefit of the levered platform. In some embodiments, any amount of the rotational output force may be instead directed to function as a motor.

[0054] In some embodiments, the tops of medium containers 1200 1201 are open, and are filled with a medium when lowered into a medium reservoir, with all electronic components maintained safely away from exposure to the medium, where when the medium containers are elevated after being filled, and then the medium flows from the elevated medium container, either to an elevated reservoir, or through drop pipe 1305, where pressure can then be created using gravity, for the medium to be provided to pipes for delivery to endpoints, eliminating the need for a pump. For example, a water utility, rather than incurring the costs of pumping, can elevate the water from a reservoir and deliver it to customers by using gravity provided water pressure.

[0055] **In some embodiments, including one complete embodiment, obtain the desired turbine and generator.** In some embodiments, including one complete embodiment, a generator is added. In some embodiments, which may include one complete embodiment, the generator selected outputs the volts, amperes, and hertz of the desired final output, with a converter utilized of the appropriate specifications to power internal embodiment components. In some embodiments, including one complete embodiment, a turbine is added. In some embodiments, including one complete embodiment, the turbine utilized is a low head high flow turbine. In some embodiments, any turbine may be used that provides adequate rotational force to the generator axle and or motor axle in conjunction with the provided medium flow. In some embodiments, when used as a motor, a generator may be left out, with the power for the unit components provided by an external source.

[0056] In some embodiments, including one complete embodiment, in reference to FIG. 1, turbine 1300 is a Kaplan turbine, which roughly resembles a fan blade on an axle, and is designed to provide rotational force with 6 feet (1.82 meters) of water drop, which may be purchased in a self-enclosed unit that accepts piping, or may have a blade of a diameter just less than the interior diameter of the previously determined pipe diameter of 10 inches (0.25 meters), where the turbine blade edges may be sanded so as to not touch the pipe, and the turbine is appropriately oriented in the pipe to support optimal output.

[0057] In some embodiments, including one complete embodiment, in reference to FIG. 1, generator 1500 has a rated output of 10,000 watts when provided 1,800 revolutions per minute and 13.3 horsepower, where the horsepower specified requires force of 38.8 pounds (17.6 kilograms) ((13.3 = n pounds of force * 1800 revolutions per minute))/5252 horsepower constant). In some embodiments, where a motor is principally desired, a generator may be used with a rated output providing only that required to provide power to the unit components, while allowing the rest of the rotational force to be provided to rotate the axle to function as a motor.

[0058] In some embodiments, including one complete embodiment, where the speed of the axle rotated by the turbine is not adequate to rotate the generator at or near optimal speed, or rotate the motor axle to achieve the desired speed, gears may be attached to connect the force from the turbine axle to the generator axle selected utilizing gear ratios, which increase the speed of rotation while proportionately reducing the force based on the ratio of teeth on interconnected gears. In some embodiments, including one complete embodiment, to achieve the previously determined requirement of a 5 times speed increase, gear 1401 with 25 teeth may be attached to the axle of turbine 1400 that interconnects with gear 1501 with 5 teeth on generator 1500. In some embodiments, a gear box may be purchased that provides the desired increase in speed. In some embodiments, other means of controlling speed may be used.

[0059] **In some embodiments, including one complete embodiment, construct a support structure.** In some embodiments, including one complete embodiment, a support structure for unit components is made of a material able to support the weight of the filled medium tanks and forces enacted by the raising and lowering force providing devices, for example, utilizing high grade steel. In some embodiments, including one complete embodiment, the medium containers are positioned to provide optimal flow to the turbine, where the height of the support structure center beam, allows medium containers to become full and empty on each trip, and is therefore approximately half of the combined height of each medium container, plus the water drop pipe, and intermediary pipes, while the width of the structure is approximately the width of a medium container, plus the width of the both steel beams, while the length of the support structure is approximately the length of each end tank, plus if center tanks are used, the length of one center tank since the center tanks are vertically on top of each other, plus the length of the pipes, from the end tanks to the center medium containers, which may be of a length that minimizes the height the tanks need to be raised. To optimize component fit, the exact sizes of the support structure may be determined at the time of construction, after available components have been purchased and or assembled. In some embodiments, many of the embodiment components may be 3D printed, which may include the support structure, tanks, pipes, enclosure, and force providing devices.

[0060] In some embodiments, including one complete embodiment, in reference to FIG. 1, the support structure consists of 12 beams, connected together by means which may include welding, or bolting together through bolt holes drilled in appropriate locations, where 4 base beams 1000 1001 1002 1003 are assembled in a rectangle, with beams 1004 1005 extending from either side of the center of the width of the base, in order to support the liquid container see saw platform beams 1006 1007 1008 assembled in a rectangular formation, with beams 1005 1006 having holes drilled through the top center, which may contain bearings, and beams 1007 1008 having holes drilled through the middle center, to allow beam 1005 to be connected to beam 1006 with bolt 1009, and beam 1008 to be connected to beam 1006 with bolt 1010, with bolts of adequate thickness and strength to support the weight of the medium filled containers, to allow the see saw formation to operate back and forth, which provides the efficiency benefit of partial leverage, and with one side providing an optional extended lever, and with beams 1011 1012 1013 1014 extending vertically from beams 1006 1008 in a formation that prevents the left medium container from shifting forward and backward, and beams 1015 1016 positioned to prevent the left medium container from shifting laterally, and beams 1017 1018 1019 1020 extending vertically from beams 1006 1008 in a formation that holds right medium container in place, and beams 1021 1022 extending vertically from support beam 1006 1008 in a formation that prevents the right medium container from shifting laterally.

[0061] In some embodiments, including one complete embodiment, in reference to FIG. 1, the force from the force providing device piston is connected to a control arm, which operates the platform back and forth through the full required range of motion, where the control arm is constructed of 4 high strength beams 1023 1024 1025 1026, where beam 1023 is coupled to piston of cylinder 1100 by means which may include welding or having a hole drilled through it to be bolted to the piston's bolt port, beam 1024 bolted to beam 1023 and extending down to beam 1025, having a length of approximately the height of the piston, with beam 1025 bolted to the bottom of beam 1024 and welded to the bottom of beam 1007, with components connected and of adequate dimensions and spacing to allow the piston to control the platform through its full range of motion. In some embodiments, a force providing device and associated control arm may be included on each side of the embodiment. In some embodiments, which may include one complete embodiment, the control arm may be omitted and the medium containers are instead directly raised and lowered by the force providing devices. In some embodiments, the force providing devices may be mounted horizontally from the base to the platform, and extend outward and upward. In some embodiments, one or more force providing devices operate, instead of or in addition to tanks, a traditional water wheel structure to circulate a substance.

[0062] In some embodiments, which may include one complete embodiment, where a see saw formation is not used, tanks may each be mounted on top of a raising and lowering force providing device, which is automated as provided for in this disclosure. In some embodiments, where space constraints are less important and increased system efficiency is desired, an extended lever may be used, and or may be on each side of the structure, with each side of the support platform operated by its own additional set of force devices and controllers.

[0063] In embodiments where the tanks and or support platform transition height exceeds the height of an available force providing device, or accelerated cycle time is desired, force providing devices may be stacked, where corresponding control components as explained in this disclosure are repeated as needed, and where hydraulics or pneumatics are utilized, manual pump handles may be connected by a metal bar in a manner that allows them to be operable by one motor, where stacked force providing devices, if they utilize single acting cylinders, are stacked on each side of the tank support platform, operating in opposition, where when one set of cylinder pistons is raising the medium containers, the other set of cylinders has their release valves held by timer operated motors as provided for in this disclosure.

[0064] In some embodiments, the tanks may be raised and lowered by extended levers on one or both sides that are powered by one or more electric motor(s), with the efficiency provided by the levers, reducing the energy required by the electric motor(s). In some embodiments, the see saw platform formation may rotate to allow for the top tank to move vertically above the bottom tank.

[0065] **In some embodiments, including one complete embodiment, select and mount one or more force providing devices, including but not limited to hydraulic, pneumatic, mechanical leverage, motorized mechanical leverage, and or functionally equivalent device(s), where such a force providing device may be in the form of a unified unit or connected components.** Commodity hydraulics and pneumatics are available at the time of disclosure that may each provide force of up to 2,000,000 pounds (907,000 kilograms), and may be powered by means including an electric motor, which may be in the unit or in an external pump, which compresses a substance such as a liquid, gas, or air, to provide force to the hydraulics, pneumatics, or functional equivalents. Commodity hand operable hydraulic bottle jacks are available at the time of disclosure with a 14 inch (35.6 millimeters) piston extension length and are rated to provide 100,000 pounds (45,360 kilograms) of force at roughly 1 inch (2.5 centimeters) every 5 seconds. Commodity hand operable hydraulic pumps and cylinders are available at the time of disclosure that may provide force of up to 190,000 pounds (86,180 kilograms), where the piston may extend and retract at a rate of around 1 inch (2.5 centimeters) per 5 seconds. A double acting cylinder in conjunction with a corresponding double acting pump provides both push and pull force. Commodity mechanical leverage devices such as a screw jack are simply constructed and easily scalable devices, that at the time of disclosure may provide up to approximately 8,000 pounds (3,630 kilograms) of force each. In some embodiments, electric motors may operate in conjunction with some means of leverage, including the leverage provided by the see saw platform with optional extended levers. In some embodiments, a hydraulic motor is powered by a hydraulic pump, or interconnected hydraulic pumps, which raises and lowers the medium containers by means which may include rotating a chain suspended above and below while being connected to the support platform, or circulates the medium by means which may include being the motor in a medium pump. In some embodiments, including one complete embodiment, force providing devices are targeted for utilization that along with other system components consume less energy than is produced by the generator.

[0066] In some embodiments, including one complete embodiment, in reference to FIG. 1, double acting hand operable hydraulic pump 1101 operates rear flange mounted double acting hydraulic cylinder 1100, whose piston is bolted to a platform control handle 1023, with the directional valve of pump 1101 connected to and operated by linear actuator 1102, and motor 1105 supported by support beam 1106, with axle of motor 1105 welded to one end of extension rod 1104, with extension rod 1104 bolted through a drilled hole on the other end to rotatable steel cuff 1103, created by slicing a piece of metal pipe of sufficient diameter and strength to allow the handle of pump 1101 to move a full cycle and bolted to rod 1104 in such a position that extension rod 1104 has little or no friction when completing full cycles of the handle of pump 1101, with directional valve linear actuator 1102 operated by repeat cycle timers or equivalents, as described in a later step, where if more appropriate the directional valves may be instead operated directly with electric motors. In some embodiments, including one complete embodiment, the force providing devices selected provide a minimum of the force required to raise and lower the tanks and platform, and to support each tank becoming either full or empty each cycle, provide a total piston extension length adequate to move the containers fully up and down, which may be approximately calculated by adding together the heights of the tanks plus the height of the water drop pipe plus the height of intermediary pipes. In some embodiments, including one complete embodiment, the force providing devices selected provide approximately 10,000 pounds (4,535 kilograms) of force and a piston extension length of approximately 120 inches (3 meters). In some embodiments, which may include one complete embodiment, the motor to operate the handle of the hand operable force providing device pump may provide force of 225 pounds (102 kilograms) while consuming a peak of 55 watts (12 volts at 4.6 amperes), which is dramatically more force than required to operate the pump handle to power the embodiment. In some embodiments, cylinder 1100 and manual pump 1101 are replaced with ones that provide 100,000 pounds (45,360 kilograms) of force, while using the same handle motor to operate the hydraulic pump, where the tanks may hold the corresponding volume of water of up to approximately 11,990 gallons (100,000 pounds of lift / 8.34 pounds per gallon) (82,718 liters), thus nearly eliminating power consumption in the system relative to output. In some embodiments, which may include one complete embodiment, handle motor 1105 is replaced by a linear actuator connected to the handle, operated with an additional repeat cycle timer, connected in a manner similar to the later described repeat cycle timers, for the linear actuator to operate the handle in continuous back and forth cycles. In some embodiments, force providing devices may be used on either side of the platform. In some embodiments, to provide additional force, force providing devices may operate in parallel.

[0067] In some embodiments, including one complete embodiment when reduced embodiment power consumption is desired, rather than directly using an electric motor to operate pump handle(s), instead force providing device(s) including but not limited to hydraulic, pneumatic, mechanical leverage, motorized mechanical leverage, and or functional equivalents, instead provide the force required to operate the pump handle. In some embodiments, including one complete embodiment when reduced embodiment power consumption is desired, in reference to FIG. 2, which utilizes some components previously labeled in FIG. 1, hydraulic cylinder 2000 is powered by hydraulic pump 2001 with directional valve of pump 2001 controlled by linear actuator 2100 powered by and wired to repeat cycle timers 2200 2201 or relay board 1801 in a similar manner as disclosed for other pump directional valve linear actuators, and timed to ensure back and forth motion of the piston of cylinder 2000, with repeat cycle timers 2200 2201 or relay board 1801 wired to battery 1602 through power switch 1600, with the handle of pump 2001 operated by motor 1105, for piston of cylinder 2000 to operate the handle of pump 1101, where when such a force providing device is providing the amount of force to operate one or more other force providing device(s), and that force providing device is in turn operated by a motor providing the amount of force required to operate it, energy consumption in the system is further reduced, where either elevating supports are added under the pumps and cylinders operated by piston of hydraulic cylinder 2000 with appropriate adjustments to impacted components, or cylinder 2000 may be mounted with additional appropriate supports in an inverted position above the handles it will operate, so that the cylinders corresponding to the operated handles rest on the existing support structure. In some embodiments, including one complete embodiment, the speed of the force providing device operating the handles may be n times slower than the speed of the motor previously operating the handles, thus the output calculations may utilize cycle power consumption increased n times to compensate, and or embodiment components including medium container sizes and corresponding force providing devices may be adjusted to compensate for the speed reduction, where n may be best obtained by timing the force providing device operating cycle once implemented. For example, in reference to FIG. 2, if hydraulic cylinder 2000 and corresponding hydraulic pump 2001 are provide 10 pounds (4.5 kilograms) of force, and pump 2001 has its handle powered by motor 1105 providing 1 pounds (0.45 kilograms) of force while consuming 10 watts, with the hydraulics in turn providing the 10 pounds (4.5 kilograms) of force to operate the handle of the primary pump, the energy consumption if running continuously, would be per day approximately 864,000 watts (10 watts per second * 86,400 seconds per day), while the output per day of the generator is 5,975,000 watts (10,000 watts per second * 86,400 seconds per day * (8.3 second cycle run time / (120 second force providing device transition * 10 times slower))), or if center medium tanks are used that prevent medium transference interruption, the output per day of the generator is 864,000,000 watts (10,000 watts per second * 86,400 seconds per day), thus operating with a near perfect input output efficiency ratio. In some embodiments, which may include one complete embodiment, to reduce the force required to power the force providing device pump handle, a more powerful force providing device than is required to circulate the medium may be used. In some embodiments, which may include one complete embodiment, in reference to FIG. 2, containing some components in FIG. 1, hydraulic cylinder 2000 or hydraulic motor with a corresponding automated manual hydraulic pump 2001 or a bottle jack providing 100,000 pounds (45,360 kilograms) of force, automated through wiring of corresponding linear actuators or motors wired to repeat cycle timers and or relay boards in the manner provided in this disclosure, provides force to operate the handles of not only pump 1101 but also the handles of up to roughly 10,000 additional force providing pumps (100,000 pounds of output force / 10 pounds of input force required) with corresponding cylinders or hydraulic or pneumatic motors, or bottle jacks, providing 100,000 pounds (45,360 kilograms) of force each, with the force from the pistons interconnected by means which may include metal bars, with appropriately added and adjusted supporting components, with those pumps directional valves also operated by linear actuators and repeat cycle timers wired in a manner similar to that described in this disclosure, or where all directional valves are interconnected and operated by another automated piston operated on an appropriately timed cycle, providing roughly 1,000,000,000 pounds (10,000 hydraulics * 100,000 pounds of output force each) (454,000,000 kilograms) of force to lift a container holding 119,904,000 gallons of water (1,000,000,000 pounds of lift / 8.34 pounds per gallon), using the energy consumed by a single relatively small electric motor consuming 10 watts and supporting directional valve linear actuators and repeat cycle timers, utilizing roughly 12,000 watts (10 watts consumed per second * 120 lift cycle seconds * 10 times slow down when operating hydraulics with hydraulics) to lift the water which will flow for 11.5 days (119,904,000 gallons / 120 gallons per second) providing 9,992,000,000 watts (10,000 watts per second * 999,200 seconds), thus operating with a near perfect input output efficiency ratio. In another example of input output efficiency, hydraulics provided 10 pounds (4.5 kilograms) of input force, to provide 100,000 pounds (45,360 kilograms) of output force, over a pump handle cycle time of 25 seconds when using hydraulics and 1 second when using an electric motor, layered 3 times, where the pistons of a layer operate the handles of the following layer until the last layer provides output force, not including the first hydraulics as a layer, provides for a total of 100 million output providing units ((100,000 pounds of output force / 10 pounds of input force) ^ layers) which provide 10 trillion pounds of output force (100,000,000 units providing output force * 100,000 pounds of output force per unit), completing a cycle in a little over 10 minutes ((25 seconds ^ layers) + 1 second), all while utilizing only 10 pounds (4.5 kilograms) of input force. Archimedes is recorded as having stated "Give me a lever and a place to stand and I will move Earth." Utilizing this system, braced against a celestial body of appropriate mass and trajectory, Archimedes could have moved Earth with the force of his hand. Such an embodiment may be made possible by an additional law of physics discovered by the inventor, where layered leverage provides efficiency gains as a result of gains in layer output force (total output = (unit output force / unit input force) ^ layers) exceeding gains in layer cycle time (total cycle time = unit cycle time ^ layers). In some embodiments, including one complete embodiment, if any utilized commodity component, such as the force providing devices and or operating motors, doesn't perform with the force or speed expected, units of that component may simply be added. In some embodiments, such layered force providing device arrangements may be built directly into other force providing devices. Therefore, embodiments can provide for a self-powered generator and motor that can easily be brought up to any level of output, allowing for clean continuous self-powered electricity generation and fuel free propulsion, all at a cost that is effectively zero when amortized over time.

[0068] In some embodiments, including one complete embodiment, to determine the force providing cylinders(s) and pump(s) to select corresponding to the determined specifications, the manufacturer's product guide is used. In some embodiments, to determine the force providing cylinders(s) and pump(s) to select corresponding to the determined specifications, calculations may be made, where a cylinder may be selected based on factors including the previously determined piston extension length and being rated to support the previously determined force, while a corresponding pump may be selected based on its output of pounds (kilograms) of pressure per square inch (centimeter) which provides force, flow in gallons (liters) per minute which provides speed, and reservoir gallons (liters) which must adequately fill the cylinder, where the internal area of the cylinder to determine pump gallons (liters) required may be calculated as the constant Pi of 3.14 multiplied by the diameter of the piston multiplied by the extension length of the piston, while the speed of the piston in inches (centimeters) per minute may be calculated as the internal area of the piston divided by the gallons (liters) per minute provided by the pump, and pounds (kilograms) of pressure per square inch (centimeter) required of the pump may be determined by setting the pounds of output force required equal to piston diameter multiplied by the constant Pi 3.14 multiplied by pounds (kilograms) of pressure per square inch (centimeter), where if such calculations call for more powerful pumps and cylinders than can be operated in a net positive energy system, pumps and cylinders may instead be used that provide required force but at a slower transition speed so as to consume less energy than produced in the system. For example, to provide 10,000 pounds (4,535 kilograms) of force moving at 1 inch per second over 12 inches (30 centimeters), using a cylinder with a piston of the corresponding extension length of 12 inches (30 centimeters) with a 2 inch (5 centimeter) diameter and thus an internal area of 0.33 gallons (2 inch diameter * 3.14 Pi * 12 inches) (1.23 liters), requires a pump with a corresponding tank size of 0.33 gallons (1.23 liters) that provides flow of 3.96 gallons per minute (0.33 gallons * (1 inch piston extension per second * 12 inches)) (15 liters) at 1,592 pounds of pressure per square inch (10,000 pounds of required output force = (3.14 Pi * 2 inch piston diameter) * pounds of pressure per square inch)) (722 kilograms of pressure per 6.54 square centimeters). In another example, to provide 100,000 pounds (45,350 kilograms) of force otherwise utilizing the same specifications requires 15,923 (7,220 kilograms) pounds of pressure per square inch (100,000 pounds of required output force = (3.14 pi * 2 inch piston diameter) * pounds of pressure per square inch)). However, in another example, to reduce the pounds (kilograms) of input pressure per square inch (centimeter) required in the previous example by 10 times, the diameter of the piston is increased 10 times, resulting in a requirement of 1,592 pounds of pressure per square inch (100,000 pounds of required output force = (3.14 pi * 20 inch piston diameter) * pounds of pressure per square inch)) (722 kilograms of pressure per 6.45 square centimeters). In some embodiments, which may include one complete embodiment, the gain in efficiency provided by certain force providing device configurations, which may include reducing input force required relative to output force, may contribute to energy being captured in an embodiment in excess of that consumed by the embodiment.

[0069] In some embodiments, which may include one complete embodiment, double acting hydraulic cylinder 1100 is instead, in reference to FIG. 3, double acting hydraulic cylinder 3000, operated by electric double acting hydraulic pump 3100, which is operated by pump switch 3101, operated by motor 3200 which has welded to its axle rod 3201 to press the buttons on pump switch 3200, with the switch operating motor secured to the switch by metal clamp 3202, with motor 3200 operated by repeat cycle timers disclosed in a later step, or with the pump switch cut off and the operating wires connected to repeat cycle timers in a similar manner, to raise and lower the liquid container support platforms on a cycle that allows each tank to become approximately full and empty on each cycle.

[0070] In some embodiments, other means of leverage, including hydraulic, pneumatic, or mechanical, controlled electronically or manually, may be used. In some embodiments, commodity hydraulic jacks, which may be operated electronically, may operate the medium container platform and corresponding tanks. In some embodiments, the force providing devices rest on the steel support beams of the support structure. In some embodiments, a screw jack may be used. In some embodiments, a hydraulic or pneumatic screw jack may be used, where the motor on the traditional screw jack is replaced with a hydraulic or pneumatic device operating in its place. In some embodiments, a hydraulic motor may operate the platform by means which may include a linked chain providing force to the platform handle through a pulley system. In some embodiments, a hydraulic motor powered medium pump may provide force to circulate a medium past the turbine.

[0071] **In some embodiments, including one complete embodiment, secure medium containers to the support platform and connect piping between medium containers and around turbine.** In some embodiments, including one complete embodiment, the pipes may be made of the same material as the tanks, and connected and secured in place through means which may include welding or industrial sealant. In some embodiments, including one complete embodiment, in reference to FIG. 1, tanks 1200 1201 are placed on platform beams 1006 1008, pipe 1300 is connected from the top of the tank 1200 to the bottom of pipe connector 1308, and pipe 1301 is connected from the bottom of tank 1200 to the top of pipe connector 1302, and pipe 1303 is connected to the top of tank 1201 and the bottom of pipe connector 1308, pipe 1303 is connected from the bottom of tank 1201 to the top of pipe 1302, where drop pipe 1305 provides the medium drop to create pressure to elbow joint 1306, which connects to pipe 1307, which encloses the turbine, with pipe 1307 connected to base support 1014, with pipe 1307 connected to tee joint pipe connector 1308, to drop a medium into the medium tanks, with pipe connector 1308 having a hole drilled through, sealed by elastomer seal 1318, through which the axle of turbine 1400 passes through and has mounted to it gear 1401, which may be a worm gear if the turbine axle is perpendicular to the generator axle, which is connects with complimentary gear 1501, with a teeth ratio to provide the previously determined increase in speed and proportional reduction of force, with gear 5015 coupled by means which may include welding to the axle of generator 1500, where because a Kaplan turbine is oriented to have a vertically extending axle, generator 1500 may mounted to the support structure vertically above the turbine, ensuring that the generator axle rotates in the designed direction for electricity generation, and if not the generator and or other relevant components are repositioned accordingly, with corresponding changes to pipe positions. In some embodiments, which may include one complete embodiment, the center pipes and generator are fixed in place to the support structure, with flexible hoses between the pipe connectors 1302 and 1308, allowing the center components to remain fixed in place as the outer pipes move. In some embodiments, which may include one complete embodiment, to provide greater efficiency by removing the friction of an elbow joint, turbine 1400 is oriented vertically in pipe 1305 rather than horizontally, with a corresponding adjustment to positioning of generator 1500. In some embodiments, which may include one complete embodiment, water pipe 1305 provides a greater drop height to increase pressure on the turbine and thus increase output, with supporting components adjusted accordingly.

[0072] **In some embodiments, including one complete embodiment, connect a battery and power switch.** In some embodiments, including one complete embodiment, a battery which may be in the form of uninterruptable power supply, allows the unit to start and provide output and power components during tank transitions, where the battery receives power from the generator. In some embodiments, including one complete embodiment, in reference to FIG. 1, battery 1600 allows the unit to turn on when a medium isn't flowing through the system, where the battery provides the current type, voltage, amperes, and hertz required to provide power to repeat cycle timers 1700 1701, the pump handle motor 1105, and pump directional valve linear actuator 1102, where the output wires of generator 1500 are connected to the battery to recharge it, and may be connected through an overcharge controller, and or an electrical converter, and or a resistor, if there is a mismatch of current type, voltage, amperes, or hertz, or the strength of the electrical output from the generator would damage the embodiment components, where the battery is attached to support structure by means which provide for replaceability, such as metal brackets coupled through mounting holes with corresponding holes drilled in the support structure, where power switch 1600 turns the unit on or off, by opening and closing the circuit between the battery 1602, the electronically controlled pump motor 1105, and the repeat cycle timers 1700 1701 which power pump valve linear actuator 1102, where the power switch is externally accessible to the operator, and may be attached to the unit, by means including bolts or screws.

[0073] **In some embodiments, including one complete embodiment, implement a timer based control system.** A repeat cycle timer is a commodity component that's used to operate electronics, such as a sprinkler system, by turning the system on at a fixed interval, running it for a fixed interval, turning the system off for a fixed interval, and looping the cycle. The instructions for configuring the timers, when purchased from a quality supplier, will be in the manual accompanying the timers. For example, some timers may have positive and negative terminals for each the power source and the device, along with a knob to set the seconds on per cycle, and a knob to set the seconds off per cycle. In some embodiments, the functional equivalent of a repeat cycle timer may be used, including a computer controlled relay board operated by computer commands to open and close the circuits to produce the same effect, and or to turn the units on and off to meet peak demand at specific times or based on current consumption. In some embodiments, including one complete embodiment, motors including those driving linear actuators may be utilized that have two wires and are operable in both directions, driven by for example a reversible brushed direct current motor, though other types of motors may be utilized which may have additional wires to be wired in the manner corresponding to their accompanying instructions.

[0074] In some embodiments, including one complete embodiment, the timers are set to operate the hydraulics to raise for the previously determined cycle, hold for previously determined medium container emptying time, then lower for the previously determined cycle time, hold for the previously determined medium container emptying time, and repeat, but are to be adjusted based on the performance of the purchased components by observing embodiment operation.

[0075] In some embodiments, including one complete embodiment, in reference to FIG. 1, repeat cycle timers 1700 1701 are selected to support powered components and operate pump directional valve linear actuator 1102, in a timed back and forth cycle timed to allow the medium containers to fully empty and fill each transition, thereby ensuring continuous motion is provided in the system, where repeat cycle timers 1700 1701 open and close circuits that cause the pump directional valve to raise and lower the piston of cylinder 1100, on the previously determined interval, and for the previously determined interval, with positive wire of linear actuator 1102 attached to both repeat cycle timer 1700 positive terminal, and repeat cycle timer 1701 negative terminal, and negative wire of linear actuator 1102 attached to repeat cycle timer 1700 negative terminal and repeat cycle timer 1701 positive terminal, with repeat cycle timers 1700 and 1701 each connected to the corresponding positive and negative terminals of battery 1602. In some embodiments, where an electronic hydraulic pump is used, the pump switch is either operated by linear actuator 1102, or the switch is removed and its operating wires are directly connected to repeat cycle timers in the manner previously described to allow the medium tanks to empty and fill during each transition. The repeat cycle timers may be attached to a support structure beam 1000 using bolts or straps, which makes them replaceable, or any other method as long as it doesn't damage the timers. In some embodiments, wiring to repeat cycle timers or functional equivalents may be done in any manner that achieves the desired effect.

[0076] In some embodiments, the equivalent of a repeat cycle timer may be used, where a computer controlled relay board 1801 is wired in the same manner as the commodity repeat cycle timers, and a computer 1800 is embedded in the unit which uses software to operate a computer controlled relay board, with a connected motor, to perform the functions of a repeat cycle timer, with the code to operate the computer controlled relay board later disclosed to write operating software.

[0077] In some embodiments, the force providing devices may be manually operated, and repeat cycle timer(s) or equivalent may be excluded.

[0078] **In some embodiments, including one complete embodiment, connect an electrical power output connector and converter if required.** In some embodiments, including one complete embodiment, in reference to FIG. 1, a commodity power connector 1601 may be added between the generator and be externally accessible to the user, which may be in the form of an outlet or positive and negative terminals. In some embodiments, where the generator does not provide output as desired including current type, amperes, voltage, and hertz, a commodity power converter providing appropriate adjustments may be wired between the generator and the power output connector 1601.

[0079] **In some embodiments, which may include one complete embodiment, when implementing a motor, optionally attach a valve and optionally attach a linear actuator to control flow rate to control motor output.** In some embodiments, which may include one complete embodiment, in reference to FIG. 1, to modify flow control for motor output, the valve on the input flow pipe of the turbine may be controlled by a linear actuator 1802 wired to relay board 1801 to allow it to be reversible, where linear actuator positive wire is wired to first relay positive terminal and second relay negative terminal, and linear actuator negative wire is wired to first relay negative terminal and second relay negative terminal, where the motor may not be allowed to restrict flow beyond the level required to power the generator, and where a motor may be most applicable when the embodiment is stationary and its weight is unimportant.

[0080] **In some embodiments, which may include one complete embodiment, create software for unit operation and or to control an array of units.** In embodiments where a computer is embedded in the unit, a computer may be used that runs custom software on startup, and may be connected to a network accessible relay board. The software code to operate the computer connected relay board, may resemble that described below. In some embodiments, the computer may be a Raspberry Pi, connected to an Ethernet controlled relay board through and Ethernet crossover cable, with the computer set on startup to run software developed in the Python language from the disclosed description, where the software is installed by connecting to the computer through telnet or secure shell, then at the command prompt typing "nano run.py" and adding and saving the software code, then at the command prompt typing "nano /etc/rc.local" and adding and saving the line "python /$location/run.py", where $location is the path to the directory containing the previously created software file.

[0081] The software code to operate the computer connected relay board, in some embodiments, which may include one complete embodiment, provides functionality comprising:
> import the library for connecting to the relay;
> import the library for accessing system resources;
> create variables holding values for relay on off values;
> create variables to hold the state of each relay including those controlling motors for the force providing devices(s) and the flow valve;
> create a variable holding the system cycle seconds;
> create variables to hold the relay board connection, IP address, username, and password.
> create a function to update the relay board;
>> establish a connection to the relay, if it has not been initialized, or has been dropped
>> creating a string sequence of relay states;
>> send the states to the relay board.

> create a function to listen for user input;
>> read the current key pressed;
>> if the key pressed is the forward arrow, set the flow valve backward state to off and forward state to on;
>> if the key pressed is the backward arrow, set the flow valve forward state to off and backward state to on;
>> if the key pressed is the spacebar, set the flow valve backward state to off.
> define a function to run on script execution;
>> retrieve the current system time;
>> perform a continuous loop;
>>> create a variable holding seconds elapsed as the time initialized minus the current system time;
>>> if the remainder is zero, when seconds elapsed are divided by force providing device cycle seconds, transition the force providing device(s) by changing corresponding relay state(s);
>>> call the function to see if the user has provided any commands to the unit;
>>> call the function to update the relay with current changes.

[0082] In some embodiments, where a large number of generator units are being operated concurrently, to turn the units on and off at specific times and or based on current power consumption, network accessible relay boards may be installed in the units, and wired in place of or in addition to the power switch to be able turn the units on and off, and may be controlled by a computer running software provide functionality comprising:
> import a library for accessing the relays;
> import a library for system resource access;
> initialize and set variables holding relay on and off values;
> create an array of IP addresses of unit on off relay boards;
> initialize an array of unit relay board connections;
> create an array of unit output watts;
> create an array of unit on off states;
> create variables holding the unit relay boards username(s) and password(s);
> create variables to hold the total watts available across all units, the current watts being consumed, and whether or not time based watts are to be used;
> create an array of pairs of times and time desired watts;
> create a variable indicating whether or not unit power consumption meters are to be used;
> create and array of power consumption meter IP addresses;

> create variables holding the power consumption meters username(s) and password(s);

> create variables holding the minimum and maximum power consumption to be allowed before switching units on or off.

> create a function to send a command to a unit at a specified index in the unit IP address array;

>> create a connection to the unit on off relay board, at the IP address at the provided index in the unit IP address array, if the connection has not been initialized, or had been dropped;

>> create a variable holding the command string to be sent to the unit based on whether the unit is to be turned on or off;

>> send the command the unit relay;

>> wait for the command to go through, then disengage all on off relays.

> define a function to run on script execution;

>> iterate through each unit IP address and call the function to switch the unit on;

>> run a continuous loop;

>>> create variables to hold a unit index iterator, an output display message, and the current action;

>>> proceed if power consumption meters are being used;

>>>> create variables to hold average and total consumption of unit power, and add up total possible power output;

>>>> loop through each power consumption meter IP address;

>>>>> retrieve the current power consumption number;

>>>>> calculate the average consumption;

>>>>> add the current consumption to the total;

>>>>> add the unit maximum capacity to the total consumption capacity;

>>>>> increment the index.

>>>> calculate the consumption percentage as the consumption total divided by the total unit capacities;

>>>> create a message to display the states the currently consumed watts and the current watt capacity;

>>>> if the average consumption is greater than the maximum consumption level before more unit should be turned on, then set the unit action equal to on;

>>>> if the average consumption is less than the minimum consumption level before more unit should be turned off, then set the unit action equal to off;

>>>> iterate through unit states until one is found that is either off, if looking to turn a unit on, or on if looking to turn a unit off, then send the command to switch the unit state.

>>> proceed if power consumption meters are not used, and instead the total watts to be provided by the units are determined by the current time;
>>>> iterate through each time watts pair, and if the current time is equal to the specified time, send commands to turn units on or off, until the desired level of output is produced, and display each action on the command line.
>>> sleep for a moment before looping again.

[0083] **In some embodiments, including one complete embodiment, construct and attach a protective enclosure.** In some embodiments, including one complete embodiment, in reference to FIG. 1, a protective enclosure 1900 houses all of the components, to protect the electronic components from external elements, to protect the operator from the force providing devices, tank weights, and the extraordinarily rare event of a cylinder explosion, and may be shaped to fit around the support structure, and may be made of a material such as sheets of steel, plastic, or carbon fiber, formed by means which may include fabrication or welding in the dimensions of the unit and may allow power connector 1601 and the power switch 1600 to be externally accessible to the operator.

[0084] **In some embodiments, including one complete embodiment, enjoy clean continuous self-powered energy and or propulsion.** In some embodiments, set the embodiments power switch 1600 to on, to allow the unit to run briefly to validate construction, then turn it off. In some embodiments, including one complete embodiment, if implemented as a generator, the embodiment is then ready to have an experienced licensed electrician connect it to a power grid, and turn on the unit's power switch to on, or if implemented as a motor, be appropriately connected in an engine compartment.

CLAIMS

What is claimed is:

1. A medium circulator able to function as a motor and or power an electricity generator, with the invention comprising:
   a means for holding a transferrable medium;
   a turbine or functional equivalent;
   force providing device(s) including but not limited to hydraulic, pneumatic, mechanical leverage, motorized mechanical leverage, and or functional equivalents;

said force providing device(s) able to be powered or operated, directly or indirectly, by a generator and or by another source;

said force providing device(s) able to provide for the flow of a medium past said turbine which in turn is able to provide rotational force to either or both an electricity generator axle and or an axle to function as a motor.

2. Further comprising claim 1, said force providing device(s) operated by force providing device(s) including but not limited to hydraulic, pneumatic, motorized mechanical leverage, and or functional equivalents to optimize the input output efficiency.

3. A method performed by an apparatus comprising:

providing electricity which directly or indirectly powers force providing device(s) including hydraulic, pneumatic, mechanical leverage, motorized mechanical leverage, and functional equivalents;

transferring a transferrable medium to rotate a turbine;

said turbine directly or indirectly transferring rotational force to rotate a generator axle and or function as a motor;

said generator providing electricity directly or indirectly to power or operate directly or indirectly force providing devices or additional force providing devices operating said force providing devices;

4. Further comprising claim 1, said force providing device(s) operated by force providing device(s) including but not limited to hydraulic, pneumatic, mechanical leverage, motorized mechanical leverage, and or functional equivalents to optimize the input output efficiency.

5. A method for constructing an apparatus comprising:

obtaining medium container(s) and or conduit(s), a turbine or functional equivalent, and force providing devices including but not limited to hydraulic, pneumatic, mechanical leverage, motorized mechanical leverage, and functional equivalents;

ensuring the attachment to a support structure of said force providing device(s), medium container(s) and or conduit(s), and a turbine;

6. Further comprising claim 1, force providing device(s) including but not limited to hydraulic, pneumatic, motorized mechanical leverage, and or functional equivalents, operatively coupled to said force providing device(s), to optimize the input output efficiency.

7. A medium delivery system, with the invention comprising:

a means for holding a medium;

force providing device(s) including but not limited to hydraulic, pneumatics, mechanical leverage, motorized mechanical leverage, and or functional equivalents;

said force providing device(s) able to provide for the flow of said medium.

8. Further comprising claim 5, said force providing device(s) operated by force providing device(s) including but not limited to hydraulic, pneumatic, or motorized mechanical leverage, to optimize the input output efficiency.

9. A manually operable force providing device including hydraulic, pneumatic, mechanical leverage, motorized mechanical leverage, and or functional equivalents, converted to an automatic force providing device, with the invention comprising:

   a motor or motorized device able to provide the input force required by a force providing device;

   a motor or motorized device able to control the force providing device direction;

   a connection between said force providing device input force receiver and corresponding motor able to take said input force receiver through a cycle;

   a power source;

   one or more repeat cycle timer(s) or functional equivalents;

   said repeat cycle timer(s) able to be powered by said power source able to control said motor(s) to control said input force receiver and or said valve.

10. An apparatus comprising:

   force providing device(s) including but not limited to hydraulic, pneumatic, mechanical leverage, motorized mechanical leverage, and or functional equivalents able to operate one or more force providing device(s) including but not limited to hydraulic, pneumatic, mechanical leverage, motorized mechanical leverage, and or functional equivalents to improve input output efficiency.

11. A non-transitory computer-readable recording medium holding stored instructions, which when executed by one or more processing devices, cause the one or more processing devices to implement a method comprising:

   turning power producing units on and off to meet desired power output, either or both at specific times, or by reading the power consumption meter of one or more units, and if the average power being consumed is above a certain threshold, additional units are turned on, and if power being consumed is below a certain threshold, units are turned off.

FIG. 1

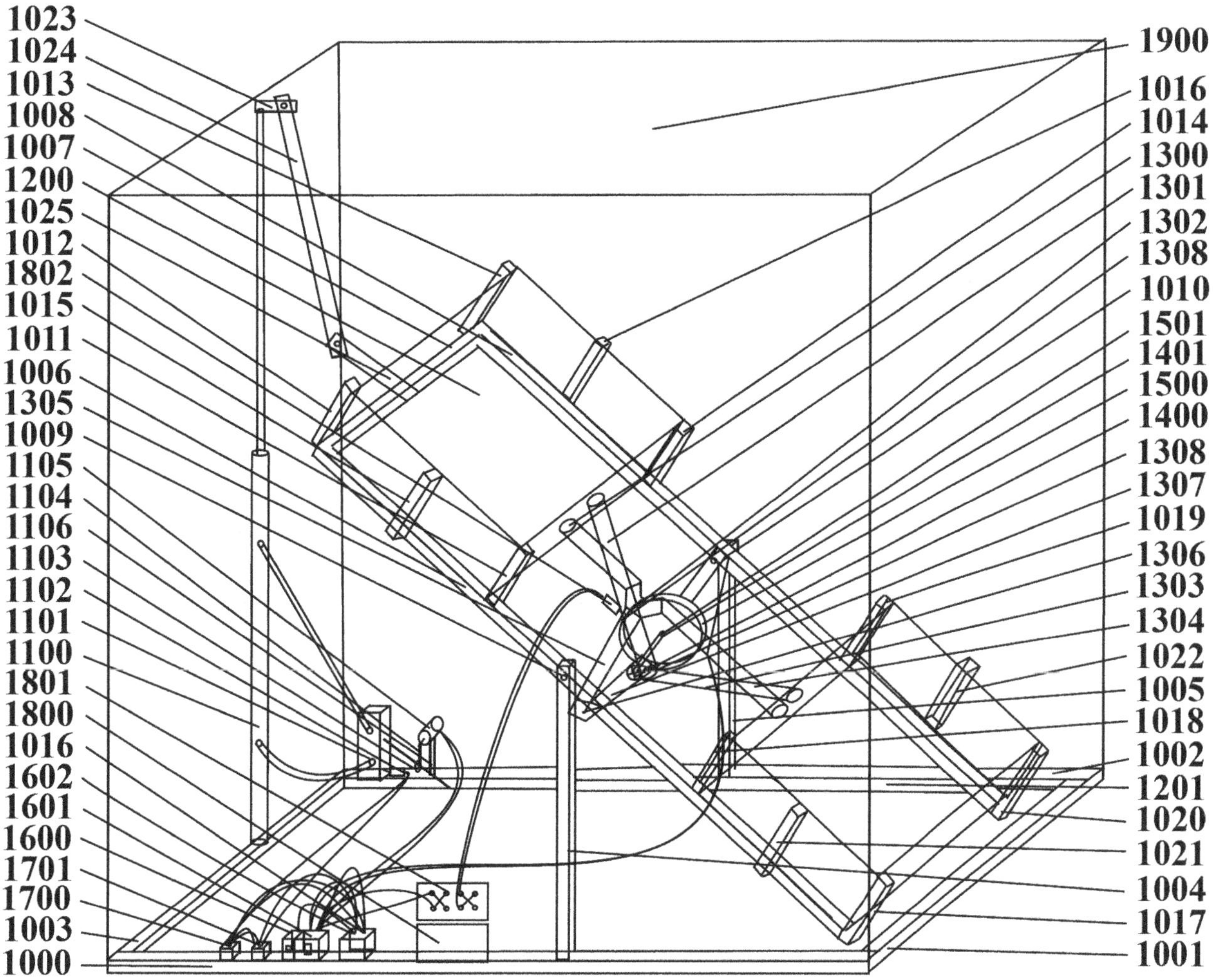

Jonathan Bannon Maher

FIG. 2

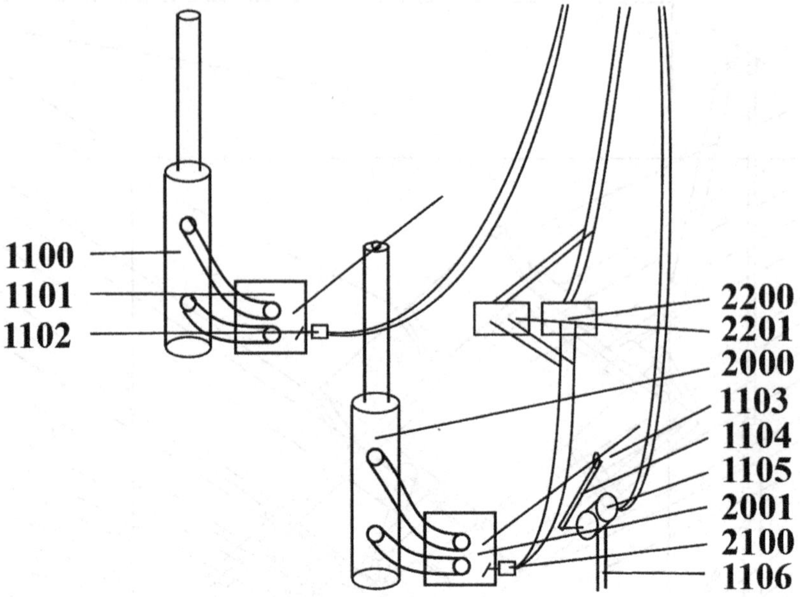

Jonathan Bannon Maher

FIG. 3

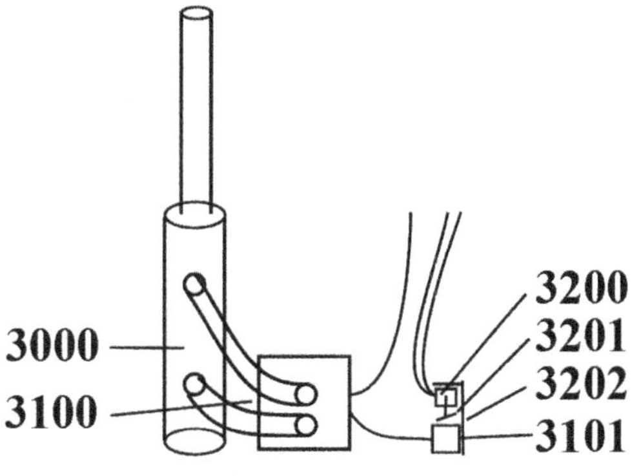

Jonathan Bannon Maher

FIG. 4

4000 — import the library for connecting to the relay

```
import telnetlib
```

4001 — import the library for system resource access

```
import sys
```

4002 — create variables holding values for relay on off values

```
off = 0
on = 1
```

4003 — create variables to hold the state of each relay including those controlling motors for the force providing devices and the flow valve

```
relay_1_lift_up_state = off
relay_2_lift_down_state = off
relay_3_flow_valve_forward_state = off
relay_4_flow_valve_backward_state = off
```

4004 — create a variable holding the system cycle seconds

```
lift_cycle_seconds = 15
```

4005 — create variables to hold the relay board connection, IP address, username, and password

```
relay = None
relay_ip = "169.254.1.1"
relay_username = "admin"
```

relay_password = "admin"

4006 – create a function to update the relay board, first establishing a connection to the relay, if it has not been initialized, or has been dropped, then creating a string sequence of relay states, and sending the states to the relay board

```
def update_relay():

    if not relay:
        relay = telnetlib.Telnet(relay_ip, 23)
        relay.read_until(b"User Name: ")
        relay.write(
            relay_username.encode('ascii') + b"\n")
        relay.read_until(b"Password: ")
        relay.write(relay_password.encode('ascii') + b"\n")

    command = str(relay_1_lift_up_state)
    command+= str(relay_2_lift_down_state)
    command+= str(relay_3_flow_valve_forward_state)
    command+= str(relay_4_flow_valve_backward_state)
    relay.write(command.encode('ascii') + b"\n")
```

4007 – create a function to listen for user input

```
def listen():
```

4008 – read the current key pressed

```
    key_pressed = ord(sys.stdin.read(1))
```

4009 – if the key pressed is the forward arrow, set the flow valve backward state to off and forward state to on

```
    if key_pressed == 39:
```

relay_4_flow_valve_backward_state == off
relay_3_flow_valve_forward_state == on

4010 – if the key pressed is the backward arrow, set the flow valve forward state to off and backward state to on

```
if key_pressed == 37:
    relay_3_flow_valve_forward_state == off
    relay_4_flow_valve_backward_state == on
```

4011 – if the key pressed is the spacebar, set the flow valve backward states to off

```
if key_pressed == 37:
        relay_3_flow_valve_forward_state == off
        relay_4_flow_valve_backward_state == off
```

4012 – define a function to run on script execution

```
def __main__():
```

4013 – retrieve the current system time

```
time_initialized = sys.time()
```

4014 – perform a continuous loop

```
while True:
```

4015 – create a variable holding seconds elapsed as the time initialized minus the current system time

```
seconds_elapsed = time_initialized - sys.time()
```

4016 – if the remainder is zero, when seconds elapsed are divided by lift cycle seconds, transition the lifts

```
if seconds_elapsed%lift_cycle_seconds == 0:
   if relay_1_lift_up_state == on:
      relay_1_lift_up_state = off
      relay_2_lift_down_state = on
   if relay_2_lift_down_state == on:
      relay_1_lift_up_state = off
      relay_2_lift_down_state = on
```

4017 – call the function to see if the user has provided any commands to the unit

```
listen()
```

4018 – call the function to update the relay with current changes

```
update_relay()
```

FIG. 5

5000 — import a library for accessing the relays

```
import telnetlib
```

5001 — import a library for system resource access

```
import sys
```

5002 — initialize and set variables holding relay on and off values

```
off = 0
on = 1
```

5003 — create an array of IP addresses of unit on off relay boards

```
unit_ips = ["169.254.1.3", "169.254.1.4"]
```

5004 — initialize an array of unit relay board connections

```
relays = []
for unit_ip in unit_ips:
   relays.append(None)
```

5005 — create an array of unit output watts

```
unit_watts = [5000, 5000]
```

5006 — create an array of unit on off states

```
unit_states = [on, off]
```

5007 – create variables holding the default unit relay board username and password

unit_username = "admin"
unit_password = "admin"

5008 – create variables to hold the total watts available across all units, the current watts being consumed, and whether or not time based watts are to be used

watts_total = 0
watts_current = 1000000
time_watts = False

5009 – create an array of pairs of times and time desired watts

times_watts = [["7:00", 1000000],["20:00",500000]]

5010 – create a variable indicating whether or not unit power consumption meters are to be used

power_meter = True

5011 – create and array of power consumption meter IP addresses

power_meter_ips = ["169.254.1.1","169.254.1.2"]

5012 – create variables holding the power consumption meters username and password

power_meter_username = "admin"
power_meter_password = "admin"

5013 – create variables holding the minimum and maximum power consumption to be allowed before switching units on or off

```
consumption_minimum = 0.5
consumption_maximum = 0.9
```

5014 – create a function to send a command to a unit at a specified index in the unit IP address array

```
def send_command(unit_index, command):
```

5015 – create a connection to the unit on off relay board, at the IP address at the provided index in the unit IP address array, if the connection has not been initialized, or had been dropped

```
    if not relays[unit_index]:
        relays[unit_index] =
            telnetlib.Telnet(unit_ips[unit_index], 23)
        relays[unit_index]
            .read_until(b"User Name: ")
        relays[unit_index]
            .write(unit_username.encode('ascii') + b"\n")
        relays[unit_index]
            .read_until(b"Password: ")
        relays[unit_index]
            .write(unit_password.encode('ascii') + b"\n")
```

5016 – create a variable holding the command string to be sent to the unit based on whether the unit is to be turned on or off

```
    command_string = "00"
    if command == on:
        command_string = "01"
    if command == off:
        command_string = "10"
```

5017 – send the command the unit relay

```
relays[unit_index]
    .write(command_string.encode('ascii') + b"\n")
```

5018 – wait for the command to go through, then turn off all relays

```
time.sleep(1)
command_string = "00"
relays[unit_index]
    .write(command_string.encode('ascii') + b"\n")
```

5019 – define a function to run on script exectution

```
def __main__():
```

5020 – iterate through each unit IP address and call the function to switch the unit on

```
    index = 0
    for unit_ip in unit_ips:
        send_command(index, on)
        index += 1
```

5021 – run a continuous loop

```
    while True:
```

5022 – create variables to hold a unit index iterator, an output display message, and the current action

```
        index = 0
        message = ""
```

```
    action = None
```

5023 – proceed if power consumption meters are being used

```
    if power_meter:
```

5024 – create variables to hold average and total consumption of unit power, and add up total possible power output

```
        consumption_average = 0
        consumption_total = 0
        consumption_capacity = 0
```

5025 – loop through the each power consumption meter IP address

```
        while index < len(power_meter_ips):
```

5026 – retrieve the current power consumption number

```
            power_meter_address =
                "http://" +
                power_meter_ips[index]+
                "/?username=" +
                power_meter_username +
                "&password=" +
                power_meter_password +
                "&command=consumption"
            power_meter_consumption =
                urllib.open(power_meter_address).read()
```

5027 – calculate the average consumption

```
consumption_average +=
   power_meter_consumption/unit_watts
```

5028 – add the current consumption to the total

```
consumption_total +=
   power_meter_consumption
```

5029 – add the unit maximum capacity to the total consumption capacity

```
consumption_capacity +=
   unit_watts[index]
```

5030 – increment the index

```
index = index + 1
```

5031 – calculate the consumption percentage as the consumption total divided by the total unit capacities

```
consumption_percentage =
   consumption_total/consumption_capacity
```

5032 – create a message to display the states the currently consumed watts and the current watt capacity

```
message = str(consumption_total) + " of " +
      str(consumption_capacity) +
      " watts consumed"
```

5033 — if the average consumption is greater than the maximum consumption level before more unit should be turned on, then set the unit action equal to on

```
if consumption_percentage >
    consumption_maximum:
    action = on
```

5034 — if the average consumption is less than the minimum consumption level before more unit should be turned off, then set the unit action equal to off

```
if consumption_average <
    consumption_minimum:
    action = off
```

5035 — iterate through unit states until one is found that is either off, if looking to turn a unit on, or on if looking to turn a unit off, then send the command to switch the unit state

```
index = 0
unit_found = False
if action:
    for unit_state in unit_states:
        if not unit_found:
            if (action == on and
                unit_state == off) or
                (action == off and
                unit_state == on):
                send_command(index, action)
                unit_found = True
                unit_state[index] == "on"
                state = "on"
                if action == off: state = "off"
```

```
            message += ", unit turned " + state
          index += 1
```

5036 — proceed if power consumption meters are not used, and instead the total watts to be provided by the units are determined by the current time

```
if not power_meter:
```

5037 — iterate through each time watts pair, and if the current time is equal to the specified time, send commands to turn units on or off, until the desired level of output is produced, and display each action on the command line

```
  for time_watt in time_watts:
    if time == time_watt[0]:
      desired_watts = time_watt[1]
      transition_complete = False
      index = 0
      for unit_state in unit_states:
        while not transition_complete:
          if watts_current > time_watt[1]:
            action = off
            watts_current -= unit_watts
            message = "unit turned off"
          if watts_current < time_watt[1]:
            action = on
            watts_current += unit_watts
            message = "unit turned on"
          if (action == on and
            unit_state == off) or
             (action == off and
            unit_state == on):
              send_command(index, action)
              unit_state[index] == "on"
              state = "on"
```

```
            if action == off: state = "off"
            message +=
               ", unit turned " + state
            index += 1
         action = None
      print message
      message = ""
```

5038 — sleep for a moment before looping again

```
time.sleep(1)
```

SELF-POWERED MOTOR AND GENERATOR

INVENTOR JONATHAN BANNON MAHER

TECHNICAL FIELD

[0001] Embodiments of the invention relate to the fields of motors, generators, physics, engineering, and programming.

ABSTRACT

[0002] Systems, methods, apparatuses, and in some embodiments computer programs encoded on a computer storage medium, provide for clean continuous portable self-powered energy generation and propulsion, consistent with the laws of physics, by in some embodiments, including one complete embodiment, transferring force from force providing devices not limited to but including hydraulics and or pneumatics and or mechanical leverage and or motorized mechanical leverage, to provide rotational force to power an electricity generator, and or function as a motor, where energy may be captured in excess of that consumed as a result of the differential between input force required and output force provided by certain force providing device configurations. The invention permanently solves global warming, provides reduced cost of living and cost of goods to alleviate poverty, provides unlimited clean energy for evaporated water purification and atmospheric carbon dioxide splitting, and eliminates the need for every other method of energy production, including nuclear technology – thus reducing nuclear weapons technology proliferation.

BACKGROUND

[0003] The majority of the proceeds from the licensing of this patent will be going to causes that support the well being of humanity, and your support in ensuring the patent is forever in every way as strong as possible, will be providing a service to all the world.

[0004] This section is intended to introduce the reader to various aspects of the art that may be related to various aspects of the present techniques, which are described and or claimed. This discussion is believed to be helpful in providing the reader with background information to facilitate a better understanding of the various aspects of the

present disclosure. Accordingly, it is understood that these statements are to be read in this light, and not a citation of any prior art.

[0005] Patent filings on structurally differentiated and fundamentally deficient disclosures may exist that may attempt to claim any invention that is self-powered, based on previously publicly known failed attempts to build such devices, however any such disclosures do not enable the purported inventions, with overly broad claims not supported by the disclosure that also fail to distinctly claim the invention, and any such patent filings are inherently invalidated by prior public disclosure, the enablement requirement, and the claims support requirement.

[0006] Known and proposed energy production, transmission, and storage systems have some or all of the following deficiencies:

[0007] 1. External fuel source required: an external fuel source is utilized such as oil, gas, coal, wind, sun, water currents, geothermal heat, hydrogen, or uranium, where the cost of providing fuel in the form of electricity or gasoline to a vehicle over its useful life potentially exceeds the cost of the vehicle, and about a quarter of airline costs are from fuel.

[0008] 2. Environmentally unfriendly: nuclear energy production, including fission, fusion, and cold (LENR), results in toxic waste and or materials, geothermal often circulates contaminants from the ground, fracking creates toxic water, hydroelectric dams decompose organic matter producing the potent global warming gas methane, solar panel manufacturing often releases toxic byproducts including greenhouse gases far more potent and long lived than those from fossil fuels, hydrogen takes substantially more energy to produce and transport than it provides, while clean energy sources may utilize slowly degrading flammable toxic batteries to store energy for when the sun is not shining, wind is not blowing, or water is not adequately flowing.

[0009] 3. Intermittent: wind, solar, and traditional water energy systems provide variable output, with average output often found to be around 20% of rated output, as a result of environmental conditions, meaning a 5 kilowatt system typically produces average output of only 1 kilowatt.

[0010] 4. Not portable: solar panels can only function in the sun, wind turbines in wind, water turbines in water currents, nuclear in a stable highly controlled environment, and carbon with the aid of an emissions pipe.

[0011] 5. Extremely expensive transmission costs: power lines are required for all forms of nuclear energy, including fission, fusion and cold, for farms of solar, wind, and water, and for fossil fuels plants, while pipelines are generally required for oil and gas, including natural gas, which is principally methane, a greenhouse gas more than twenty times more potent than carbon dioxide that may be leaked during extraction, transport, and consumption. Power line and fossil fuel pipe line cost of installation per 1 mile (1.6 kilometers) has been found to be up to around 20 times the average annual income in the United States, and lines must be replaced every 30 to 50 years, so in the United States alone, with 300 thousand miles (480 thousand kilometers) of power lines, and 200 thousand miles (320 thousand kilometers) of pipelines, replacement would require an expenditure around 25 times the national debt, passed on to consumers, and dragging down the economy, with every other developed nation in a similar situation. Power lines require environmental destruction during installation, leave visible blight, are forever vulnerable to cyber attacks, transmit power from central sites that are inherently more prone to failure and blackout than a decentralized system, dissipate power during transmission, with high voltage power lines having health consequences for those living nearby.

[0012] 6. High initial costs: in addition to previously cited expense of power lines, fossil fuel pipes, and ongoing fuel costs, clean energy farms require an allocation of land, and degrading batteries requiring periodic replacement, which is why tax credits are often required for clean energy systems to be affordable. In addition to those factors, comparing a 1 kilowatt rated clean energy system to a 1 kilowatt rated traditional system, may require multiplying the cost of the clean energy system by approximately 10 times, 5 times to account for enough electricity generation to be stored for the equivalent continuous output, and 5 times for the battery storage.

[0013] 7. Vulnerable to weather: wind turbines, hydroelectric dams, and solar panels, can be made ineffective by environmental conditions such as freezing temperatures, snow, and rain, and similar to power lines, may be taken down by extreme weather and lightning strikes.

[0014] 8. Vulnerable to black outs: power transmitted over power lines creates vulnerability for critical facilities such as hospitals and data centers.

[0015] 9. Vulnerable to cyber attack: utility scale energy systems often require a hackable computer to operate, and are therefore forever vulnerable to computer viruses able to take down and or destroy nuclear power plants and the electrical grid, even if such systems aren't connected to the Internet, as demonstrated by the Stuxnet virus.

[0016] 10. Causes deaths: plants, rigs, and pipes, for current and proposed forms of nuclear, hydrogen, gas, and oil energy can explode, and coal mines can collapse, while wildlife is killed by wind, ocean, wave, and river turbines, hydroelectric dams, solar condensers, solar panel and battery manufacturing byproducts, and nuclear waste.

[0017] 11. Encourage nuclear weapon proliferation: fission, fusion, and cold (LENR) nuclear energy are or can be one step from weaponizable, while hydrogen can be obtained by a terrorist at a hydrogen fuel station to create a powerful compressed hydrogen explosion, as verified by reviewing a video of a balloon filled with hydrogen being lit on fire. Fusion is particularly disturbing, as a fusion weapon could be created with the power of an exploding star, able to take out the planet, and resulting in an extinction level event for humans, a scenario even more likely when considering increasingly autonomous – and therefore inevitably hackable by individuals – weapons control.

BRIEF DESCRIPTION OF THE ILLUSTRATIONS

[0018] Illustrations are presented by way of example, and not by way of limitation, where some embodiments may not contain all components, may contain additional components, and may contain functionally similar components.

[0019] FIG. 1 is an illustration of an example of an embodiment, which utilizes the force from hydraulics and or pneumatics and or mechanical leverage and or motorized mechanical leverage to provide rotational force to power an electricity generator and or function as a motor.

[0020] FIG. 2, containing some components in FIG. 1, is an illustration of an example of an embodiment of an automated hydraulic pump and cylinder whose piston operates other hydraulic pump handles to operate their corresponding hydraulic cylinder pistons to improve system input output efficiency.

[0021] FIG. 3 is an illustration of an example of an embodiment of a traditional electric hydraulic pump and cylinder which in some embodiments may be used in place of the converted manual to automatic hydraulic pump and cylinder in FIG. 1.

[0022] FIG. 4 is an illustration of an example of an embodiment of hydraulic bottle jacks stacked in opposition between gear racks and automated to provide back and forth force for use by an embodiment which in some embodiments may be used in place of the converted manual to automatic hydraulic pump and cylinder in FIG. 1.

[0023] FIG. 5 is an example of an embodiment of a structure to transfer force through a medium.

[0024] FIG. 6 is an example of an embodiment of a pendulum structure to transfer force.

DETAILED DESCRIPTION

[0025] The disclosure is related to the field of clean continuous portable self-powered energy and propulsion. It is understood that any reference to a person skilled in the art, recognizes that at the time of filing, there is no one else skilled in the art of this particular field, or a closely related field. Given the extraordinary nature of the disclosure, regardless of how full, clear, concise and exact the disclosure in enabling the production and use of embodiments of the disclosure, what could be construed to be undue experimentation during production and use, is simply the ordinary effort required in the assembly and use of an embodiment of such a disclosure.

[0026] It is understood that, as in any engineering or design project, the development of any actual implementation will include numerous implementation specific decisions made to achieve the developers' specific goals, such as compliance with business related and system related constraints, which may vary from one implementation to another. It is understood that such a development effort might be complex and time consuming, but is nevertheless a routine undertaking of design, fabrication, and manufacture for those skilled in the art having the benefit of this disclosure. The disclosed steps may be read as prefaced by "In some embodiments, including one complete embodiment, ", may be executed or performed in other orders or sequences, and are not limited to the order and sequence shown and described, which are provided to enable ease in constructing an embodiment, and along with each components of each step, may be removed, modified, combined, or rearranged, and other steps and or step components may be added, without departing from the scope of this disclosure and or invention. Although embodiments of the invention have been described and illustrated in the disclosed implementations, it is understood that the present disclosed subject matter, including apparatuses, methods, specification, and illustrations, has been made only by way of example, not by way of limitation, and the methods and apparatuses may be used in

other systems, and that numerous changes and optimizations in the details of implementation of the invention and or embodiment are made without such modifications departing from the spirit and scope of this disclosure and or embodiments of the invention. Although the disclosure has been shown and described with respect to one or more embodiments, features of the disclosed embodiments can be combined and rearranged in various ways, and changes including equivalent alterations, substitutions, modifications, and additional efficiencies will of course occur to someone of ordinary skill in the art without departing from the spirit and scope of this disclosure and or invention. In particular regard to the various functions performed by the described components, the terms used to describe such components are intended to correspond, unless otherwise indicated, to any component which performs the specified function of the described component, or is functionally equivalent to the described component, even though not structurally equivalent to the disclosed structure which performs the function in the implementations described in this disclosure. In addition, while a particular feature of the disclosure may have been provided with respect to only one of several embodiments, such feature may be combined with one or more other features of other embodiments as may be desired and advantageous for any given or particular application. In some instances, well-known circuits, structures and techniques have not been shown in detail in order not to obscure the understanding of this disclosure. Articles in this disclosure such as "a" "an" and "the" may allow for both singular and plural forms. Verbs in this disclosure such as "is" may be read as "may be". Conjunctions in this disclosure such as "or" as used herein may be interpreted as inclusive or meaning any one or any combination, where "A, B or C" means "any of the following: A; B; C; A and B; A and C; B and C; A, B and C". Relational terms in this disclosure, for example first and second, top and bottom, left and right, are to distinguish one entity or action from another, and may not necessarily require or imply a relationship, or order between, such entities or actions. The disclosure includes the best mode contemplated by the inventor, a completely described specific embodiment, along with optional components and alternative embodiments to best suit the implementer, measurements in imperial and metric units to support universal understanding, and dramatically exceeds claims support requirements and enablement requirements by allowing for selection and or construction of the required components to be carried out easily, quickly, and routinely by persons of ordinary skill in the art, who are provided the additional benefit of utilizing readily available commodity components whenever possible. The present disclosure includes material protected by copyrights, and the owner of the copyrights hereby reserves all rights, but with authorization for publication as required by government patent offices. Various embodiments of the present invention may provide all, some or none of the disclosed technical advantages.

[0027] The computer code descriptions disclosed, in order to provide comprehensive enabling disclosure, rather than utilizing flow charts, which according to Patent Cooperation Treaty 11.11a are prohibited from containing "text matter, except a single word or words, when absolutely indispensable, such as... a few short catchwords indispensable for understanding", are provided in a text only format where the number of arrows preceding a line indicate logical block level, semicolons indicate a new segment of a logical block, and periods indicate the closure of one or more logical blocks. It is understood that any computer code representations in this disclosure are merely illustrative, rather than restrictive. While code may be written in nearly any computer language, including Java and C++, the illustrative computer code descriptions were derived from code written the Python language, which may be run through the Python interpreter, with appropriate supportive libraries, which at the time of disclosure, may run on nearly any computer, for example one with an Intel or AMD processor, running a current version of Linux, Windows, or Mac OS. All code components may read as if prefaced by "In some embodiments, including one complete embodiment, ". In some embodiments, functionality may be modified, rearranged, excluded, and added. To provide more fundamental computer system details, in some embodiments, the functionality associated with the disclosed computer code descriptions may be referred to as a script, module, software, software application, or code, and can be written in any form of language, including compiled, interpreted, declarative, or procedural, able to be deployed in any form suitable for use in a computing environment, including as an independent or integrated program, module, component, or subroutine, for execution by the computer system, implemented on one or more independent or integrated computers, utilizing a central processing unit in the form of one or more general or special purpose microprocessors, in conjunction with digital electronic circuitry, which may include special purpose logic circuitry such as a field programmable gate array or application specific integrated circuit, with the computer controlled by and operatively coupled to tangibly embodied software and or firmware, which may include code that creates an environment for code execution, including individual or combined use of processor firmware, a protocol stack, a database management system, and an operating system, where such software and or firmware may exist in one or more parts in memory on one or more computers, and is encoded on one or more tangible non transitory software carriers, such as individual or combined use of a random or serial access device or substrate, a semiconductor memory device, transient or persistent random access memory, a magnetic, magnetic optical, or optical disk, or encoded on an artificially generated transmitted signal, for example, optical, electrical, or electromagnetic, transmitted using a sending and a receiving apparatus, where the interaction between the user and the software may be implemented by operatively coupling, to the local implementing computer, or a local computer connected to one or more remote computers through a local or wide area network, a display device which may implement liquid crystals or light emitting diodes, a keyboard, and a pointing device.

[0028] The inventor retains absolutely no liability for any implementation of this invention, and the invention is implemented exclusively at the risk and liability of the implementer.

[0029] Calculations, formulas, and specific units are not in any way restrictive, are not be relied upon, are not required as presented to produce an embodiment of the invention, are provided exclusively as a courtesy to enhance enablement for those resizing components and or constructing alternative embodiments, and may contain inaccurate assumptions easily modified during practice, with all calculations utilized to select and estimate components and unit output being rough estimates that may vary greatly based on factors that include the type and quality of purchased and or manufactured components and embodiment construction, where embodiments may be constructed utilizing an effectively endless range of output and component configurations and selection processes.

[0030] In some embodiments, for quality control purposes, all components may be manufactured from scratch.

[0031] Embodiments of the invention provide some or all of the following benefits over previously discussed predecessors:

[0032] 1. Self-powered for free output: embodiments provide the first energy and motor system in the history of the known universe to not require an external fuel source. Embodiments can produce endless energy, and provide endless transportation range, until there is a system failure, and therefore may potentially produce energy and propulsion until gravity driven orbital drift causes the Earth to be consumed by the Sun in a few billion years – assuming the units and humans are still on Earth. Embodiments reduce transportation costs by allowing implementing vehicles to operate without fuel, allowing for effectively free endless transportation range after purchase, and additionally allow for travel by supersonic jets and flying cars, which have been impractical principally as a result of fuel costs.

[0033] 2. Clean: the manufacture and use of embodiments produces no notable harmful environmental byproducts, nor the potential for deaths associated with predecessors.

[0034] 3. Continuous: embodiments produce electricity and propulsion that is continuous and stable.

[0035] 4. Portable: embodiments are able to function as well in a basement closet as in a car.

[0036] 5. Cyber attack proof: embodiments are self-contained thus require no hackable computer to operate and are therefore immune to computer viruses.

[0037] 6. Blackout proof: embodiments are designed to be kept indoors and on-site, and are thus ideal for critical facilities such as hospitals and data centers that can't afford a blackout from failed power lines or plants.

[0038] 7. Inexpensive: embodiments can be manufactured and operated at the lowest cost total cost possible, because they don't require fuel, installation and maintenance of power lines, an allocation of land, or degrading batteries.

[0039] 8. Weatherproof: embodiments are self-contained for indoor use and therefore aren't vulnerable to environmental factors such as freezing temperatures, snow, rain, lightning strikes, or extreme weather events.

[0040] 9. Eliminates energy output storage: because additional embodiments of this system can be utilized at peak times with limited cost, storage of energy is no longer relevant, for either utilities or homes, even for peak output needs.

[0041] 10. Eliminates expensive power lines and fossil fuel pipe lines: because embodiments are designed to be kept on site, and any number of units can be utilized to meet peak power needs, power and fossil fuel transmission lines and their associated costs are now rendered irrelevant, thus substantially unburdening all economies globally of associated costs. Land currently holding power lines, as well as arrays of solar panels, wind turbines, and hydroelectric turbines, can be reclaimed to reduce visual pollution and make space for a growing population. Roofs of solar panels can be removed to allow for roof tiles that reflect heat to maintain a cool house in summer. No thinking person will ever want, nor could a functional government allow, any type of nuclear reactor –

fission, fusion, cold (LENR) – in a car or home, leaving only now irrelevant power lines for transmission, thereby making those sources wholly irrelevant.

[0042] 11. Potentially profitable: in a standard home use scenario, embodiments may be the only way for a unit owner to make a profit from selling energy back to the utility at wholesale rates.

[0043] 12. Alleviates poverty: because energy is effectively free after embodiment purchase, and the purchase price is less per unit of output than other energy systems, embodiments reduce the cost of living and the cost of goods for every person on Earth, thus reducing poverty.

[0044] 13. Powers water purification and pumping: 1 in 10 people live without access to clean water, while climate change driven droughts fuel conflicts. Because embodiments make energy nearly free over their useful lives, the energy intensive nature of evaporated water purification — which removes nearly every contaminant with a higher boiling point than water, and potentially all others can be removed with a standard carbon filter and ultraviolet light – is no longer a barrier, nor is pumping, thus embodiments provide for the global resolution of clean water needs for individual consumption and agriculture.

[0045] 14. Powers reduced water consumption: effectively eliminates the energy cost of operating electricity powered showers that require only a cold water pipe, and recirculate, filter to potentially cleaner than direct from pipe, and heat water, to provide exact continuous temperature and pressure control, as well as powering low voltage electric showerheads that mix air with water to provide the effect of the same output using dramatically less water.

[0046] 15. Powers atmosphere cleaning: because an existing specialized laser can disassociate atmospheric carbon dioxide molecules into carbon molecules and oxygen molecules, and because embodiments can provide effectively unlimited and continuous clean energy, a power source is now available to reduce carbon dioxide in the air, and potentially other greenhouse gasses, if corresponding devices are developed.

[0047] 16. Reduces nuclear and hydrogen weapon proliferation: because embodiments eliminate any need for any type of nuclear energy, including fission, fusion, and cold (LENR), each of which is or may be weaponizable and one step from nuclear weapons technology, and also eliminate the need for hydrogen, which a terrorist can obtain at a hydrogen fuel station to create a powerful compressed hydrogen explosion verifiable by watching a video of a balloon filled with hydrogen being lit on fire, I have provided us all a fundamentally safer world.

[0048] 17. Therefore, a few applications of embodiments include powering: all transportation vehicles including automobiles, trains, jets, spaceships, cargo ships, cruise ships, tugboats, boats, submarines, hover boards, jet packs, including in vertical take off and landing configurations; powering electrical grids as well as homes, offices, factories, hospitals, and data centers that would like to disconnect from external power sources to end their recurring bill, be permanently immune from blackouts, use clean energy, and save money; televisions, washing machines, dishwashers, showers, and water pumps; portable consumer electronics, such as phones and laptops, through an internally installed miniaturized embodiment eliminating the need to recharge; personal rapid transport; home hydroponic production systems, including light, temperature control, and nutrient water circulation; spaceship electro magnetic ion drives using a fraction of the fuel of traditional rockets; high intensity laser powered solar sails, with the laser powered by a large number of these energy units on the surface of any space based body with a limited atmosphere; space tourism; video streaming planetary sampling probes journeying an unlimited number of years into the universe; space colonies; and inter galactic travel.

[0049] Additionally, some embodiments may make use of an embodiment of a custom torque converter which allows a turbine or functional equivalent connected to the motor to spin at a high consistent speed, in a transferrable medium filled container, with an opposing moveable turbine or functional equivalent connected to an output axle, with the rotational force of the output axle determined by the proximity of the turbines.

[0050] **In some embodiments, including one complete embodiment, determine desired output of generator and or motor, then identify corresponding force providing device(s), not limited to but including hydraulic, pneumatic, mechanical leverage, motorized mechanical leverage, or functionally equivalent device(s), where such a force providing device may be in the form of a unified unit or connected components, and acquire generator and force providing device(s).** In some embodiments, because the size of the unit is driven primarily by the generator and force producing device(s), and their size is driven primarily by their output, the desired unit output is first decided, then corresponding components purchased, with the embodiment built around those components. In some embodiments, which may include one complete embodiment, the generator selected outputs the volts, amperes, and hertz of the desired final output, with a converter utilized of the appropriate specifications to power internal embodiment components. In some embodiments, where an existing generator is to be powered as part of the embodiment, for example, when replacing a wind or water turbine to power an existing generator, the force providing device(s), gears, and other components may be selected that correspond to the force and speed required to power such a generator, and where the rotational force provided by the embodiment is connected by means which may include an axle coupler or a gear mounted on the axle of the generator that matches a gear on the axle of the unit. In some embodiments, when used as a motor, a generator may be left out, with the power for the unit provided by an external source.

[0051] Commodity hydraulics and pneumatics are available at the time of disclosure that may each provide force up to 2,000,000 pounds (907,000 kilograms), and may be powered by means including an electric motor, which may be in the unit or in an external pump, which compresses a substance such as a liquid or gas to provide force to the hydraulics, pneumatics, or functional equivalents. Commodity hand operable hydraulic bottle jacks are available at the time of disclosure that may provide 100,000 pounds (45,400 kilograms) of force with a 14 inch (35.6 centimeter) piston extension length moving at roughly 1 inch (2.5 centimeters) every 5 seconds. Commodity hand operable hydraulic pumps and cylinders are available at the time of disclosure that may provide force of up to 190,000 pounds (86,200 kilograms). A double acting cylinder in conjunction with a double acting pump provides both push and pull force. Commodity mechanical leverage devices such as a screw jack are available at the time of disclosure that may provide up to around 8,000 pounds (3,630 kilograms) of force each, and may implement an electric motor to utilize the leveraged force provided by certain mechanical structures. In some embodiments, a hydraulic screw jack may be used, where the motor on the traditional screw jack is replaced with a hydraulic device operating in its place. In some embodiments, rotational force may be provided by one or more hydraulic motor(s).

[0052] In some embodiments, which may include one complete embodiment, given horsepower and revolutions per minute are often specified as the measurement of input required to power a generator, and the measure of the output of a motor, to determine the pounds of input force to be provided by the force providing devices, in order to provide a specific output horsepower at a specific number of revolutions per minute, horsepower is equal to pounds of force multiplied by revolutions per minute with the result divided by the horsepower constant of 5252. In some embodiments, which may include one complete embodiment, to convert pounds of force from the previous calculation to pound feet of force, as used in the standard calculation of horsepower, a gear may be placed on the generator and or motor axle, which is coupled to the force providing gear immediately before it, where each of those gears have a radius of 1 foot, while maintaining all gear ratios in the system. In some embodiments, after determining the number of pounds of force and revolutions per minute required of the generator and or motor, a calculation may be made to determine the number of times the speed of the hydraulic cylinder piston(s) will have to be increased to provide that output speed in revolution per minute, which can be done utilizing gears which increase speed in proportion to a reduction in force corresponding to the gears' teeth ratio, and then multiplying the determined speed differential by the force required by the generator and or to act as a motor, to identify the required original input force. In some embodiments, including one complete embodiment when including a generator and providing the functionality of a motor, once the horsepower and revolutions per minute have been implemented as required by the generator, an additional gear may be added that interconnects with a gear providing rotational force to the generator axle, where such a gear has a teeth ratio that provides for a change in speed to match the desired revolutions per minute to function as a motor, while more powerful hydraulics may be used to offset the reduction in force resulting from the change in speed, where the output force required by the motor is multiplied by the total geared speed differential in the system, and added to the force required by the generator after it is multiplied by its corresponding speed differential, to determine the required initial input force.

[0053] In some embodiments, including one complete embodiment, a 10,000 watt output commodity generator is purchased, which may require 1800 revolutions per minute at 13.3 horsepower, while a commodity manual hydraulic pump and cylinder may be used to provide force moving at approximately 1 inch every 5 seconds, which is 15 inches (38 centimeters) per minute, that when rotating a 1 inch (2.5 centimeters) diameter gear with a circumference of 3.14 inches (8 centimeters), provides 4.8 revolutions per minute, where the optimal rotation of the axle of the 10,000 watt generator is provided at 1,800 revolutions per minute with 13.3 horsepower, resulting in a ratio of the piston speed to desired generator axle speed of 1:375, meaning to provide the required 1,800 revolutions per minute, the force will have to be passed through gears having a total teeth ratio of 1:375, while the force is also decreased by a factor of 375, where the horsepower specified requires force of 38.8 pounds ((13.3 = n pounds of force * 1800 revolutions per minute))/5252 horsepower constant) (17.6 kilograms), and therefore the force providing device to rotate the 10,000 watt generator for maximum output is to provide force of around 15,550 pounds (7,030 kilograms) – 38.8 pounds (17.6 kilograms) of force required by the generator multiplied by the 375 times speed differential. For validation through comparative reference, a wind turbine typically provides optimal rotational force at around 20 revolutions per minute which is then passed through gears to increase speed while reducing force to power a generator, so for example, the previously determined 15,500 pounds of force at 4.8 revolutions per minute converted to 20 revolutions per minute, would provide rotational force of approximately 3,720 pounds (15,500 pounds of force / (20 revolutions per minute / 4.8 revolutions per minute)) (1,687 kilograms). For further validation, the set of factors required to fully validate the embodiment as providing a self-powered generator, are the principle that gears will increase speed in proportion to reduction of force, along with the speed and power consumption of the input force, and the speed and force required to produce maximum output by the generator. For further validation, using an example to assess power consumption, the previously calculated required input force of 15,550 pounds may be around twice the force provided by a commodity electric hydraulic jack used to raise and lower a car, which may consume a maximum of 180 watts (12 volts x 15 amps) before blowing the car outlet's fuse, thus the total consumption by a pair of such commodity electric hydraulic car jacks to provide required force consumes 360 watts, deducted from the 10,000 watt generator output, resulting in a dramatically net positive energy production system.

[0054] In some embodiments, including one complete embodiment, where a motor is implemented, given the average car engine may provide about 250 horsepower at a maximum of 7,000 revolutions per minute, the previously provided 1,800 revolutions per minute may be increased about 4 times utilizing gears with a teeth ratio of 1:4, and by solving for pounds of force required for 250 horsepower finds around 343 pounds (156 kilograms) of force are required ((250 = n pounds of force * 7200 revolutions per minute))/5252 horsepower constant), therefore with a total speed increase ratio of 1500 (375 times generator geared speed increase * 4 times motor geared speed increase) and corresponding force decrease, output requires an initial force of 514,500 pounds (343 pounds of force * 1,500 times force reduction) (234,000 kilograms), plus about 15,500 pounds (7,260 kilograms) for the generator that powers the hydraulics, for a total input force of around 530,000 pounds (241,000 kilograms).

[0055] In some embodiments, including one complete embodiment, to determine the force providing cylinders(s) and pump(s) to select corresponding to the determined specifications, the manufacturer's product guide is used. In some embodiments, to determine the force providing cylinders(s) and pump(s) to select corresponding to desired specifications, calculations may be made, where a cylinder may be selected based on factors including being rated to support the previously determined force, while a corresponding pump may be selected based on its output of pounds (kilograms) of pressure per square inch (centimeter) which provides force, flow in gallons (liters) per minute which provides speed, and reservoir gallons (liters) which must adequately fill the cylinder, where the internal area of the cylinder to determine pump gallons (liters) required may be calculated as the constant Pi of 3.14 multiplied by the diameter of the piston multiplied by the extension length of the piston, while the speed of the piston in inches (centimeters) per minute may be calculated as the internal area of the piston divided by the gallons (liters) per minute provided by the pump, and pounds (kilograms) of pressure per square inch (centimeter) required of the pump may be determined by setting the pounds of output force required equal to piston diameter multiplied by the constant Pi 3.14 multiplied by pounds (kilograms) of pressure per square inch (centimeter), where if such calculations call for a more powerful pumps and cylinders than can be operated in a net positive energy system, pumps and cylinders may instead be used that provide required force but at a slower transition speed so as to consume less energy than produced in the system. For example, to provide 10,000 pounds (4,535 kilograms) of force moving at 1 inch per second over 12 inches (30 centimeters), using a cylinder with a piston of the corresponding extension length of 12 inches (30 centimeters) with a 2 inch (5 centimeter) diameter and thus an internal area of 0.33 gallons (2 inch diameter * 3.14 Pi * 12 inches) (1.23 liters), requires a hydraulic pump with a corresponding tank size of 0.33 gallons (1.23 liters) that provides flow of 3.96 gallons per minute (0.33 gallons * (1 inch piston extension per second * 12 inches)) (15 liters) at 1,592 pounds of pressure per square inch (10,000 pounds of required output force = (3.14 Pi * 2 inch piston diameter) * pounds of pressure per square inch)) (10.98 megapascals). In another example, to provide 100,000 pounds (45,350 kilograms) of force otherwise utilizing the same specifications requires 15,923 pounds of pressure per square inch (100,000 pounds of required output force = (3.14 pi * 2 inch piston diameter) * pounds of pressure per square inch)) (109.76 megapascals). However, in another example, to reduce the input pounds (kilograms) of pressure per square inch (centimeter) required in the previous example by 10 times, the diameter of the piston is increased 10 times, resulting in a requirement of 1,592 pounds of pressure per square inch (100,000 pounds of required output force = (3.14 pi * 20 inch piston diameter) * pounds of pressure per square inch)) (10.98 megapascals). In some embodiments, which may include one complete embodiment, the gain in efficiency provided by certain force providing device configurations, which may include reducing input force required relative to output force, may contribute to energy being captured in an embodiment in excess of that consumed by the embodiment.

[0056] In some embodiments, including one complete embodiment, in reference to FIG. 1, the pistons of rear flange mounted double acting hydraulic cylinders 1100 1102 1104, are coupled to gear beam 1200, by means which may include welding or bolting, and are powered by commodity double acting hand operable manual hydraulic pumps 1101 1103 1105, converted to run automatically, where motors control the pump handles and directional valves, and motor 1402 is supported by beam 1405, axle of motor 1402 is coupled to one end of rod 1403, by means which may include welding, with rod 1403 coupled in a manner that allows it to rotate beam 1411, which has affixed to it, by means which may include welding, cuffs 1412 1413 1414, where the cuffs may be created by slicing pieces of metal pipe of sufficient diameter and strength to allow the handles of their corresponding hydraulic pumps to move full cycles to allow all hydraulic pump handle cuffs to work together to operate the hydraulic pump handles through their range of motion, where motor 1402 is positioned and connected in such a way that when provided power it continuously takes the pump handles through their range of motion, while linear actuators 1408 1409 1410 are coupled to the directional valves of hydraulic pumps 1101 1103 1105, with the coupling method determined after reviewing the directional valves of the acquired pumps, or if more appropriate the directional valves may be instead operated directly with electric motors, with linear actuators 1408 1409 1410 wired to later described repeat cycle timers or functional equivalents to change the direction of directional valves on hydraulic pumps 1100 1002 1004 on a timed loop to maintain effectively continuous motion of the pistons of hydraulic cylinders 1100 1102 1104. In some embodiments, which may include one complete embodiment, pistons of hydraulic cylinders 1100 1102 1104 each have an extension length of up to 1 foot (0.3 meters) and the previously identified diameter, with each pair of pump and cylinder providing 190,000 pounds (86,200 kilograms) of force, providing a total of 570,000 pounds (258,500 kilograms) of force, exceeding the previously determined required 514,500 pounds (241,300 kilograms) of force, and the motor to operate the pump handles may consume at peak load 120 watts (12 volts at 10 amperes), providing force of 225 pounds (102 kilograms), substantially more than is required to operate the hydraulic pump handles to continuously power the embodiment. In some embodiments, which may include one complete embodiment, in reference to FIG. 1, instead of handle motor 1103, a linear actuator is coupled to operate the pump handles in back and forth cycles through beam 1411, controlled with a pair of repeat cycle timers in a nearly identical configuration as described for the pump directional valve linear actuators and corresponding repeat cycle timers.

[0057] In some embodiments, which may include one complete embodiment, in reference to FIG. 1, instead of hydraulic cylinders 1100 1102 1104 and pumps 1101

1103 1105, in reference to FIG. 4, pairs of single acting hydraulic bottle jacks 1100 1102, where the number of pairs is determined based on providing force that meets or exceeds the required force in the embodiment, are attached to and mounted in opposition on either side of gear rack 1200 1201 1202 4000, where gear rack bottom 1200 and gear rack top 4100 are extended to accommodate the additional pistons, with the inverted hydraulic cylinders mounted by means which may include welding to metal support beam 4100 held in place by metal support beam 4101 extending from the base support structure, where to provide force in both directions, the hydraulic jack being pushed upon by the hydraulic jack currently providing force, has its release valve opened the entire time so it doesn't resist, and such hydraulics are operated in an automated manner using the components disclosed to convert a manual hydraulic pump to an automated hydraulic pump, where operating each jack handle is a motor, such as handle motor 1402, providing rotational force through the rod and cuff previously described for converting a manual hydraulic pump to an automatic hydraulic pump, and connected to each release valve is a linear actuator, such as directional valve linear actuator 1412, or if more appropriate the valves may be instead operated directly with electric motors, which are powered by a connection to repeat cycle timers 1400 1401 or relay board 1701 that are timed to operate the linear actuators or motors to ensure continuous back and forth force to the gear rack 2100 1201 1202 4100. In some embodiments, which may include one complete embodiment, six pairs of hydraulic bottle jacks 1100 1102, with each bottle jack rated to provide force of 100,000 pounds (45,400 kilograms), provide for total bi-directional force of 600,000 pounds (272,160 kilograms), which exceeds the previously calculated requirement of 530,500 pounds (241,300 kilograms) of bi-directional force.

[0058] In some embodiments, including one complete embodiment when reduced embodiment power consumption is desired, rather than directly utilizing an electric motor to operate pump handles, force providing device(s) including but not limited to hydraulic, pneumatic, mechanical leverage, motorized mechanical leverage, and or functional equivalents, instead provide the force required to operate the pump handles. In some embodiments, including one complete embodiment when reduced embodiment power consumption is desired, in reference to FIG. 2, which utilizes some components previously labeled in FIG. 1, hydraulic cylinder 2000 is powered by hydraulic pump 2001 with directional valve of hydraulic pump 2001 controlled by linear actuator 2100 powered by and wired to repeat cycle timers 2200 2201 or relay board 1701 in a similar manner as disclosed for other hydraulic pump directional valve linear actuators, and timed to ensure continuous back and forth motion of piston of hydraulic cylinder 2000, with repeat cycle timers 2200 2201 or relay board 1701 wired to power switch 1501, with the handle of hydraulic pump 2001 operated by motor 1408, for the piston of hydraulic cylinder 2000 to operate the handles of the selected hydraulics such as hydraulic pumps 1101 1103 1105, or hydraulic bottle jacks in FIG. 4, where when such a force providing device is providing the amount of force to operate one or more other force providing device(s), and that force providing device is in turn powered by a motor providing the amount of force required to operate it, the energy consumption in the system is further reduced, where either elevating supports are added under the pumps and cylinders operated by piston of hydraulic cylinder 2000 with appropriate adjustments to impacted components, or hydraulic cylinder 2000 is mounted with additional appropriate supports in an inverted position above the handles it will operate so that the cylinders corresponding to the handles rest on the previously detailed support structure. If the speed of the hydraulic pistons operating the handles is n times slower than the speed of the motor previously operating the handles, the force and gear ratios may be adjusted by a factor of n to compensate, with n being best obtained by timing the relevant hydraulics operating cycle. For example, in reference to FIG. 2, if piston of hydraulic cylinder 2000 and corresponding hydraulic pump 2001 are pump handle operating hydraulics providing 1,000 pounds (453 kilograms) of force, where hydraulic pump 2001 has its handle powered by motor 1408 providing 10 pounds (4.5 kilograms) of force while consuming 60 watts, while the handle operating hydraulic piston in turn provides 1000 pounds (45 kilograms) of force to operate the pump handles of the thirty 190,000 pounds (86,200 kilograms) of force providing hydraulic cylinders or hydraulic motors, with a corresponding 30 pumps used instead the 3 pumps, to compensate for a factor of 10 reduction in speed, while a gear ratio change increases speed 10 times, the approximate energy consumption would be a continuous 60 watts, while the output would be either a continuous 633,000 watts (10,000 watts previously calculated as provided per 8,000 pounds of force) or 250 horsepower, thus operating with a near perfect input output efficiency ratio. In some embodiments, which may include one complete embodiment, in reference to FIG. 2 derived from previously labeled components in FIG. 1, a piston of a single hydraulic cylinder 2000 or hydraulic motor with a corresponding automated manual hydraulic pump 2001 providing 100,000 pounds (45,360 kilograms) of force, automated through wiring of corresponding motors which may be in the form of linear actuators wired to repeat cycle timers and or relay boards in the manner provided in this disclosure, piston of hydraulic cylinder 2000 provides force to operate the handles of not only the previously identified hydraulics but also to the handles of roughly 10,000 force providing pumps (100,000 pounds of output force / 10 pounds of input force required) with corresponding cylinders or hydraulic or pneumatic motors, or bottle jacks, providing 100,000 pounds (45,360 kilograms) of force each, with the force from the hydraulic cylinder pistons interconnected by means which may include metal bars, with appropriately adjusted and replicated supporting components, with those pumps directional valves also operated by linear actuators and repeat cycle timers in a manner similar to that provided for in this disclosure, or where all directional valves are interconnected and operated by another automated hydraulic cylinder piston on an appropriately timed cycle, providing roughly 1,000,000,000 pounds (10,000 hydraulics * 100,000 pounds of output force each) (454,000,000 kilograms) of continuous force using the limited watts consumed by the hydraulic pump directional valve linear actuators and repeat cycle timers, and the 110 watts of energy consumed by relatively small hydraulic pump handle electric motor 1408, to produce continuous output up to roughly either 62,500,000 watts (1,000,000,000 pounds of force / 10 times speed reduction / 16,000 pounds of force previously calculated as required per 10,000 watts) or 48,450 horsepower (1,000,000,000 pounds of force / 10 times speed reduction / 516,000 pounds of force previously calculated as required per 250 horsepower), while operating with a near perfect input output efficiency ratio. In another example of input output efficiency, a hydraulic device, provided 10 pounds (4.5 kilograms) of input force, to provide 100,000 pounds (45,360 kilograms) of output force, over a pump handle cycle time of 25 seconds when using hydraulics and 1 second when using an electric motor, layered 3 times, where the pistons of a layer operate the handles of the following layer until the last layer provides output force, not including the first hydraulic device as a layer, provides for a total of 100 million output providing units ((100,000 pounds of output force / 10 pounds of input force) ^ layers) which provide 10 trillion pounds of output force (100,000,000 units providing output force * 100,000 pounds of output force per unit), completing a cycle in a little over 10 minutes ((25 seconds ^ layers) + 1 second), all while utilizing only 10 pounds (4.5 kilograms) of input force. Archimedes is recorded as having stated "Give me a lever and a place to stand and I will move Earth." Utilizing this system, braced against a celestial body of appropriate mass and trajectory, Archimedes could have moved Earth with the force of his hand. Such an embodiment may be made possible by an additional law of physics discovered by the inventor, where layered leverage provides efficiency gains as a result of gains in layer output force (total output = (unit output force / unit input force) ^ layers) exceeding gains in layer cycle time (total cycle time = unit cycle time ^ layers). In some embodiments, including one complete embodiment, if any utilized commodity component, such as the force providing devices and or operating motors, doesn't perform with the force or speed expected, simply add units of that component and or adjust gear ratios. In some embodiments, such layered force providing device arrangements may be built directly into other force providing devices. Therefore, embodiments can provide for a self-powered generator and or motor that can easily be brought up to any level of output, allowing for clean continuous electricity generation and the fastest possible and fuel free transport, all at a cost that is effectively zero when amortized over time.

[0059] In some embodiments, which may include one complete embodiment, in reference to FIG. 1, double acting manual hydraulic pumps 1101 1103 1105 are replaced by, in reference to FIG. 3, a double acting electric hydraulic pump 3001 matched with hydraulic cylinder 3000 or hydraulic motor to support the desired speed and force. In some embodiments, which may include one complete embodiment if an electric double acting hydraulic pump is used, the amperage drawn by the electric hydraulic pump may be a function of the load that the electric motor is indirectly raising and lowering. For example, where the pump operates its cylinder piston at one inch per second, and therefore requires 5 times less force than the previously cited manual hydraulic pumps, where gear ratios are adjusted accordingly for the speed increase, the correspondingly reduced required force of 106,100 pounds (530,500 pounds / 5 times speed increase) (49,000 kilograms), may be provided by a 12 volt pump, which at full load provides 2,900 pounds of pressure per square inch (20 megapascals) at 13 gallons per minute, while according to a manufacturer, drawing 795 amperes for a total maximum consumption of 9,540 watts.

[0060] In some embodiments, force is instead provided by screw jacks to provide force in the system. In some embodiments, which may include one complete embodiment, force providing devices offer more force than is required, to reduce input force required, and increase system efficiency. In some embodiments, variable pressure hydraulic pumps may be used to precisely control the force provided by the hydraulics or pneumatics. In some embodiments, pairs of alternating force providing devices may be used to generate motion in the system. In some embodiments, other means of mechanical or hydraulic or pneumatic leverage, controlled electronically or manually, may be used. In some embodiments, multiple force providing devices may be used to support the maximum output capacity of the generator, or for redundancy to enhance reliability. In some embodiments, a traditional steam or car engine crankshaft may transfer rotational force from the hydraulics to the gears.

[0061] In some embodiments, when designed for small devices, such as portable consumer electronics, micro hydraulics, micro pneumatics, or miniaturized mechanical leverage devices such as a miniaturized screw jack, or functional equivalents, may be used. The simplicity of scissor screws jacks compared to hydraulics, may make them easier to miniaturize through three dimensional printing, where a screw jack may be constructed with components including a top and bottom that connect the gear beam and the support base, two blocks that allow a threaded screw to rotate through them, and 8 identical supports, two connecting to the top and the left screw block, two connecting to the bottom and the left screw block, two connecting from the top to the right screw block, two connecting from the bottom to the right screw jack, with a reversible motor connected to the screw, and the reversible motor connected to timers in the manner later described, to endlessly extend and retract the screw jack. Given the continuous movement of the screw jack, it may be best to have the screw in the screw jack rotate through a threaded block with a center that contains a reservoir of lubricant, so that each stroke of the screw jack is lubricated to allow for greater durability, where the screw and the threads it passes through may be made out of a durable metal such as high grade steel or titanium.

[0062] **In some embodiments, including one complete embodiment, obtain gears with corresponding gear racks and axles, to provide unidirectional force at an increased speed.** Gears increase speed in proportion to a decrease in force, as a ratio of teeth per gear pair. Ratchet gears are designed to allow force to be applied in a single direction, by using a structure called a pawl, which disengages from gear pickup points when the gear is not rotated in a specific direction. A gear rack is effectively a flattened gear, in the shape of a beam, with teeth that allow it to use linear force to engage and rotate a gear. Depending on the application, and the pounds (kilograms) of pressure per square inch (centimeter) provided to the gears, automotive or aerospace quality gears may be used, or gears may be custom manufactured including with extended widths to distribute pressure, and where gear types may include simple traditional spur gears, helical gears that may be smoother at high speeds, or others.

[0063] In some embodiments, including one complete embodiment, gear racks, gears, and supporting components are chosen or designed that are of adequate thickness at every point, and are rated and or produced of sufficiently high grade material, so as to never deform under applied forces, where the gear racks height may be determined by the components, and other dimensions may be determined to support the width of the teeth of the gears being powered. In some embodiments, including one complete embodiment, the specified gears and supporting components are made of a material such as high strength steel, which may have a yield strength of around 80,000 pounds of pressure per square inch (550 megapascals). In some embodiments, a metal such as titanium may be used, which may have a yield strength rating of 128,000 pounds of pressure per square inch (880 megapascals), and is substantially lighter than steel, but may be more brittle, and may currently trade for many times more, due to difficulties in extracting the element despite its abundance. In some embodiments, any materials may be used that provide adequate resilience.

[0064] In some embodiments, including one complete embodiment, in reference to FIG. 1, a dual gear rack is created by welding two gear racks 1201 1202 to a steel support beam 1200, where the gear racks are those designed to support the diameter and number of teeth of gear rack gears 1203 1204, where all components are of an adequate width to sustain the previously determined forces without deforming, with the gear racks welded to the support base 1200 in such a way that they precisely and snugly connect with the left sides of gears 1203 1204, with the steel support beam 1200 secured to pistons of hydraulic cylinders 1100 1102 1104 through welding or corresponding bolt ports and bolt holes in the support beam, to transfer the force provided.

[0065] In some embodiments, including one complete embodiment, in reference to FIG. 1, gear rack 1200 1201 1202 is aligned to operate gear rack gears 1203 1204, gear rack gear 1203 is coupled back to back to ratchet gear 1205, by means which may include welding, gear rack gear 1204 is coupled back to back to ratchet gear 1206, by means which may include welding, ratchet gear 1205 engages support gear 1209 through pawl 1207 bolted in place through a drilled hole, and ratchet gear 1206 engages support gear 1210 through pawl 1208 bolted in place through a drilled hole, with the pawls allowing the gears to engage in only one direction, resulting in bidirectional rotational force converted to unidirectional rotational force, where support gears 1209 1210 and have an increase in teeth relative the gears providing them force in order to increase speed, and interlock with each other and with gear 1211, which is coupled back to back to ratchet gear 1212, by means which may include welding, and engages through pawl 1213 bolted in place through a drilled hole in weighted steel momentum wheel 1214, which may be omitted or may weigh any amount including that corresponding to the force required by the generator, or the generator and the motor, to maintain continuity during force providing device transition, and is coupled back to back, by means which may include welding, to gear 1215, which provides a teeth ratio relative to the gear providing it force in order to further increase speed, and interlocks with generator axle gear 1216, with the the ratios of teeth of gear pairs selected to achieve the desired speed increases at each point to provide the final desired unidirectional speed.

[0066] In some embodiments, including one complete embodiment, in reference to FIG. 1, to convert the input force to unidirectional force at the previously determined 375 times speed increase, gear rack gears 1203 1204 each have 5 teeth, ratchet gears 1205 1206 each have 5 teeth, support gears 1209 1210 each have 100 teeth, where 5 teeth of the previous gear in proportion to 100 teeth on this gear provides a speed increase of 20 times, gear 1211 has 5 teeth, ratchet gear 1212 has 5 teeth, gear 1215 has 95 teeth, where 5 teeth of the previous gear in proportion to 95 teeth on this gear provides a further speed increase of 19 times, generator axle gear 1216 has 95 teeth, which therefore together achieve unidirectional speed at around the overall 1:375 previously determined teeth ratio (1:(20*19) = 1:380), converting the input speed to around 1800 revolutions per minute. If layered force providing devices are utilized, gear ratios may be additionally adjusted, and or gears may be added, as previously discussed, to provide appropriately adjusted speed. In some embodiments, instead of a momentum wheel, additional force providing devices may be added and timed with the repeat cycle timers to ensure there is never a stop in force. In some embodiments, including one complete embodiment, the generator is attached in such a position that its axle is rotated in the designed direction for electricity generation, or an additional gear may be added if necessary before the generator axle gear to change the rotational direction.

[0067] In some embodiments, including one complete embodiment, in reference to FIG. 1, when implementing the system to rotate an axle to function as a motor, to additionally gear the motor axle to achieve the desired revolutions per minute, motor axle 1218 has passing through its center generator axle gear 1216 which connects to gear 1217, where the gears have a teeth ratio that achieves the revolutions per minute desired of the motor.

[0068] In some embodiments, including one complete embodiment, in reference to FIG. 1, when implementing the system to rotate an axle to function as a motor, to additionally gear the motor axle, which requires as determined by previous calculations increasing the speed by 4 times, to around the previously determined 7,000 revolutions per minute, generator axle gear 1216 has 95 teeth and gear 1217 has 24 teeth, approximately providing the previously determined 1:4 ratio of teeth, thereby rotating the motor axle 1218 at around 7,200 revolutions per minute. In some embodiments, which may include one complete embodiment, any number of gear and gear rack teeth and ratios may be used that allow for the desired force and speed to be achieved.

[0069] The rotational speed of the gear powering the generator should match the optimal speed of the generator for maximum output, and if the speed doesn't match, given the inability to control in advance the precise speed of the purchased commodity force providing devices, the gear providing rotational force may be switched to one whose teeth provide for an appropriate speed adjustment. In some embodiments, to accurately measure the output of the embodiment, a torque gauge may be used to measure pound feet of force, and a tachometer may be used to determine revolutions per minute, or alternatively a dynamometer may be used which measures pound feet of force as well as revolutions per minute, and then horsepower may be calculated as pound feet of force multiplied by revolutions per minute with the result divided by the horsepower constant 5252.

[0070] In some embodiments, where the unit is not guaranteed to operate in a continuously upright manner, such as for vertical takeoff and landing, or in a reduced or zero gravity environment, ratchet pawls are to be pushed down to compensate for gravity to ensure desired engagement, by means which may include mounting to each pawl one or more springs, opposing magnet pairs, or small electric motors.

[0071] In some embodiments, any other method of converting linear force to rotational force may be used, including other systems of gears, which may utilize gears of assorted number, size, arrangement, and teeth. In some embodiments, which may include one complete embodiment, rotational force is provided by one or more hydraulic motor(s) and corresponding pump(s), requiring no linear to rotational force conversion, whose pump operation may be automated using the electric motors, linear actuators, repeat cycle timers, layered force providing devices, and or functional equivalents, all as provided for in this disclosure.

[0072] In some embodiments, a traditional steam or car engine crank shaft 1200, may replace the gear racks 1201 1202, and be connected by means which may include bolting directly from the top of the hydraulic cylinders pistons gear rack bar 1200 to a gear added between gear rack gears 1203 1204 so it can provide force in either direction, or directly to gear 1204, where the connected gear may be adjusted in diameter to support a full piston cycle, and where if connecting to gear 1204 to ensure the system provides rotational force in only one direction gear 1214 functions as a weighted momentum wheel providing adequately weighted momentum to ensure continuous motion in a single direction, with additional initial input force added to power the momentum wheel, or may have another connected gear providing the weighted momentum after the speed has been reduced and the gear can provide the required momentum at a reduced weight, or may require the additional component or functional equivalents of a starter motor, and where the rear flange mounted hydraulic cylinder may be mounted at its base to an intermediary joint that may allow the cylinder to move with the crank shaft. In some embodiments, instead of a hydraulic piston a bi-directional hydraulic gear motor 1100 connects with and provides rotational force to a gear added between and interlocking with gear rack gears 1203 1204, operating in place of removed gear rack 1200 1201 1202, where the hydraulic motor can provide rotational force in either direction, in accordance with the flow of connected automated hydraulic pump 1101 as provided for in this disclosure, with adjustments made to gears and gear ratios to provide the speed and force required. In some embodiments, the hydraulic motor operating hydraulic pump 1101, is operated by another force providing device, as provided for in this disclosure, increasing its input output efficiency. In some embodiments, the hydraulic motor may be coupled directly to the generator axle or torque converter or motor axle or a rotational force output device such as a turbine or tire. In some embodiments, multiple hydraulic motors are interconnected to provide rotational force, by means which may include connecting the hydraulic motors through additional gears interlocking with gear rack gears 1203 1204.

[0073] In some embodiments, a gear beam may be used that has folding teeth that engage in only one direction, with individual gear teeth each attached to the beams by a bolt that passes through a gear beam hole, with each tooth extending beyond the length of its underlying support, where the teeth connect with the teeth of the gears, so when the force providing devices go up, the teeth are engaged, and when the force providing devices go back down, the teeth disengage by folding themselves in, thereby allowing alternating hydraulics or equivalents to provide the continuous application of force to the generator and or motor axle, so that each properly engages the teeth of the corresponding gear, or providing such beam structures on each side of the force providing devices in vertical opposition to capture force in both directions.

[0074] In some embodiments, in reference to FIG. 1, instead of using gear racks 1200 1201 1202, in reference to FIG. 5, force from the hydraulics is transferred through a pressurized medium such as liquid or gas 5001, in conduit or container 5000, which may be made of adequate thickness and of a material of adequate strength so as to not deform under applied forces, for example titanium or high strength steel, where the substance container has one hole connected to pipe t-joint 5100 and another hole connected to pipe t-joint 5102, with pipe 5101 connecting the two t-joints, and turbine 5200 passes through pipe 5102, and through a hole drilled in t-joint 5102, sealed by elastomer seal 5201, with the axle of turbine 5200 having a gear 5202, that connects between, in reference to FIG. 1, initial force transfer gears 1203 and 1204, where when the piston of hydraulic cylinder 1100 is connected to, in reference to FIG. 5, container piston 5002 which operates plate 5003 back and forth in container 5000, with the piston sealed in the container by elastomer seal 5001, turbine 5200 is spun back and forth.

[0075] In some embodiments, which may include one complete embodiment, in reference to FIG. 1, instead of using gear racks 1200 1201 1202, in reference to FIG. 6, force from the hydraulics is transferred through a weighted pendulum 6000, where the pendulum may be made of a heavy material, for example steel or iron, and of a weight that exerts force that meets or exceeds that required by the generator and or momentum wheel and or to function as motor, which is raised for release by hydraulic cylinder piston pendulum bar 6100 which is coupled to piston of hydraulic cylinder 1100 by means which may include welding or bolt 6101, with hydraulic cylinder mounted horizontally on an additional support beam, with piston pendulum bar 6100 moving in back and forth cycles pendulum wheel 6000, with pendulum wheel 6000 having on each side support bars 6200 6201 coupled together by bolt 6202 utilizing lubrication and or a bearing to minimize friction, with the other end of support bars 6200 6101 coupled together by bolt 6203 which also passes through pendulum weight support beam 6204, and utilizes lubrication and or a bearing to minimize friction, where pendulum bar 6100 provides rotational force to the axle generator 1300, after being passed through any necessary gears to increase speed, where the disclosed repeat cycle timers or functional equivalents are wired and timed to push piston of hydraulic cylinder 1100 into a position for release to complete a cycle, then timed to wait for the cycle to complete, then pull it into a position and release it to repeat the cycle, where the pendulum can be of any weight supported by the hydraulics, with the pendulum motion rate of decay driven by factors which may include the resistance of the generator, gravity, and rotational friction, where the heavier the pendulum the longer it will take to be stopped each cycle. In some embodiments, the pendulum speed may be driven by gravity at 9.8 meters per second per second and it may achieve a peak oscillation speed of around the 22 miles (35.4 kilometers) per hour.

[0076] In some embodiments, which may include one complete embodiment, in reference to FIG. 1, instead of using gear racks 1200 1201 1202, in reference to FIG. 7, which utilizes some components from FIG. 1, force from hydraulic motor 1100 is used to raise weight 7100, where the weight can be of any weight supported by the hydraulics, attached to cable 7101 which is connected to and wrapped around rod 7001 which is coupled directly or indirectly through gears to the axle of generator 1300 or is able to function as a motor axle, where weight 7100 may be composed of a heavy material, for example steel or iron, and of a weight that exerts force that meets or exceeds that required by the generator and or momentum wheel and or to function as motor, where axle of hydraulic motor 1100 couples and decouples with an inset on axle 7001 on a timed cycle, with weight 7100 raised and dropped in cycles, where during drop the force provided by the weight is compounded by gravity, with hydraulic motor 1100 connected to and resting on support beam 7002 by connecting to a rectangular opening on support beam 7200 which act as a track for hydraulic motor 1100 which is able to move on the track by welded bolt 7003, with axle 7001 passing through one or more support bars including support bar 7000 utilizing lubrication and or a bearing to minimize friction, with support bar 7000 being the approximate height of the desired drop, where weight 7100 may be enclosed in container 7300, which may be made of a material including that used for the weight, with container 7300 holding a medium 7301, which may be a liquid including water, which is at the bottom of the container to cushion the impact of the weight as it completes its drop, where the axle coupled to the generator may instead first be passed through gears to obtain the speed and force desired by the generator and or to function as a motor, where repeat cycle timers 7201 7202 or functional equivalents, connected in a manner similar to that for other disclosed repeat cycle timers, are wired and timed to use linear actuator 7200 to extend hydraulic motor 1100 into a position to lock with axle 7001 to raise weight 7100 and then retract linear actuator 7200 to allow the weight to drop, then timed to wait for the cycle to complete, then pull repeating the cycle. In some embodiments, the weight fall speed may be driven by gravity at 9.8 meters per second per second and it may achieve a peak speed of around the 22 miles (35.4 kilometers) per hour, and the weight obtaining the compounding benefit of gravity may be only that in excess of the force required to power the generator.

[0077] In some embodiments, any other method of transferring and or increasing input speed may be used.

[0078] **In some embodiments, including one complete embodiment, construct a support structure around, and appropriately connect, the force providing devices, gears, gear racks, axles, and generator.** In some embodiments, including one complete embodiment, in reference to FIG. 1, in order to support components, a load bearing support structure is constructed of adequate thickness and of a material of adequate strength so as to not deform under the pressures in the embodiment, for example high strength steel or titanium. In some embodiments, including one complete embodiment, in reference to FIG. 1, the structure consists of base beams 1000 1001 1002 1003, connected together in a rectangle, an inside base beam 1004 to support and connect to the hydraulic cylinders, an inside base beam 1005 to support the generator, gear support beam 1007, extending from base beam 1000 and supported by beam 1006 extending from beam 1003 and beam 1008 extending from beam 1001, and having welded to it gear support rods 1009 1010 1011, where in order to ensure the gears precisely align as previously described, the gears may be mounted on the gear support rods before the support rods are fixed into place by means which may include welding. In some embodiments, including one complete embodiment, the exact dimensions of each support structure components are matched to support the corresponding components.

[0079] In some embodiments, the method of connecting the beams and rods may be bolting or welding or any other means that may secure the beams. In some embodiments, the support structure, or parts thereof, may consist of solid components created rather than connected, by means that may include molds or 3D printing.

[0080] In some embodiments, including one complete embodiment, in reference to FIG. 1, generator 1300 is attached firmly to the support structure beam 1007, by means which may include its built in bolt mounts, with corresponding holes drilled in the support beam, or by welding, at a location the ensures generator support gear 1215 is securely engaged with generator gear 1216 mounted on the generator axle.

[0081] In some embodiments, in reference to FIG. 1, if the potential for even the smallest discontinuity in force can't be allowed, gears 1203 1204 may each have their own force providing device and beam, operating in opposition on the appropriate sides of gears 1203 1204, with the force providing devices precisely timed to overlap to prevent any discontinuity in force.

[0082] **In some embodiments, including one complete embodiment, implement a timer based control system.** A repeat cycle timer is a commodity device used to open and close power circuits to operate systems such as a water sprinkler, turning it on for a fixed interval, turning if off for a fixed interval, and then repeating the cycle. The instructions for configuring the timers, when purchased from a quality supplier, will be in the manual accompanying the timers. For example, some timers may have positive and negative terminals for each the power source and the device, along with a knob to set the seconds on per cycle, and a knob to set the seconds off per cycle. In some embodiments, including one complete embodiment, motors including those driving linear actuators may be utilized that have two wires and are operable in both directions, driven by a reversible brushed direct current motor, though other types of motors may be utilized which may have additional wires to be wired in the manner corresponding to their accompanying instructions.

[0083] In some embodiments, including one complete embodiment, in reference to FIG. 1, repeat cycle timers 1400 1401 are set to run on intervals timed to allow the force providing devices attached gear teeth beams to move approximately one length in a back and forth cycle while minimizing any discontinuity in force during directional transition, with the repeat cycle timer settings implemented as explained by the manufacturer in included instructions. The repeat cycle timers are attached to support beam 1000, by means which may include bolts through drilled corresponding bolt holes, straps, or any other method that doesn't damage the timers. In some embodiments, including one complete embodiment, in reference to FIG. 1, repeat cycle timers 1400 1401 operate linear actuators 1408 1409 1410 affixed respectively to the directional valves of hydraulic pumps 1101 1103 1105, where positive wire of each linear actuator is attached to both repeat cycle timer 1400 positive terminal, and repeat cycle timer 1401 negative terminal, and negative wire of each linear actuator is wired to repeat cycle timer 1400 negative terminal and repeat cycle timer 1401 positive terminal, with repeat cycle timers 1401 and 1402 each wired to the corresponding positive and negative terminals of on off switch 1501 which provides power through a connection to battery 1500, in order to operate the hydraulic pump directional valves as required by the embodiment. In some embodiments, where an electric force providing device pump is used, repeat cycle timers 1400 1401 open and close circuits that are wired to and cause, in reference to FIG. 3, electric pump 3100 operating switch motor 3200, or the exposed wires from the operating switch, to raise and lower the piston of cylinder 3000, on a fixed interval, and for a fixed interval, as required by the embodiment. In some embodiments, which may include one complete embodiment, in reference to FIG. 4, where force providing devices are mounted in opposition to each other, to provide both push and pull force for continuous motion, components including repeat cycle timers and wiring are repeated with appropriately adjusted timing.

[0084] In some embodiments, which may include one complete embodiment, the functional equivalent of a repeat cycle timer may instead be used, including a relay board 1701 controlled by computer 1700, operated by computer software that opens and closes the circuits as previously described, to produce the same effect, and or to turn the units on and off to meet peak demand at specific times or based on current consumption, with code and components utilized as disclosed in the later step describing creation of unit operation software code.

[0085] In some embodiments, the repeat cycle timer may be excluded if the hydraulics are to be manually operated.

[0086] **In some embodiments, including one complete embodiment, to transfer electricity to power embodiment components, wire the generator through any necessary adapter or converter to the battery and an on off power switch.** In some embodiments, including one complete embodiment, in reference to FIG. 1, battery 1500 allows the unit to start by providing initial power to pumps handle motor 1402 and repeat cycle timers 1400 1401 which power the pumps' directional valve linear actuators, where the battery has the capacity to simultaneously power the required components. In some embodiments, including one complete embodiment, in reference to FIG. 1, output wires of generator 1300 are connected to on off power switch 1501 and battery 1500, and if there is a mismatch of current type, amperes, volts, or hertz, or the strength of the electrical output from the generator would damage the repeat cycle timers, hydraulic pump operating motor, and or battery, they may be wired to pass through a corresponding electrical converter and or overcharge controller. In some embodiments, including one complete embodiment, in reference to FIG. 1, in order for the battery to be replaceable, it may be attached to support beam 1000 by means including metal brackets bolted through corresponding drilled holes. In some embodiments, including one complete embodiment, in reference to FIG. 1, on off power switch 1501 is externally accessible to the operator, and may be coupled to the unit, by means including welding or bolts.

[0087] **In some embodiments, including one complete embodiment, attach power output connector.** In some embodiments, including one complete embodiment, in reference to FIG. 1, the output wires of generator 1500 are connected to power connector 1502, which provides the outlet or wire connectors that provide electricity to the end user, and is externally accessible after any protective enclosure is mounted. In an embodiment where the generator does not provide output as desired including current type, amperes, volts, and hertz, generator 1500 may be wired to a power converter which is then wired to output connector 1502.

[0088] **In some embodiments, including one complete embodiment, if implementing the unit as a motor, optionally mount Continuously Variable Transmission or Torque Converter, or construct and mount a Tube Torque Converter.** A traditional torque converter spins an engine connected impeller opposing an axle connected turbine, inside a container filled with fluid, allowing the output axle and the motor axle to operate at different speeds, engaged by the force transferred through the fluid, however, at high levels of force, traditional torque converter seals may fail. A Continuously Variable Transmission, is a commodity system that transfers force from an engine to a motor axle while allowing an engine and the motor axle to operate at different speeds, by varying the circumference of opposing wheels connected through a belt, though these devices may provide uneven engagement. In some embodiments, which may include one complete embodiment, a traditional torque converter or Continuously variable transmission may be used. In some embodiments, a hydraulic motor is used in conjunction with variable force hydraulic pumps and a torque converter may not be required.

[0089] In some embodiments, which may include one complete embodiment, a new type of torque converter is constructed, named a Tube Torque Converter, which is less concerned with maintaining force transference efficiency than existing systems, because embodiments where it is implemented, may not utilize an external fuel source and can therefore be easily made more powerful to compensate in proportion to the loss of efficiency, where embodiments allows the motor's connected force transference device to spin at very high speeds while fixed in place, with an opposing force transference device operating at any level of speed, determined by moving the opposing force transference device back in forth in a transferrable medium filled container, of adequate dimensions so as to not generate compressive force that damages seals, therefore providing completely smooth acceleration with fewer parts and at a lower construction and maintenance cost than traditional torque converters and gear trains, and without the jump of gear shifting or traditional Continuously Variable transmission.

[0090] In some embodiments, including one complete embodiment, in reference to FIG. 1, the torque converter is produced by having motor axle 1218 spin an attached high strength fixed pitch propeller turbine 1602, or another fluid rotational force transference device such as an impeller, at a constant speed, inside high pressure rated pipe 1600, where the additional diameter of the container relative to the turbine allows for continuous flow around the turbine, with turbines each set to not come closer on all sides of the pipe and the compartment seal caps than would cause to damaging pressure from accumulating on the elastomer seals, with the pipe containing commodity torque converter fluid 1601, with a high strength fixed pitch propeller turbine 1603, or another fluid rotational force transference device such as an impeller, attached to axle 1604, which spins inside the pipe at the rate controlled by how close it is to opposing turbine 1602, with engagement controller 1607 controlling axle 1604 as a steel rod with a hole that fits the axle secured to push and pull the axle by a divot fitting the engagement controller in the axle or by metal holders on either side surrounding the axle, with the pipe 1600 having two threaded ends over which threaded end caps of the same thickness as the pipe will be placed, with holes in each cap for the axles sealed by commodity heavy duty torque converter elastomer seals 1605 1606, attach the turbine component to the axle such that it is able to move back and forth, with two caps, and a hole of the diameter of the axles, made in each cap, with the axles passing through. In some embodiments, which may include one complete embodiment, pipe 1600 has pressure reduction chambers on either side of the pipe, through which the axle passes on either side, also filled with torque converter fluid, created by sealing the inner compartment with two circular high strength plates whose diameter matches the diameter of the inside of the pipe, and whose center hole diameter matches that of the axles, is cut from an additional pair of caps using a welder or metal cutting machine, which are welded into place, to form a physical barrier to reduce the force on the elastomer seals. In some embodiments, which may include one complete embodiment, fixed pitch propeller turbines 1602 1603 have a 12 inch diameter and are made of high strength steel, and high strength steel pipe 1600 is of 24 inch diameter by 36 inch length 0.5 inch thick, where the distance from the turbines to the pipe may be 6 inches, with turbine pitch selected, and distances and dimensions adjusted to optimize efficiency.

[0091] In some embodiments, including one complete embodiment, in reference to FIG. 1, the engagement of this torque converter may be controlled by means including a linear actuator 1702, connected to the torque converter engagement controller 1607 handle by means which may include welding or bolting, and able to provide adequate force to smoothly transition the engagement controller 1607 over its full range, or mounted to the support structure, or a double acting rear flange mounted hydraulic cylinder in place of linear actuator 1702 with an extension length able to take engagement controller 1607 through its full range of motion, and controlled by hydraulic pump and corresponding switch where the switch moves forward at the press of the gas pedal, and the switch moves backwards when pressed by the brake pedal, or with the switch connected to a computer controlled relay board, either directly by cutting off the switch and connecting the wires to the computer controlled relay board, or by using a motor that press the hydraulic pump switches. In some embodiments, software to control the hydraulic cylinder piston and pump engagement, or the stepper motor engagement, may be implemented as described in a later step, and may have pre-determined engagement distances for each level of output, to account for non-linear distance to force transfer. In some embodiments, which may include one compete embodiment, when the embodiment is providing force to a fixed object such as a tire, the primary unit may move back and forth rather than the output turbine of the torque converter axle providing force to the tire.

[0092] **In some embodiments, when implemented as a vehicle motor for vertical take off and landing, allow for dynamically angling of the unit with a hydraulic cylinder with hydraulic pump controller and optional high strength encasement.** In some embodiments, an optional force providing device, including but not limited to hydraulic, pneumatic, motorized mechanical leverage, or functional equivalent, is attached between the support structure and a vehicle that allows the engine to be dynamically angled, for applications such as the vertical take off and landing of a car, airplane, or space ship. In some embodiments, in reference to FIG. 1, force providing device cylinder 1106 is operated by pump 1107, with cylinder 1106 mounted to the support structure and the vehicle in a manner that allows it to move the engine through a range of motion providing for vertical take off and landing as well as forward motion, where electronically controlled pump switch 1108 either has mounted on it reversible motor 1406 with button pressing rod 1405 held in place by metal clamp 1407, or has its switch cut off with the control wires wired to replay board 1701 for back and forth control as provided for in this disclosure in a similar manner as reversible motor 1406. If a cylinder is added, it can be connected to a support structure beam 1000 with the other end connected to the vehicle, and support structure beam 1002 attached to the vehicle through a jointed bracket, to allow for the motor to rotate in a manner that allows for vertical takeoff and landing. In some embodiments, where extreme force is applied, and used in life or death situations, such as airplanes, cylinders may be enclosed in high strength encasement made of a material such as titanium in the exceptionally unlikely event of a pressurized cylinder explosion, and embodiments may be redundant, in case of a failure. In embodiments where fluid utilized by the pump and piston would be exposed to freezing temperatures, the fluid must be prevented from freezing by means which may include thermal insulation and or an anti-freeze agent added to the fluid.

[0093] **In some embodiments, including one complete embodiment, optionally write software and implement supporting components to control unit operation and or an array of units.** In some embodiments, which may include one complete embodiment, where computer control of the unit is desired, to operate the force providing devices, and or to operate the torque converter, and or to operate the unit angling for vertical take off and landing, software may be written and supporting components implemented, where a computer may be embedded in the unit, running custom software on startup, and may be connected through a network connection to a network accessible relay board. In some embodiments, where one or more variable pressure force providing device pumps are used, pressure levels corresponding to output levels may be stored for use by the software which sets the pumps operating levels by controlling relay board attached motors, including in the form of standard, stepper, linear actuator, or hydraulic, that control pressure level valves and or handles according to desired output. In some embodiments, where a tube torque converter is used, force transference distances corresponding to output levels may be stored for use by the software which sets the distance corresponding to the desired output level by operating a motor, including in the form of standard, stepper, linear actuator, or hydraulic, connected to the torque converter engagement handle. The software code to operate the computer connected relay board, may resemble that disclosed below. In some embodiments, the computer may be a Raspberry Pi, connected through an Ethernet controlled relay board through an Ethernet crossover cable, that is set on startup to run software developed in the Python language from the disclosed description, where the software is installed by connecting to the computer through telnet or secure shell, then at the command prompt typing "nano run.py" and adding and saving the software code, then at the command prompt typing "nano /etc/rc.local" and adding and saving the line "python /$location/run.py", where $location is the path to the directory containing the previously created software file.

[0094] In some embodiments, the software code for unit control, provides functionality comprising:
> import a library for connecting to the relay;
> import a library for system resource access;
> create variables holding relay on off values;
> initialize variables for each relay state, including force providing device up and down, torque converter forward and backward, and unit forward and backward;
> create a variable holding force providing device cycle seconds;
> create variables holding the relay board connection, IP address, username, and password.
> create a function to update the relay board;
>> initialize a connection to the relay board, if the connection has not been initialized, or has been dropped
>> create a string holding the states of the relays;
>> send the current relay states to the relay board.
> create a function that when called listens for computer input from the operator;
>> read in any user key press;
>> if the key pressed is forward, set the torque converter control device relay state to forward;
>> if the key pressed is backward, set the torque converter control device relay state to backward;
>> if the key pressed is up, set the unit position control device relay state to forward;
>> if the key pressed is down, set the unit position control device relay state to down;
>> if the key pressed is the spacebar, set all torque converter and unit positioning relay states to off.
> create the function called on program start;
>> record the program initialization time;
>> run a continuous loop;
>>> if the seconds elapsed are divisible without remainder by the force providing device cycle seconds switch the force providing device direction by changing corresponding relay states;

>>> check for user input;
>>> call the function to send the current relay states to the relay board.

[0095] In some embodiments, where a large number of generator units are being operated concurrently, to turn the units on and off at specific times and or based on current power consumption, network accessible relay boards may be installed in the units, and wired in place of or in addition to the power switch to be able turn the units on and off, and may be controlled by a computer running software provide functionality comprising:
> import a library for accessing the relays;
> import a library for system resource access;
> initialize and set variables holding relay on and off values;
> create an array of IP addresses of unit on off relay boards;
> initialize an array of unit relay board connections;
> create an array of unit output watts;
> create an array of unit on off states;
> create variables holding the default unit relay board username and password;
> create variables to hold the total watts available across all units, the current watts being consumed, and whether or not time based watts are to be used;
> create an array of pairs of times and time desired watts;
> create a variable indicating whether or not unit power consumption meters are to be used;
> create and array of power consumption meter IP addresses;
> create variables holding the power consumption meter's username and password;
> create variables holding the minimum and maximum power consumption to be allowed before switching units on or off.
> create a function to send a command to a unit at a specified index in the unit IP address array;
>> create a connection to the unit on off relay board, at the IP address at the provided index in the unit IP address array, if the connection has not been initialized, or had been dropped;
>> create a variable holding the command string to be sent to the unit based on whether the unit is to be turned on or off;
>> send the command the unit relay;
>> wait for the command to go through, then turn off all relays.
> define a function to run on script execution;
>> iterate through each unit IP address and call the function to switch the unit on;
>> run a continuous loop;

>>> create variables to hold a unit index iterator, an output display message, and the current action;
>>> proceed if power consumption meters are being used;
>>>> create variables to hold average and total consumption of unit power, and add up total possible power output;
>>>> loop through each power consumption meter IP address;
>>>>> retrieve the current power consumption number;
>>>>> calculate the average consumption;
>>>>> add the current consumption to the total;
>>>>> add the unit maximum capacity to the total consumption capacity;
>>>>> increment the index;
>>>> calculate the consumption percentage as the consumption total divided by the total unit capacities;
>>>> create a message to display the states the currently consumed watts and the current watt capacity;
>>>> if the average consumption is greater than the maximum consumption level before more unit should be turned on, then set the unit action equal to on;
>>>> if the average consumption is less than the minimum consumption level before more unit should be turned off, then set the unit action equal to off;
>>>> iterate through unit states until one is found that is either off, if looking to turn a unit on, or on if looking to turn a unit off, then send the command to switch the unit state;
>>> proceed if power consumption meters are not used, and instead the total watts to be provided by the units are determined by the current time;
>>>> iterate through each time watts pair, and if the current time is equal to the specified time, send commands to turn units on or off, until the desired level of output is produced, and display each action on the command line;
>>> sleep for a moment before looping again.

[0096] **In some embodiments, including one complete embodiment, attach protective enclosure.** In some embodiments, including one complete embodiment, in reference to FIG. 1, enclosure 1900 protects the electronic components from external elements and the operator from the utilized forces as well as the extraordinarily rare event of a cylinder explosion, where the enclosure is shaped to fit around the support structure and all components, and may be made of sheets a strong lightweight environmentally resistant material, for example high strength steel, aluminum, or carbon fiber, formed by means which may include fabrication or welding of sheets in the dimensions of the unit, with the enclosure coupled to the support structure by means which may include bolts through corresponding drilled holes, where there may be holes in the enclosure allowing the power connector 1502 and power switch 1501 to be externally accessible to the operator. In some embodiments, many of the embodiment components may be 3d printed from a single model, including the support structure, protective enclosure, gears and supports, and potentially the force providing device(s).

[0097] **In some embodiments, including one complete embodiment, enjoy clean continuous self-powered energy and or propulsion.** In some embodiments, turn the unit's power switch to on, and allow the unit to run for a full cycle to validate structural stability, and once the unit is verified as correctly constructed, turn it off. In some embodiments, including one complete embodiment, if implemented as a generator, the embodiment is then ready to have an experienced licensed electrician connect it to a power grid, and turn on the unit's power switch to on, or if implemented as a motor, be appropriately connected in an engine compartment.

CLAIMS

What is claimed is:

1. An apparatus capable of powering a generator and or functioning as a motor, with the invention comprising:
   force providing devices(s) including but not limited to hydraulic, pneumatic, mechanical leverage, motorized mechanical leverage, and or functional equivalents;
   said force providing device(s) able to be powered or operated, directly or indirectly, by a generator and or another source;
   said force providing device(s) force able to be transferred directly or indirectly by means that provide rotational force to said generator axle and or to function as a motor.
2. Further comprising claim 1, gears, a crankshaft, or functional equivalent(s), able to convert multidirectional force to unidirectional rotational force.
3. Further comprising claim 1, a weighted structure, which may be of any shape, including circular or spherical, attached to an axle, able to maintain momentum.
4. Further comprising claim 1, said force providing device(s) able to be operated by force providing device(s) including but not limited to hydraulic, pneumatic, mechanical leverage, motorized mechanical leverage, and or functional equivalents to optimize input output efficiency.
5. A method performed by an apparatus comprising:

transmitting electricity to operate directly or indirectly one or more force providing device(s) including but not limited to hydraulic, pneumatic, mechanical leverage, motorized mechanical leverage, and functional equivalents;

said force said force providing device(s) force transferred directly or indirectly by means that provide rotational force to a generator axle and or to function as a motor;

transmitting electricity from the generator directly or indirectly to operate directly or indirectly said force providing devices;

6. Further comprising claim 5, converting multidirectional force to unidirectional rotational force through gears, a crankshaft, or functional equivalent, with supporting components.
7. Further comprising claim 5, a weighted structure, which may be of any shape, including circular or spherical, attached to an axle, maintaining momentum.
8. Further comprising claim 5, said force providing device(s) operated by force device(s) including but not limited to hydraulic, pneumatic, mechanical leverage, motorized mechanical leverage, and or functional equivalents to optimize the input output efficiency ratio.
9. A method for constructing an apparatus comprising:

   obtaining a power source including but not limited to a generator, repeat cycle timers or functional equivalents, force providing device(s) including but not limited to hydraulic, pneumatics, mechanical leverage, motorized mechanical leverage, and or functional equivalents;

   ensuring the attachment to a support structure of said force providing device(s), said repeat cycle timers or functional equivalents, and said power source;
10. Further comprising claim 9, attaching directly or indirectly gears, a crankshaft, or functional equivalent with supporting components to convert multidirectional force to unidirectional rotational force.
11. Further comprising claim 9, attaching a weighted structure, which may be of any shape, including circular or spherical, to an axle, to maintain momentum.
12. Further comprising claim 9, operatively coupling said force providing device(s) to force providing device(s) including but not limited to hydraulic, pneumatic, mechanical leverage, motorized mechanical leverage, and or functional equivalents to optimize the input output efficiency ratio.
13. A force transference device, with the invention comprising:

   a turbine or functional equivalent;

   a means for holding a transferrable medium;

individual or combined implementation of force providing device(s) including hydraulic, pneumatic, mechanical leverage, motorized mechanical leverage, and functional equivalents;
said force providing devices able to cause the flow of a medium through said means for holding a transferrable medium past said turbine.

14. A force transference device, with the invention comprising:
a pendulum;
a generator axle or motor axle;
individual or combined implementation of force providing device(s) including hydraulic, pneumatic, mechanical leverage, motorized mechanical leverage, and functional equivalents;
said pendulum connected directly or indirectly to said generator axle or motor axle;
said pendulum able to be operated by said force providing device(s).

15. A method performed by an apparatus, with the invention comprising:
a pendulum moved for release by individual or combined implementation of force providing device(s) including hydraulic, pneumatic, mechanical leverage, motorized mechanical leverage, and functional equivalents;
a pendulum transferring force to a generator axle or motor axle;
said pendulum connected directly or indirectly to said generator axle or motor axle;
said pendulum able to be operated by said force providing device(s).

16. A force transference device, with the invention comprising:
a weight;
an axle;
individual or combined implementation of force providing device(s) including hydraulic, pneumatic, mechanical leverage, motorized mechanical leverage, and functional equivalents;
said weight able to couple directly or indirectly to said axle;
said weight able to be raised by said force providing device(s);
said weight able to rotate said axle.

17. A method performed by an apparatus, with the invention comprising:
a weight moved for release by individual or combined implementation of force providing device(s) including hydraulic, pneumatic, mechanical leverage, motorized mechanical leverage, and functional equivalents;
said weight transferring force to rotate an axle.

18. A torque controller, with the invention comprising:
a pair of opposing rotational force transference devices;

a container capable of enclosing a transferrable medium and said rotational force transference devices;

said rotational force transference devices adjustable in proximity.

19. A manually operable force providing device including hydraulic, pneumatic, mechanical leverage, motorized mechanical leverage, and or functional equivalents, converted to an automatic force providing device, with the invention comprising:

    a motor or motorized device able to provide the input force required by a force providing device;

    a motor or motorized device able to control the force providing device direction;

    a connection between said force providing device input force receiver and corresponding motor able to take said input force receiver through a cycle;

    a power source;

    one or more repeat cycle timer(s) or functional equivalents;

    said repeat cycle timer(s) able to be powered by said power source able to control said motor(s) to control said input force receiver and or said valve.

20. An apparatus comprising:

    one or more force providing device(s) including but not limited to hydraulic, pneumatic, mechanical leverage, motorized mechanical leverage, and or functional equivalents able to operate one or more force providing device(s) including but not limited to hydraulic, pneumatic, mechanical leverage, motorized mechanical leverage, and or functional equivalents to improve input output efficiency.

21. A non-transitory computer-readable recording medium holding stored instructions, which when executed by one or more processing devices, cause the one or more processing devices to implement a method comprising:

    turning power producing units on and off to meet desired power output, either or both at specific times, or by reading the power consumption meter of one or more units, and if the average power being consumed is above a certain threshold, additional units are turned on, and if power being consumed is below a certain threshold, units are turned off.

22. A non-transitory computer-readable recording medium holding stored instructions, which when executed by one or more processing devices, cause the one or more processing devices to implement a method comprising:

    adjusting engagement of a torque converter controller or hydraulic pressure controller(s), utilizing a computer controlled motor or motorized device, adjusted according to user input and or stored engagement to output levels, to control output.

FIG. 1

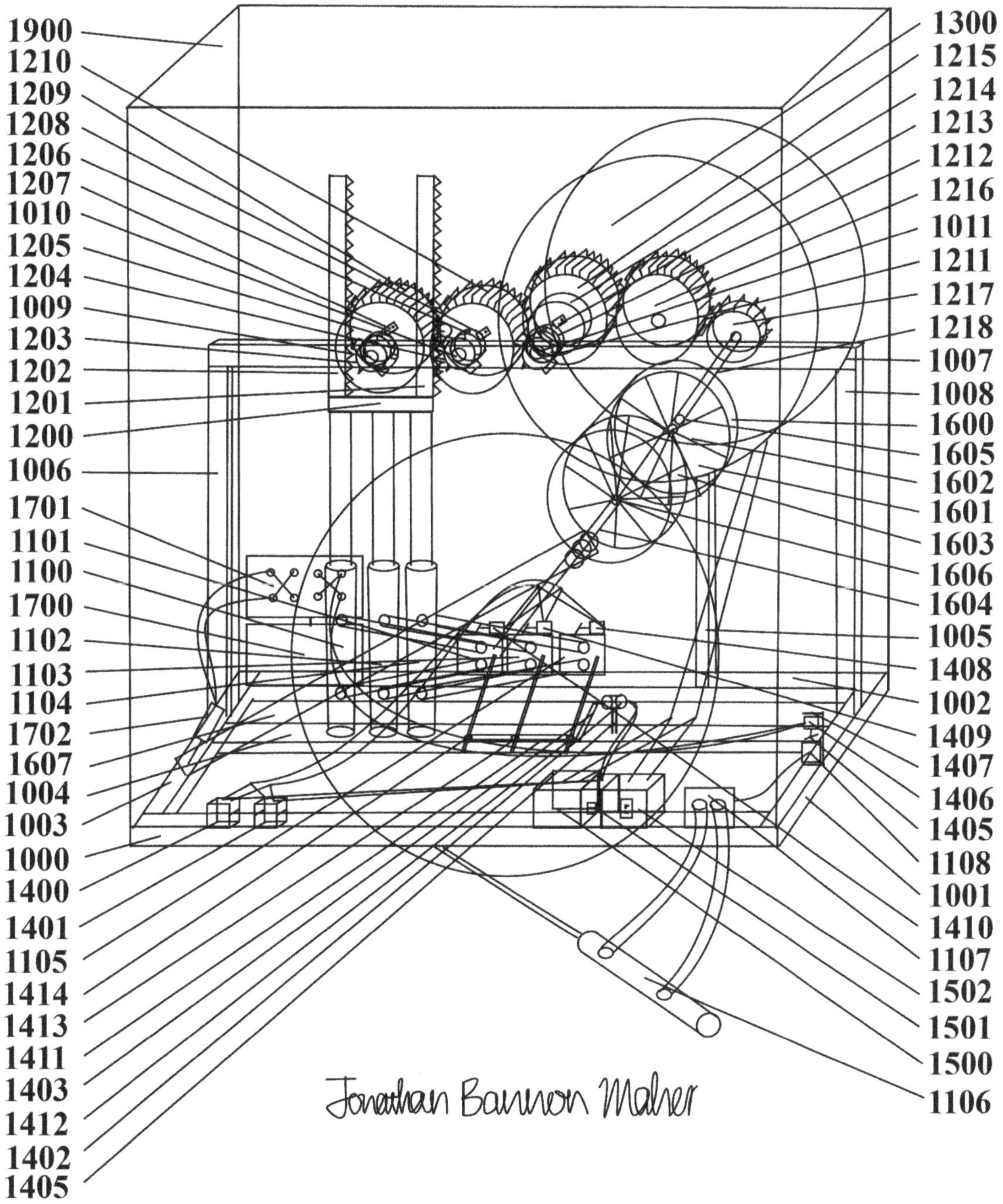
1900
1210
1209
1208
1206
1207
1010
1205
1204
1009
1203
1202
1201
1200
1006
1701
1101
1100
1700
1102
1103
1104
1702
1607
1004
1003
1000
1400
1401
1105
1414
1413
1411
1403
1412
1402
1405
1300
1215
1214
1213
1212
1216
1011
1211
1217
1218
1007
1008
1600
1605
1602
1601
1603
1606
1604
1005
1408
1002
1409
1407
1406
1405
1108
1001
1410
1107
1502
1501
1500
1106

FIG. 2

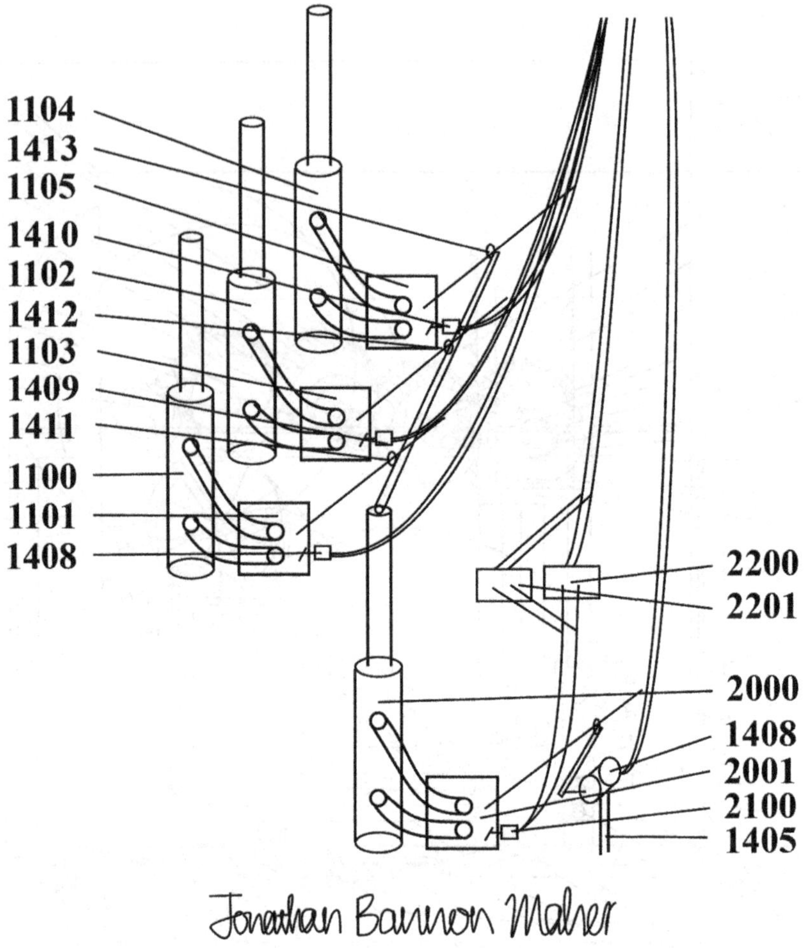
1104
1413
1105
1410
1102
1412
1103
1409
1411
1100
1101
1408
2200
2201
2000
1408
2001
2100
1405
Jonathan Bannon Maher

FIG. 3

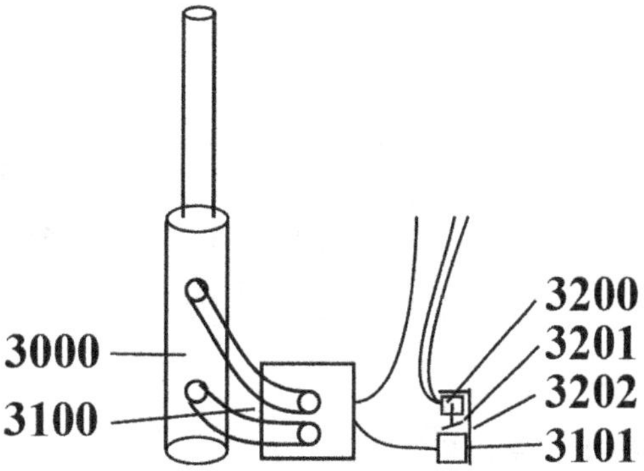

Jonathan Bannon Maher

FIG. 4

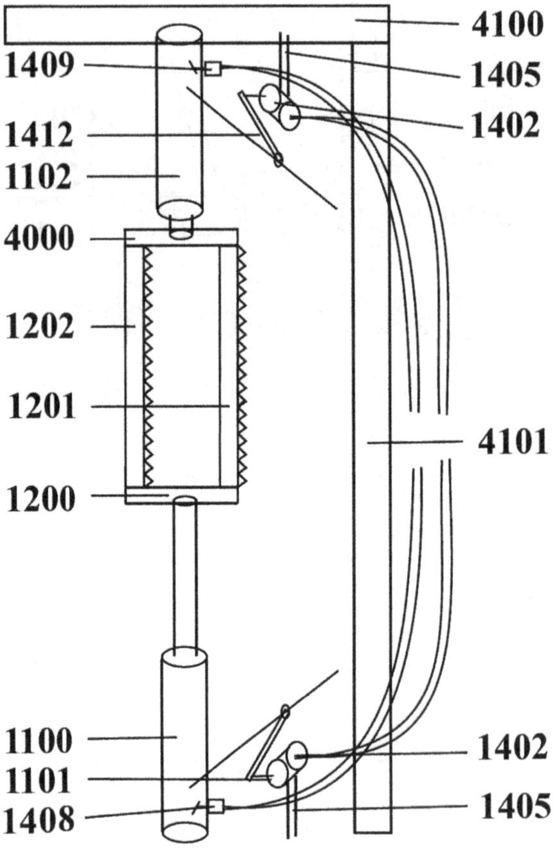

Jonathan Bannon Maher

FIG. 5

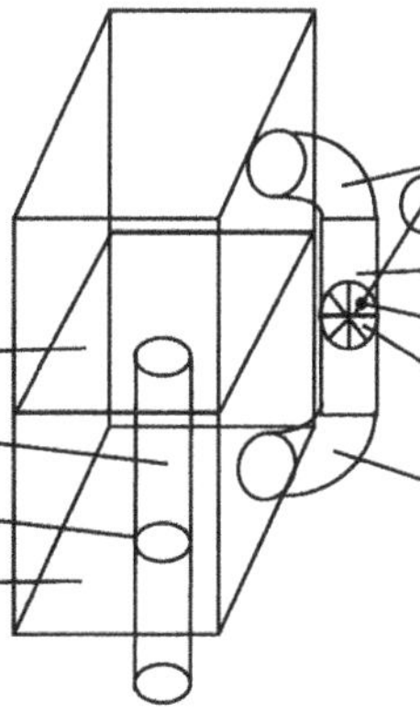

Jonathan Bannon Maher

FIG. 6

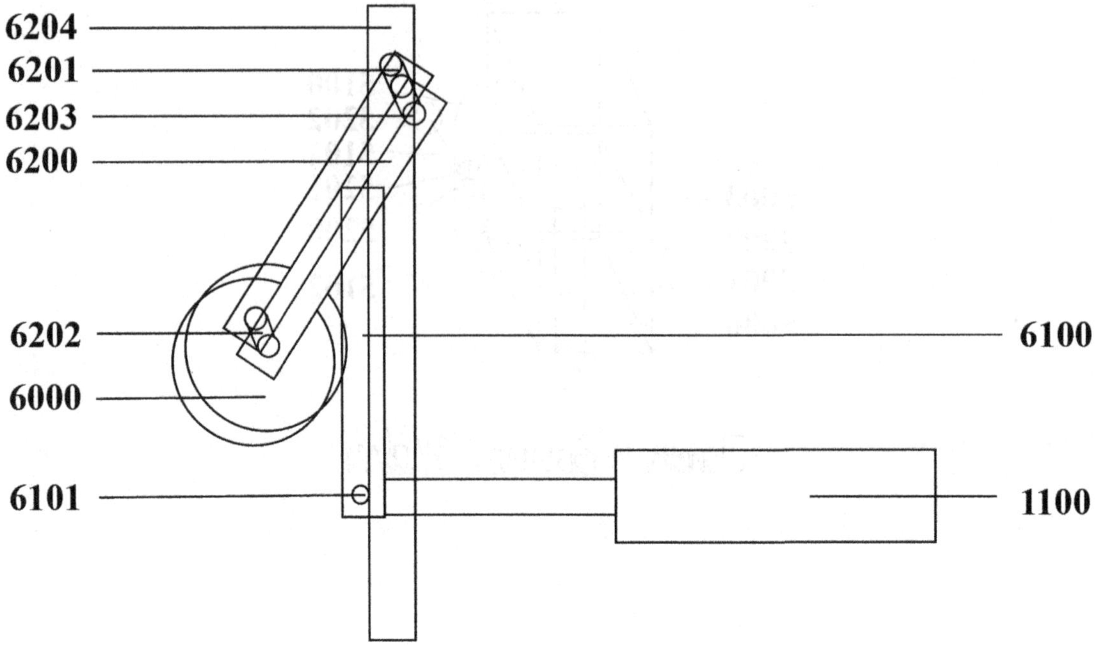

Jonathan Bannon Maher

FIG. 7

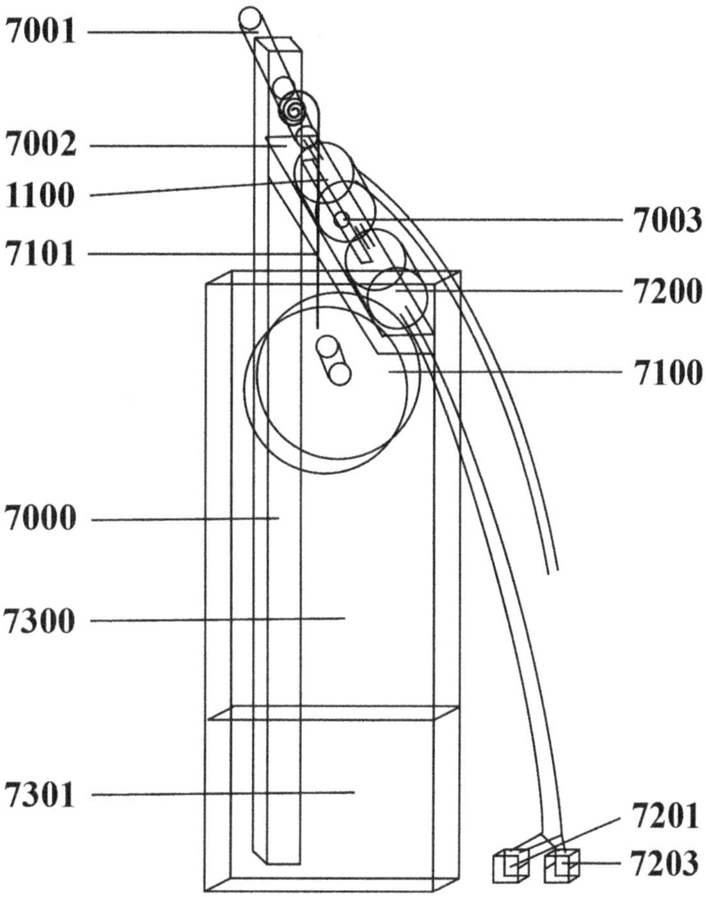

Jonathan Bannon Maher

FIG. 8

8000 — import a library for connecting to the relay

```
import telnetlib
```

8001 — import a library for system resource access

```
import sys
```

8002 — create variables holding relay on off values

```
off = 0
on = 1
```

8003 — initialize variables for each relay state, including lift up and down, torque converter forward and backward, and unit forward and backward

```
relay_1_lift_up_state = off
relay_2_lift_down_state = off
relay_3_torque_forward_state = off
relay_4_torque_backward_state = off
relay_5_unit_forward_state = off
relay_6_unit_backward_state = off
```

8004 — create a variable holding force providing device(s) cycle seconds

```
lift_cycle_seconds = 5.0
```

8005 — create variables holding the relay board connection, IP address, username, and password

```
relay = None
relay_ip = "169.254.1.1"
relay_username = "admin"
relay_password = "admin"
```

8006 – create a function that when called, initializes a connection to the relay board, if the connection has not been initialized, or has been dropped, then sends the current state of each relay to the relay board

```
def update_relay ():

    if not relay:
        relay = telnetlib.Telnet(replay_ip, 23)
        relay.read_until(b"User Name: ")
        relay.write(relay_username.encode('ascii') + b"\n")
        relay.read_until(b"Password: ")
        relay.write(relay_password.encode('ascii') + b"\n")

    command = str(relay_1_lift_up_state)
    command+= str(relay_2_lift_down_state)
    command+= str(relay_3_torque_forward_state)
    command+= str(relay_4_torque_backward_state)
    command+= str(relay_5_unit_forward_state)
    command+= str(relay_6_unit_backward_state)
    relay.write(command.encode('ascii') + b"\n")
```

8007 – create a function that when called listens for computer input from the operator

```
def listen():
```

8008 – read in any user key press

```
    key_pressed = ord(sys.stdin.read(1))
```

8009 — if the key pressed is forward, set the torque converter control device relay state to forward

```
if key_pressed == 39:
    relay_4_torque_backward_state == off
    relay_3_torque_forward_state == on
```

8010 — if the key pressed is backward, set the torque converter control device relay state to backward

```
if key_pressed == 37:
    relay_3_torque_forward_state == off
    relay_4_torque_backward_state == on
```

8011 — if the key pressed is up, set the unit position control device relay state to forward

```
if key_pressed == 38:
    relay_6_unit_backward_state == off
    relay_5_unit_forward_state == on
```

8012 — if the key pressed is down, set the unit position control device relay state to down

```
if key_pressed == 40:
    relay_5_unit_forward_state == off
    relay_6_unit_backward_state == on
```

8013 — if the key pressed is the spacebar, set all torque converter and unit positioning relay states to off

```
if key_pressed == 32:
    relay_3_torque_forward_state == off
    relay_4_torque_backward_state == off
```

```
    relay_5_unit_forward_state == off
    relay_6_unit_backward_state == off
```

8014 – create the function called on program start

```
def __main__():
```

8015 – record the program initialization time

```
  time_initialized = sys.time()
```

8016 – continuously check if the seconds elapsed are divisible without remainder by the force providing device cycle seconds, and if so switch the lift direction, then check for user input and call the function to send the current relay states to the relay board

```
  while True:
    seconds_elapsed = time_initialized – sys.time()
    if seconds_elapsed%lift_cycle_seconds == 0:
      if relay_1_lift_up_state == on:
        relay_1_lift_up_state = off
        relay_2_lift_down_state = on
      if relay_1_lift_up_state == off:
        relay_1_lift_up_state = on
        relay_2_lift_down_state = off
    listen()
    update_relay()
```

FIG. 9

9000 — import a library for accessing the relays

```
import telnetlib
```

9001 — import a library for system resource access

```
import sys
```

9002 — initialize and set variables holding relay on and off values

```
off = 0
on = 1
```

9003 — create an array of IP addresses of unit on off relay boards

```
unit_ips = ["169.254.1.3", "169.254.1.4"]
```

9004 — initialize an array of unit relay board connections

```
relays = []
for unit_ip in unit_ips:
    relays.append(None)
```

9005 — create an array of unit output watts

```
unit_watts = [5000, 5000]
```

9006 — create an array of unit on off states

```
unit_states = [on, off]
```

9007 – create variables holding the default unit relay board username and password

```
unit_username = "admin"
unit_password = "admin"
```

9008 – create variables to hold the total watts available across all units, the current watts being consumed, and whether or not time based watts are to be used

```
watts_total = 0
watts_current = 1000000
time_watts = False
```

9009 – create an array of pairs of times and time desired watts

```
times_watts = [["7:00", 1000000],["20:00",500000]]
```

9010 – create a variable indicating whether or not unit power consumption meters are to be used

```
power_meter = True
```

9011 – create and array of power consumption meter IP addresses

```
power_meter_ips = ["169.254.1.1","169.254.1.2"]
```

9012 – create variables holding the power consumption meters username and password

```
power_meter_username = "admin"
power_meter_password = "admin"
```

9013 – create variables holding the minimum and maximum power consumption to be allowed before switching units on or off

```
consumption_minimum = 0.5
consumption_maximum = 0.9
```

9014 – create a function to send a command to a unit at a specified index in the unit IP address array

```
def send_command(unit_index, command):
```

9015 – create a connection to the unit on off relay board, at the IP address at the provided index in the unit IP address array, if the connection has not been initialized, or had been dropped

```
  if not relays[unit_index]:
    relays[unit_index] =
      telnetlib.Telnet(unit_ips[unit_index], 23)
    relays[unit_index]
      .read_until(b"User Name: ")
    relays[unit_index]
      .write(unit_username.encode('ascii') + b"\n")
    relays[unit_index]
      .read_until(b"Password: ")
    relays[unit_index]
      .write(unit_password.encode('ascii') + b"\n")
```

9016 – create a variable holding the command string to be sent to the unit based on whether the unit is to be turned on or off

```
  command_string = "00"
  if command == on:
    command_string = "01"
  if command == off:
    command_string = "10"
```

9017 – send the command the the unit relay

```
relays[unit_index]
    .write(command_string.encode('ascii') + b"\n")
```

9018 – wait for the command to go through, then turn off all relays

```
time.sleep(1)
command_string = "00"
relays[unit_index]
    .write(command_string.encode('ascii') + b"\n")
```

9019 – define a function to run on script exectution

```
def main():
```

9020 – iterate through each unit IP address and call the function to switch the unit on

```
    index = 0
    for unit_ip in unit_ips:
        send_command(index, on)
        index += 1
```

9021 – run a continuous loop

```
    while True:
```

9022 – create variables to hold a unit index iterator, an output display message, and the current action

```
        index = 0
        message = ""
```

```
action = None
```

9023 – proceed if power consumption meters are being used

```
if power_meter:
```

9024 – create variables to hold average and total consumption of unit power, and add up total possible power output

```
consumption_average = 0
consumption_total = 0
consumption_capacity = 0
```

9025 – loop through the each power consumption meter IP address

```
while index < len(power_meter_ips):
```

9026 – retrieve the current power consumption number

```
power_meter_address =
    "http://" +
    power_meter_ips[index]+
    "/?username=" +
    power_meter_username +
    "&password=" +
    power_meter_password +
    "&command=consumption"
power_meter_consumption =
    urllib.open(power_meter_address).read()
```

9027 – calculate the average consumption

```
consumption_average +=
   power_meter_consumption/unit_watts
```

9028 — add the current consumption to the total

```
consumption_total +=
   power_meter_consumption
```

9029 — add the unit maximum capacity to the total consumption capacity

```
consumption_capacity +=
   unit_watts[index]
```

9030 — increment the index

```
index = index + 1
```

9031 — calculate the consumption percentage as the consumption total divided by the total unit capacities

```
consumption_percentage =
   consumption_total/consumption_capacity
```

9032 — create a message to display the states the currently consumed watts and the current watt capacity

```
message = str(consumption_total) + " of " +
      str(consumption_capacity) +
      " watts consumed"
```

9033 — if the average consumption is greater than the maximum consumption level before more unit should be turned on, then set the unit action equal to on

```
if consumption_percentage >
    consumption_maximum:
    action = on
```

9034 — if the average consumption is less than the minimum consumption level before more unit should be turned off, then set the unit action equal to off

```
if consumption_average <
    consumption_minimum:
    action = off
```

9035 — iterate through unit states until one is found that is either off, if looking to turn a unit on, or on if looking to turn a unit off, then send the command to switch the unit state

```
index = 0
unit_found = False
if action:
    for unit_state in unit_states:
        if not unit_found:
            if (action == on and
                unit_state == off) or
                (action == off and
                unit_state == on):
                send_command(index, action)
                unit_found = True
                unit_state[index] == "on"
                state = "on"
                if action == off: state = "off"
```

```
            message += ", unit turned " + state
        index += 1
```

9036 — proceed if power consumption meters are not used, and instead the total watts to be provided by the units are determined by the current time

```
if not power_meter:
```

9037 — iterate through each time watts pair, and if the current time is equal to the specified time, send commands to turn units on or off, until the desired level of output is produced, and display each action on the command line

```
    for time_watt in time_watts:
        if time == time_watt[0]:
            desired_watts = time_watt[1]
            transition_complete = False
            index = 0
            for unit_state in unit_states:
                while not transition_complete:
                    if watts_current > time_watt[1]:
                        action = off
                        watts_current -= unit_watts
                        message = "unit turned off"
                    if watts_current < time_watt[1]:
                        action = on
                        watts_current += unit_watts
                        message = "unit turned on"
                    if (action == on and
                        unit_state == off) or
                         (action == off and
                        unit_state == on):
                            send_command(index, action)
                            unit_state[index] == "on"
                            state = "on"
```

```
            if action == off: state = "off"
            message +=
               ", unit turned " + state
            index += 1
         action = None
      print message
      message = ""
```

9038 — sleep for a moment before looping again

```
time.sleep(1)
```

SELF-POWERED MOTOR AND GENERATOR

INVENTOR JONATHAN BANNON MAHER

TECHNICAL FIELD

[0001] Embodiments of the invention relate to the fields of motors, generators, physics, engineering, and programming.

ABSTRACT

[0002] Systems, methods, apparatuses, and in some embodiments computer programs encoded on a computer storage medium, provide for clean continuous portable self-powered energy generation and propulsion, consistent with the laws of physics, by in some embodiments, including one complete embodiment, utilizing buoyancy which causes an object in a fluid to rise if the weight of the object is less than to the weight of the volume of the fluid it displaces, where after a buoyant weight ascends to the top of a fluid container, it is pushed into an open air compartment attached to a rotatable chain, and then descends under the force of gravity, providing rotational force through the chain to power an electricity generator, and or rotate an axle to provide the propulsion of a traditional motor, with the buoyant weight reentering the fluid container through a compartment at the bottom, where to not have to displace the weight of all the fluid in the container, the bottom compartment is first sealed off from the top compartment of the liquid container, and the buoyant weight, with the assistance of a support weight, uses gravity when entering the bottom compartment, through a sealable entry door designed to prevent or minimize leakage, by entering at an angle to push fluid in the bottom compartment through a pipe back to the top of the fluid container, then the bottom compartment is unsealed to allow the buoyant weight to ascend to repeat the cycle, where energy is captured in excess of that consumed as a result of the differential between the lack of force required for the buoyant weight to ascend and the energy captured from the force provided by gravity as the buoyant weight descend. The invention permanently solves global warming, provides reduced cost of living to alleviate poverty, provides unlimited clean energy for evaporated water purification and atmospheric carbon dioxide splitting, and eliminates the need for every other method of energy production, including nuclear technology – thus reducing nuclear weapons technology proliferation.

BACKGROUND

[0003] The majority of the proceeds from the licensing of this patent will be going to causes that support the well being of humanity, and your support in ensuring the patent is forever in every way as strong as possible, will be providing a service to all the world.

[0004] This section is intended to introduce the reader to various aspects of the art that may be related to various aspects of the present techniques, which are described and or claimed. This discussion is believed to be helpful in providing the reader with background information to facilitate a better understanding of the various aspects of the present disclosure. Accordingly, it is understood that these statements are to be read in this light, and not a citation of any prior art.

[0005] Patent filings on structurally differentiated and fundamentally deficient disclosures may exist that may attempt to claim any invention that is self-powered, based on previously publicly known failed attempts to build such devices, however any such disclosures do not enable the purported inventions, with overly broad claims not supported by the disclosure that also fail to distinctly claim the invention, and any such patent filings are inherently invalidated by prior public disclosure, the enablement requirement, and the claims support requirement.

[0006] Known and proposed energy production, transmission, and storage systems have some or all of the following deficiencies:

[0007] 1. External fuel source required: an external fuel source is utilized such as oil, gas, coal, wind, sun, water currents, geothermal heat, hydrogen, or uranium, where the cost of providing fuel in the form of electricity or gasoline to a vehicle over its useful life potentially exceeds the cost of the vehicle, and about a quarter of airline costs are from fuel.

[0008] 2. Environmentally unfriendly: nuclear energy production, including fission, fusion, and cold (LENR), results in toxic waste and or materials, geothermal often circulates contaminants from the ground, fracking creates toxic water, hydroelectric dams decompose organic matter producing the potent global warming gas methane, solar panel manufacturing often releases toxic byproducts including greenhouse gases far more potent and long lived than those from fossil fuels, hydrogen takes substantially more energy to produce and transport than it provides, while clean energy sources may utilize slowly degrading flammable toxic batteries to store energy for when the sun is not shining, wind is not blowing, or water is not adequately flowing.

[0009] 3. Intermittent: wind, solar, and traditional water energy systems provide variable output, with average output often found to be around 20% of rated output, as a result of environmental conditions, meaning a 5 kilowatt system typically produces average output of only 1 kilowatt.

[0010] 4. Not portable: solar panels can only function in the sun, wind turbines in wind, water turbines in water currents, nuclear in a stable highly controlled environment, and carbon with the aid of an emissions pipe.

[0011] 5. Extremely expensive transmission costs: power lines are required for all forms of nuclear energy, including fission, fusion and cold, for farms of solar, wind, and water, and for fossil fuels plants, while pipelines are generally required for oil and gas, including natural gas, which is principally methane, a greenhouse gas more than twenty times more potent than carbon dioxide that may be leaked during extraction, transport, and consumption. Power line and fossil fuel pipe line cost of installation per 1 mile (1.6 kilometers) has been found to be up to around 20 times the average annual income in the United States, and lines must be replaced every 30 to 50 years, so in the United States alone, with 300 thousand miles (480 thousand kilometers) of power lines, and 200 thousand miles (320 thousand kilometers) of pipelines, replacement would require an expenditure around 25 times the national debt, passed on to consumers, and dragging down the economy, with every other developed nation in a similar situation. Power lines require environmental destruction during installation, leave visible blight, are forever vulnerable to cyber attacks, transmit power from central sites that are inherently more prone to failure and blackout than a decentralized system, dissipate power during transmission, with high voltage power lines having health consequences for those living nearby.

[0012] 6. High initial costs: in addition to previously cited expense of power lines, fossil fuel pipes, and ongoing fuel costs, clean energy farms require an allocation of land, and degrading batteries requiring periodic replacement, which is why tax credits are often required for clean energy systems to be affordable. In addition to those factors, comparing a 1 kilowatt rated clean energy system to a 1 kilowatt rated traditional system, may require multiplying the cost of the clean energy system by approximately 10 times, 5 times to account for enough electricity generation to be stored for the equivalent continuous output, and 5 times for the battery storage.

[0013] 7. Vulnerable to weather: wind turbines, hydroelectric dams, and solar panels, can be made ineffective by environmental conditions such as freezing temperatures, snow, and rain, and similar to power lines, may be taken down by extreme weather and lightning strikes.

[0014] 8. Vulnerable to black outs: power transmitted over power lines creates vulnerability for critical facilities such as hospitals and data centers.

[0015] 9. Vulnerable to cyber attack: utility scale energy systems often require a hackable computer to operate, and are therefore forever vulnerable to computer viruses able to take down and or destroy nuclear power plants and the electrical grid, even if such systems aren't connected to the Internet, as demonstrated by the Stuxnet virus.

[0016] 10. Causes deaths: plants, rigs, and pipes, for current and proposed forms of nuclear, hydrogen, gas, and oil energy can explode, and coal mines can collapse, while wildlife is killed by wind, ocean, wave, and river turbines, hydroelectric dams, solar condensers, solar panel and battery manufacturing byproducts, and nuclear waste.

[0017] 11. Encourage nuclear weapon proliferation: fission, fusion, and cold (LENR) nuclear energy are or can be one step from weaponizable, while hydrogen can be obtained by a terrorist at a hydrogen fuel station to create a powerful compressed hydrogen explosion, as verified by reviewing a video of a balloon filled with hydrogen being lit on fire. Fusion is particularly disturbing, as a fusion weapon could be created with the power of an exploding star, able to take out the planet, and resulting in an extinction level event for humans, a scenario even more likely when considering increasingly autonomous – and therefore inevitably hackable by individuals – weapons control.

BRIEF DESCRIPTION OF THE DRAWINGS

[0018] Illustrations are presented by way of example, and not by way of limitation, where some embodiments may not contain all components, may contain additional components, and may contain functionally similar components.

[0019] FIG. 1 is an illustration of an example of embodiment of the invention, which utilizes the buoyancy of a fluid to lift a buoyant weight, which upon reaching the top is pushed into to an open air compartment where it descends to rotate a chain connected to provide rotational force to an electricity generator axle to generate electricity and or to rotate or function as a motor axle, and with the buoyant weight then entering the bottom of the liquid container to repeat the cycle.

[0020] FIG. 1A is an illustration of an example of an embodiment of a segment of the compartmented chain utilized in FIG. 1.

[0021] FIG. 2 is an illustration of an example of embodiment of the threaded circular disc partition system which may be used to seal the bottom compartment.

DETAILED DESCRIPTION

[0022] The disclosure is related to the field of clean continuous portable self-powered energy and propulsion. It is understood that any reference to a person skilled in the art, recognizes that at the time of filing, there is no one else skilled in the art of this particular field, or a closely related field. Given the extraordinary nature of the disclosure, regardless of how full, clear, concise and exact the disclosure in enabling the production and use of embodiments of the disclosure, what could be construed to be undue experimentation during production and use, is simply the ordinary effort required in the assembly and use of an embodiment of such a disclosure.

[0023] It is understood that, as in any engineering or design project, the development of any actual implementation will include numerous implementation specific decisions made to achieve the developers' specific goals, such as compliance with business related and system related constraints, which may vary from one implementation to another. It is understood that such a development effort might be complex and time consuming, but is nevertheless a routine undertaking of design, fabrication, and manufacture for those skilled in the art having the benefit of this disclosure. The disclosed steps may be read as prefaced by "In some embodiments, including one complete embodiment, ", may be executed or performed in other orders or sequences, and are not limited to the order and sequence shown and described, which are provided to enable ease in constructing an embodiment, and along with each components of each step, may be removed, modified, combined, or rearranged, and other steps and or step components may be added, without departing from the scope of this disclosure and or invention. Although embodiments of the invention have been described and illustrated in the disclosed implementations, it is understood that the present disclosed subject matter, including apparatuses, methods, specification, and illustrations, has been made only by way of example, not by way of limitation, and the methods and apparatuses may be used in

other systems, and that numerous changes and optimizations in the details of implementation of the invention and or embodiment are made without such modifications departing from the spirit and scope of this disclosure and or embodiments of the invention. Although the disclosure has been shown and described with respect to one or more embodiments, features of the disclosed embodiments can be combined and rearranged in various ways, and changes including equivalent alterations, substitutions, modifications, and additional efficiencies will of course occur to someone of ordinary skill in the art without departing from the spirit and scope of this disclosure and or invention. In particular regard to the various functions performed by the described components, the terms used to describe such components are intended to correspond, unless otherwise indicated, to any component which performs the specified function of the described component, or is functionally equivalent to the described component, even though not structurally equivalent to the disclosed structure which performs the function in the implementations described in this disclosure. In addition, while a particular feature of the disclosure may have been provided with respect to only one of several embodiments, such feature may be combined with one or more other features of other embodiments as may be desired and advantageous for any given or particular application. In some instances, well-known circuits, structures and techniques have not been shown in detail in order not to obscure the understanding of this disclosure. Articles in this disclosure such as "a" "an" and "the" may allow for both singular and plural forms. Verbs in this disclosure such as "is" may be read as "may be". Conjunctions in this disclosure such as "or" as used herein may be interpreted as inclusive or meaning any one or any combination, where "A, B or C" means "any of the following: A; B; C; A and B; A and C; B and C; A, B and C". Relational terms in this disclosure, for example first and second, top and bottom, left and right, are to distinguish one entity or action from another, and may not necessarily require or imply a relationship, or order between, such entities or actions. The disclosure includes the best mode contemplated by the inventor, a completely described specific embodiment, along with optional components and alternative embodiments to best suit the implementer, measurements in imperial and metric units to support universal understanding, and dramatically exceeds claims support requirements and enablement requirements by allowing for selection and or construction of the required components to be carried out easily, quickly, and routinely by persons of ordinary skill in the art, who are provided the additional benefit of utilizing readily available commodity components whenever possible. The present disclosure includes material protected by copyrights, and the owner of the copyrights hereby reserves all rights, but with authorization for publication as required by government patent offices. Various embodiments of the present invention may provide all, some or none of the disclosed technical advantages.

[0024] The computer code descriptions disclosed, in order to provide comprehensive enabling disclosure, rather than utilizing flow charts, which according to Patent Cooperation Treaty 11.11a are prohibited from containing "text matter, except a single word or words, when absolutely indispensable, such as... a few short catchwords indispensable for understanding", are provided in a text only format where the number of arrows preceding a line indicate logical block level, semicolons indicate a new segment of a logical block, and periods indicate the closure of one or more logical blocks. It is understood that any computer code representations in this disclosure are merely illustrative, rather than restrictive. While code may be written in nearly any computer language, including Java and C++, the illustrative computer code descriptions were derived from code written the Python language, which may be run through the Python interpreter, with appropriate supportive libraries, which at the time of disclosure, may run on nearly any computer, for example one with an Intel or AMD processor, running a current version of Linux, Windows, or Mac OS. All code components may read as if prefaced by "In some embodiments, including one complete embodiment, ". In some embodiments, functionality may be modified, rearranged, excluded, and added. To provide more fundamental computer system details, in some embodiments, the functionality associated with the disclosed computer code descriptions may be referred to as a script, module, software, software application, or code, and can be written in any form of language, including compiled, interpreted, declarative, or procedural, able to be deployed in any form suitable for use in a computing environment, including as an independent or integrated program, module, component, or subroutine, for execution by the computer system, implemented on one or more independent or integrated computers, utilizing a central processing unit in the form of one or more general or special purpose microprocessors, in conjunction with digital electronic circuitry, which may include special purpose logic circuitry such as a field programmable gate array or application specific integrated circuit, with the computer controlled by and operatively coupled to tangibly embodied software and or firmware, which may include code that creates an environment for code execution, including individual or combined use of processor firmware, a protocol stack, a database management system, and an operating system, where such software and or firmware may exist in one or more parts in memory on one or more computers, and is encoded on one or more tangible non transitory software carriers, such as individual or combined use of a random or serial access device or substrate, a semiconductor memory device, transient or persistent random access memory, a magnetic, magnetic optical, or optical disk, or encoded on an artificially generated transmitted signal, for example, optical, electrical, or electromagnetic, transmitted using a sending and a receiving apparatus, where the interaction between the user and the software may be implemented by operatively coupling, to the local implementing computer, or a local computer connected to one or more remote computers through a local or wide area network, a display device which may implement fluid crystals or light emitting diodes, a keyboard, and a pointing device.

[0025] The inventor retains absolutely no liability for any implementation of this invention, and the invention is implemented exclusively at the risk and liability of the implementer.

[0026] In some embodiments, for quality control purposes, all components may be manufactured from scratch.

[0027] Calculations, formulas, and specific units are not in any way restrictive, are not be relied upon, are not required as presented to produce an embodiment of the invention, are provided exclusively as a courtesy to enhance enablement for those resizing components and or constructing alternative embodiments, and may contain inaccurate assumptions easily modified during practice, with all calculations utilized to select and estimate components and unit output being rough estimates that may vary greatly based on factors that include the type and quality of purchased and or manufactured components and embodiment construction, where embodiments may be constructed utilizing an effectively endless range of output and component configurations and selection processes.

[0028] Embodiments of the invention provide some or all of the following benefits over previously discussed predecessors:

[0029] 1. Self-powered for free output: embodiments provide the first energy and motor system in the history of the known universe to not require an external fuel source. Embodiments can produce endless energy, and provide endless transportation range, until there is a system failure, and therefore may potentially produce energy and propulsion until gravity driven orbital drift causes the Earth to be consumed by the Sun in a few billion years – assuming the units and humans are still on Earth. Embodiments reduce transportation costs by allowing implementing vehicles to operate without fuel, allowing for effectively free endless transportation range after purchase, and additionally allow for travel by supersonic jets and flying cars, which have been impractical principally as a result of fuel costs.

[0030] 2. Clean: the manufacture and use of embodiments produces no notable harmful environmental byproducts, nor the potential for deaths associated with predecessors.

[0031] 3. Continuous: embodiments produce electricity and propulsion that is continuous and stable.

[0032] 4. Portable: embodiments are able to function as well in a basement closet as in a car.

[0033] 5. Cyber attack proof: embodiments are self-contained thus require no hackable computer to operate and are therefore immune to computer viruses.

[0034] 6. Blackout proof: embodiments are designed to be kept indoors and on-site, and are thus ideal for critical facilities such as hospitals and data centers that can't afford a blackout from failed power lines or plants.

[0035] 7. Inexpensive: embodiments can be manufactured and operated at the lowest cost total cost possible, because they don't require fuel, installation and maintenance of power lines, an allocation of land, or degrading batteries.

[0036] 8. Weatherproof: embodiments are self-contained for indoor use and therefore aren't vulnerable to environmental factors such as freezing temperatures, snow, rain, lightning strikes, or extreme weather events.

[0037] 9. Eliminates energy output storage: because additional embodiments of this system can be utilized at peak times with limited cost, storage of energy is no longer relevant, for either utilities or homes, even for peak output needs.

[0038] 10. Eliminates expensive power lines and fossil fuel pipe lines: because embodiments are designed to be kept on site, and any number of units can be utilized to meet peak power needs, power and fossil fuel transmission lines and their associated costs are now rendered irrelevant, thus substantially unburdening all economies globally of associated costs. Land currently holding power lines, as well as arrays of solar panels, wind turbines, and hydroelectric turbines, can be reclaimed to reduce visual pollution and make space for a growing population. Roofs of solar panels can be removed to allow for roof tiles that reflect heat to maintain a cool house in summer. No thinking person will ever want, nor could a functional government allow, any type of nuclear reactor –

fission, fusion, cold (LENR) – in a car or home, leaving only now irrelevant power lines for transmission, thereby making those sources wholly irrelevant.

[0039] 11. Potentially profitable: in a standard home use scenario, embodiments may be the only way for a unit owner to make a profit from selling energy back to the utility at wholesale rates.

[0040] 12. Alleviates poverty: because energy is effectively free after embodiment purchase, and the purchase price is less per unit of output than other energy systems, embodiments reduce the cost of living and the cost of goods for every person on Earth, thus reducing poverty.

[0041] 13. Powers water purification and pumping: 1 in 10 people live without access to clean water, while climate change driven droughts fuel conflicts. Because embodiments make energy nearly free over their useful lives, the energy intensive nature of evaporated water purification — which removes nearly every contaminant with a higher boiling point than water, and potentially all others can be removed with a standard carbon filter and ultraviolet light – is no longer a barrier, nor is pumping, thus embodiments provide for the global resolution of clean water needs for individual consumption and agriculture.

[0042] 14. Powers reduced water consumption: effectively eliminates the energy cost of operating electricity powered showers that require only a cold water pipe, and recirculate, filter to potentially cleaner than direct from pipe, and heat water, to provide exact continuous temperature and pressure control, as well as powering low voltage electric showerheads that mix air with water to provide the effect of the same output using dramatically less water.

[0043] 15. Powers atmosphere cleaning: because an existing specialized laser can disassociate atmospheric carbon dioxide molecules into carbon molecules and oxygen molecules, and because embodiments can provide effectively unlimited and continuous clean energy, a power source is now available to reduce carbon dioxide in the air, and potentially other greenhouse gasses, if corresponding devices are developed.

[0044] 16. Reduces nuclear and hydrogen weapon proliferation: because embodiments eliminate any need for any type of nuclear energy, including fission, fusion, and cold (LENR), each of which is or may be weaponizable and one step from nuclear weapons technology, and also eliminate the need for hydrogen, which a terrorist can obtain at a hydrogen fuel station to create a powerful compressed hydrogen explosion verifiable by watching a video of a balloon filled with hydrogen being lit on fire, I have provided us all a fundamentally safer world.

[0045] 17. Therefore, a few applications of embodiments include powering: all transportation vehicles including automobiles, trains, jets, cargo ships, cruise ships, tugboats, boats, submarines, hover boards, jet packs, including in vertical take off and landing configurations; powering electrical grids as well as homes, offices, factories, hospitals, and data centers that would like to disconnect from external power sources to end their recurring bill, be permanently immune from blackouts, use clean energy, and save money; televisions, washing machines, dishwashers, showers, and water pumps; portable consumer electronics, such as phones and laptops, through an internally installed miniaturized embodiment eliminating the need to recharge; personal rapid transport; home hydroponic production systems, including light, temperature control, and nutrient water circulation; high intensity laser powered solar sails, with the laser powered by a large number of these energy units on the surface of any space based body with a limited atmosphere; video streaming planetary sampling probes journeying an unlimited number of years into the universe, recharging by utilizing an embodiment when resting on a mass producing gravity or in conjunction with a device creating artificial gravity; space colonies.

[0046] **In some embodiments, including one complete embodiment, obtain the desired generator.** In some embodiments, including one complete embodiment, a generator is added. In some embodiments, including one complete embodiment, first the generator is selected, and or the rotational force and speed to function as a motor is determined, because those specifications determine the specifications of other components in the system, including the weight and size of the buoyant weight or weights that will be used to provide force to operate the generator and or function as a motor, where the buoyant weight in turn determines the size of the fluid tank required to elevate the buoyant weight. In some embodiments, which may include one complete embodiment, the generator selected outputs the volts, amperes, and hertz of the desired final output, with a converter utilized of the appropriate specifications to power internal embodiment components. In some embodiments, when used as a motor, a generator may be left out, with the power for the unit components provided by an external source.

[0047] In some embodiments, to provide a specific horsepower at a specific number of revolutions per minute, to provide rotational force as required by a generator axle and or to function as a motor, in order to determine the pounds of force required, horsepower is equal to pounds of force multiplied by revolutions per minute with the result divided by the horsepower constant of 5252. In some embodiments, which may include one complete embodiment, to convert pounds of force from the previous calculation to pound feet of force, as used in the standard calculation of horsepower, a gear may be placed on the generator and or motor axle, which is coupled to the force providing gear immediately before it, where each of those gears have a radius of 1 foot (0.3 meters), while maintaining all gear ratios in the system. In some embodiments, to accurately measure the rotational force and speed provided by the embodiment, a torque gauge may be used to measure pound feet of force, and a tachometer may be used to determine revolutions per minute, or alternatively a dynamometer may be used which measures pound feet of force as well as revolutions per minute, and then horsepower may be calculated as pound feet of force multiplied by revolutions per minute with the result divided by the horsepower constant 5252.

[0048] In some embodiments, including one complete embodiment, in reference to FIG. 1, generator 1500 has a rated output of 10,000 watts when provided 1,800 revolutions per minute and 13.3 horsepower, where the horsepower specified requires force of 38.8 pounds (17.6 kilograms) (13.3 = ((n pounds of force * 1800 revolutions per minute)/5252 horsepower constant)). In some embodiments, where a motor is principally desired, a generator may be used with a rated output providing only that required to provide power to the unit components, while allowing the rest of the rotational force to be provided to rotate the axle to function as a motor.

[0049] **In some embodiments, including one complete embodiment, create the buoyant weight.** It is a principle of physics that buoyancy causes an object in a fluid to rise if the weight of the object is less than the weight of the volume of the fluid it displaces. In some embodiments, including one complete embodiment, a buoyant weight is made of a weight and size appropriate to provide the force and speed required by the generator and or to act as a motor while being adequately buoyant to float to the top of the fluid container. In some embodiments, including one complete embodiment, the buoyant weight is constructed of a durable rust resistant material, and is hollow with walls of a thickness that when spread across the determined dimensions it has the identified weight while sustaining buoyancy in the fluid.

[0050] In some embodiments, including one complete embodiment, the speed provided by the buoyant weight during free fall and the desired cycle period of the buoyant weight, is used as a starting point for calculations of the desired weight of the buoyant weight, where revolutions per minute provided by the free fall of the buoyant weight, without resistance from the generator and compartmented chain, may be approximately calculated using the rate of drop provided by the gravity of the celestial body on which it operates, for example Earth which causes a mass to fall at 32 feet per second per second (9.8 meters per second per second), however given the fall force will be linear rather than compounding when rotating the compartmented chain with the resistance of the generator, "per second" rather than "per second per second" may be used, where the weight of the buoyant weight may then adjusted based on the differential between the speed during linear free fall and the speed required by the generator, where to obtain linear free fall revolutions per minute, the distance traveled is divided against the circumference of a gear connected to the compartmented chain that provides rotational force to the generator and or to function as a motor, which provides revolutions per second, which is then multiplied by 60 seconds to obtain revolutions per minute, with the revolutions per minute required by the generator divided by the resulting number, to obtain the differential, where the required generator axle input force is divided by the differential to obtain the weight of the buoyant weight, with additional weight added to offset resistance from factors including the weight of the compartmented chain, mechanical friction, and air, with the weight and corresponding adjustments best identified by and made after testing and determining the rate at which the buoyant weight descends on the compartmented chain after it has been constructed, to obtain the desired force and speed. In some embodiments, including one complete embodiment, the weight and corresponding size to maintain buoyancy of the buoyant weight and supporting components are modified to obtain the desired force and speed when the embodiment is running. In some embodiments, which may include one complete embodiment, the buoyant weight has inset segments on one or more sides that help capture any fluid that might otherwise leak out past the buoyant weight as it enters the fluid container bottom compartment. In some embodiments, multiple buoyant weights may be circulated at the same time, either in sequence or in parallel, to provide continuous or improved continuity of rotational force to the compartmented chain, where the top and bottom gears around which the compartmented chain operatively wraps have a circumference that supports the determined revolutions per minute required by the generator.

[0051] In some embodiments, including one complete embodiment, the force required by the generator – and thus the resistance it provides – is 38.8 pounds (17.6 kilograms) at a speed of 1800 revolutions per minute, therefore targeting a 5 second cycle time over a 5 foot (1.5 meter) drop, the buoyant weight should fall at 1 foot (0.3 meters) per second, and where the buoyant weight without resistance will free fall rate provided by gravity of approximately 32 feet per second (9.8 meters per second), the teeth ratio of the connected gears should provide a decrease in speed of 32 times (32 feet per second / 1 feet per second desired) resulting in a corresponding 32 times increase in force, providing for a required input weight of at least 1.2 pounds (38.8 pounds of required force / 32 times speed reduction) (0.54 kilograms), with additional weight added to offset resistance from factors including the weight of the compartmented chain, mechanical friction, and air, for say a doubling of weight to 2.4 pounds (1.08 kilograms), which is to be adjusted along with corresponding size to maintain buoyancy after observing performance to obtain the force and speed desired, where the circumference of the top and bottom gears, around which the compartmented chain operatively wraps, are to provide for approximately 56 revolutions per minute (1800 revolutions per minute required by generator / 32 times speed increase by generator attached gear), thus each gear should have a circumference of approximately 1 inch (56 revolutions per minute / 60 seconds per minute) and thus a 0.32 inch diameter (1 inch circumference / 3.14 Pi) to provide 1800 revolutions per minute at force of 38.8 pounds (17.6 kilograms).

[0052] Fluids are substances that can't resist shear applied force, and are a subset of the phases of matter that include liquids, gases, plasmas, and to an extent, plastic solids. In some embodiments, to calculate the approximate volume of fluid that must be displaced for the buoyant weight to become buoyant, and thus calculate the dimensions of the buoyant weight, the pounds (kilograms) of the fluid per gallon (liter) are used, along with the cubic feet (cubic meters) per gallon (liter), where the buoyant weight has a weight that provides the pounds of force required by the generator, and has that weight spread over a volume that displaces more than the gallons (liters) of the fluid of the same weight, where such a calculation may be determined by the pounds (kilograms) required of buoyant weight divided by the pounds (kilograms) per gallon (liter) of the fluid, and therefore the buoyant weight may be of any size greater than the gallons (liters) displaced divided by the gallons (liters) of fluid per cubic foot (cubic meter).

[0053] In some embodiments, including one complete embodiment, the fluid to be used is water, which allows the fluid containers to be easily filled on site, thus reducing shipping costs, while allowing for the fluid containers to be emptied at any time without causing environmental harm. In some embodiments, the fluid to be used is a liquid metal – for example zinc, which is over six times as heavy as water – to allow for faster unit cycles while utilizing smaller containers, with corresponding adjustments made to the buoyant weight to compensate for any change in fluid density. In some embodiments, any other fluid may be used, though when accounting for safety and durability, a non-corrosive non-toxic fluid may be best. In some embodiments, if the unit is to be run outside, in temperatures below freezing, a fluid with a lower freezing point than water may be used, or an anti-freeze agent may be added, with any necessary adjustment made to the buoyant weight to compensate for any change in fluid density. In some embodiments, any fluid or medium that allows for a buoyant weight to rise may be used.

[0054] In some embodiments, including one complete embodiment, to calculate the approximate buoyant weight dimensions, water may be used, where water has a weight of 8.34 pounds per gallon (1 kilogram per liter), 1 cubic foot of water is 7.48 gallons (1 cubic meter of water is 1000 liters), a buoyant weight having the previously calculated weight of 2.4 pounds (1.08 kilograms), has to be of a material and thickness providing that weight spread over a volume that displaces more than approximately 0.3 gallons (2.4 pounds of required of buoyant weight / 8.34 pounds per gallon) (1.14 liters), and therefore the buoyant weight must be of a size greater than approximately 70 cubic inches (0.3 gallons / 7.48 gallons per cubic foot) (1,142 cubic centimeters), where for increased ascent rate the buoyant weight displacement volume may be increased 50% beyond the minimum required size for a total of 105 cubic inches (1,713 cubic centimeters), which may be spread across any combination of dimensions. In some embodiments, including one complete embodiment, the material from which the buoyant weight is constructed is any that allows the buoyant weight to achieve the desired weight and volume supporting dimensions. In some embodiments, which may include one complete embodiment, the material to be used is stainless steel, where to account for the density of stainless steel, the buoyant weight has an air tight hollow interior, with walls of a thickness that when constructed in the desired dimensions, the desired weight is achieved. In some embodiments, which may include one complete embodiment, the material to be used is polyethylene. In some embodiments, the buoyant weight may be comprised of countless other dimensions, volumes, and materials. In some embodiments, in place of the buoyant weight, any buoyant medium may be used that can be circulated in the embodiment, not limited to but including a fluid of a lesser density than the primary fluid in the fluid container, with appropriate adjustments made to supporting components.

[0055] In some embodiments, which may include one complete embodiment, in order to maintain continuous or near continuous output, multiple buoyant weights are used in the system, in sequence or in parallel, with appropriate supporting adjustments made to the repeat cycle timers or functional equivalents, to provide continuous or near continuous weight to the compartmented chain and thus continuous or near continuous rotational force to the generator or to function as a motor.

[0056] **In some embodiments, including one complete embodiment, create the buoyant weight entry support weight.** In some embodiments, including one complete embodiment, to reduce the force required for the buoyant weight to enter the bottom of the fluid container, a bottom compartment with a sealable bottom compartment partition is created in the fluid container, which removes the weight of the fluid in the top compartment from the fluid in the bottom compartment, without requiring significant force because at the time the partition is sealed the pressure is equalized on both sides of the partition, and then because in order for the buoyant weight to be buoyant, it weighs less than the fluid it's displacing in the bottom compartment, in order for the fluid to be pushed back to the top of the fluid container through a pipe connecting the bottom compartment to the top of the fluid container, a weight is released on a timed cycle behind the buoyant weight after it has dropped in place in front of the bottom compartment entry door, with both weights at an angle so gravity pushes on the weights, where total weight provided opens the bottom compartment entry door and forces fluid back to the top, and then the support weight is retracted to its original position, where the support weight's total weight exceeds that of the combined difference between the weight of the buoyant weight and that of the weight of the fluid displaced, and the column of fluid in the pipe as it flows back to the top of the fluid container, with a modest addition accounting for the weight being applied at an angle rather than vertically, where any fluid that leaks out of the bottom fluid compartment into the fluid container support structure reservoir as the buoyant weight enters, is pumped to the top of the fluid container, and or pushed back into the bottom fluid compartment during the next entry of the buoyant weight when the buoyant weight is coupled with the support weight, by means which may include an rod and receiver or an overhang on the support weight to hold the buoyant weight down, to prevent buoyancy on the leaked water. In some embodiments, including one complete embodiment, providing for the entry of the buoyant weight into the fluid container utilizing gravity and a weight, which can be pushed and pulled into place in seconds, may provide greater efficiency than providing force through other means which would have to be maintained the entire time the weighted structure is displacing fluid. In some embodiments, a force producing device, not limited to but including hydraulic, pneumatic, mechanical leverage, or motorized mechanical leverage, instead of or in addition to the support weight, provides force that allows the buoyant weight to enter the fluid container. In some embodiments, the sealable bottom compartment partition may be omitted.

[0057] In some embodiments, including one complete embodiment, to calculate the total weight of the fluid that the buoyant weight has to displace as it enters the bottom compartment of the fluid container, the weight of the fluid in the bottom compartment is added together with the weight of the fluid in the pipe when full, where the weight of the fluid in the compartment is equal to the pounds per gallon (kilograms per liter) of fluid displaced multiplied by the gallons (liters) of fluid displaced, and the weight of the fluid in the tube when full is calculated as the radius of the pipe squared, multiplied by Pi of 3.14, multiplied by the height of the pipe, with the combined weight of the determined volume is the weight per gallon (liter) of fluid per cubic foot (cubic meter) divided by the determined volume in cubic feet (cubic meters), where the weight of the buoyant weight combined with the weighted structure must weigh more than the weight of all fluid being displaced.

[0058] In some embodiments, including one complete embodiment, the weight of the fluid to displace is calculated utilizing the previously calculated size of the buoyant weight of 105 cubic inches (1,713 cubic centimeters) which displaces 0.46 gallons of water (0.07 cubic feet * 7.48 gallons per cubic foot) (1.74 liters), where the weight of the displaced water in the bottom compartment is 3.8 pounds (0.46 gallons of water * 8.34 pounds per gallon) (1.7 kilograms), which is added to the weight of the water in the pipe when full, where the 0.5 inch (1.27 centimeters) radius of the pipe, multiplied by Pi of 3.14, multiplied by a height of 5 feet (1.5 meters), provides for fluid of 47 cubic inches (3.14 pi * (0.5 inch radius) ^ 2)) * 5 feet high), or 0.03 cubic feet (765 cubic centimeters) which is 0.23 gallons of water (0.03 cubic feet * 7.48 gallons per cubic foot) (0.87 liters), weighing 1.9 pounds (0.23 gallons of water * 8.34 pounds per gallon) (7.3 kilograms), plus the weight of the bottom compartment entry door 1206, plus any pressure from water leaked from above the bottom compartment if it is imperfectly sealed, and the reduced force resulting from angled rather than vertical force, resulting in the weight of the weighted support that will provide force to the buoyant weight at the bottom being more than 7.6 pounds ((3.8 pound weight of fluid displaced + 1.9 pounds required to displace full fluid column + 5 pounds bottom compartment entry door) – 2.4 pound weight of buoyant weight) (1.2 kilograms), for example, quadrupled to a total of 15.2 pounds (6.9 kilograms).

[0059] The set of factors required to validate the embodiment as providing a self-powered generator, are the output of the generator when continuously powered by multiple buoyant weights, and the power drawn to push then retract the support weight at the bottom, and the power drawn to push the buoyant weight off the top of the fluid container into the compartmented chain, and the power drawn to pump any leaked fluid back to the top of the fluid container. In some embodiments, to approximately calculate total output of the system, the generator output may be assumed to provide continuous output as the result of using multiple buoyant weights and thus provides a continuous or near continuous 10,000 watts, and subtracted from the output are the watts consumed by the three motors and or linear actuators used in the system, which may consume at peak load 120 watts (12 volts at 10 amperes), while providing force of 225 pounds (102 kilograms), and to provide a conservative estimate, assuming each draws continuous maximum load, results in a continuous average net output of around 9,600 watts ((10,000 watts provided by generator) – 120 watts consumed to push the support weight at the bottom) - (120 watts consumed to retract the support weight at the bottom) - (40 watts consumed to send leaked fluid back to the top of the fluid container with an electric pump) - (120 watts consumed to push the buoyant weight off at the top)), for input output efficiency of approximately 96%. In some embodiments, which may include one complete embodiment, to increase efficiency, the height of the fluid tank in conjunction with supporting components including the compartmented chain, is increased in height, for example a height 100 times greater, thus raising a buoyant weight 100 times as high, while still using no energy to lift the buoyant weight but allowing the buoyant weight to provide force for 100 times as long as it descends. In some embodiments, energy consumption may be further reduced by utilizing hydraulics, pneumatics, mechanical leverage, and or motorized mechanical leverage to provide force in the system, including in place of the motors and or linear actuators and or support weight. In some embodiments, energy may be captured in excess of that consumed, as a result of no energy being expended to elevate the buoyant weight, while capturing the energy provided by gravity as the buoyant weight falls.

[0060] **In some embodiments, including one complete embodiment, construct the fluid container and support structure.** In some embodiments, including one complete embodiment, a fluid container is constructed, providing fluid enclosure on all sides except the top of one side where the buoyant weight is pushed out, where the fluid container rests on a support structure, while the bottom of one side of the fluid container has a door that allows a buoyant weight to enter while minimizing or effectively eliminating leakage including by having the compartment sides, wall, and entry door snugly fit the buoyant weight during entry and the entry door slanted inward along with its corresponding support structure and of a weight and utilizing sealant material that ensures a seal when the entry door is at rest which may be further reinforced by the door utilizing a unidirectional spring lock or a repeat cycle timer and motor in a manner as provided for in this disclosure to operate a seal compressing latch, where a reservoir that captures any leaked fluid with the walls of the reservoir fitted to the buoyant weight to minimize leakage as the buoyant weight enters, and the entry door having extended sides on the entry side that help form a seal with the buoyant weight as it enters to minimize leakage with a corresponding inset in the liquid container reservoir to prevent catching of the buoyant weight, where limiting or preventing leakage is important as the percent of fluid leaked may be roughly tied to the percentage of loss of efficiency in the embodiment as a result of having to pump the leaked fluid to the top of the fluid container, with a sealable bottom compartment partition that is able to seal off the bottom compartment from the weight of the fluid above it to limit the force required for fluid displacement, and where a motor may in conjunction with a latch that when rotated further is able to compress the seal, so that the buoyant weight only has to displace the weight of the fluid in the bottom compartment and the fluid is returned to the top of the fluid container through the return pipe, while also minimizing the volume of fluid that can leak out, while the bottom compartment on the side opposing the buoyant weight entry door has connected to it piping that sends fluid back to the top of the container as it is displaced by the buoyant weight, where the moveable partition has a valve that is opened when a motor pulls on it, and is weighted on the bottom to be held closed by gravity and with the weight helping to maintain the seal of the sealable bottom compartment partition in conjunction with a sealant material, where when the valve is opened it allows fluid from the top of the fluid container to flow into the bottom compartment to equalize pressure and release the buoyant weight. In some embodiments, which may include one complete embodiment, to obtain a potentially perfect seal for the sealable bottom compartment partition, the sealable bottom compartment partition is a circular partition with threaded sides, with the threading pattern forming a perfect or near perfect seal when engaged but when disengaged leaves an adequate gap to allow fluid to flow into the liquid container bottom compartment, connected to the fluid container through a jointed arm connected from the center of the circular partition to the fluid container, where the arm is connected in a manner that allows rotation and vertical movement, where the center of the circular partition may have a weighted plug of limited diameter closed by gravity but that can be lifted up to release fluid into the bottom compartment with the plug having an extended bottom perimeter to prevent it from being fully removed, with the top of the fluid container bottom compartment having a complimentary threaded receptacle, to allow the threaded circular sealable bottom compartment partition to screw and unscrew to form a seal, where the threaded circular sealable bottom compartment partition has gear teeth around the top of the perimeter that are engaged for rotation by a corresponding gear powered by a motor operated by a repeat cycle timer to allow for the seal to be closed and opened on a timed cycle, with the threaded circular sealable bottom compartment partition constructed to be of a weight adequate to fall into place for rotation to form a seal, but not significantly heavier to limit the force required to raise it, and where a threaded circular sealable bottom compartment partition may also be implemented in a similar manner as a replacement for the buoyant weight entry door with appropriate supporting adjustments made to other components. In some embodiments, hydraulics and or pneumatics and or mechanical leverage devices and or motors and or functional equivalents are used to displace the fluid and its corresponding weight to allow the buoyant weight to enter. In some embodiments, the sealable bottom compartment partition may be a sliding door that seals off the bottom compartment by extending from and retracting into an additional fitted compartment in the fluid container using motors and repeat cycle timers in a manner similar to that disclosed for the other sealable bottom compartment partitions. In some embodiments, the sealable bottom compartment partition may consist of an air locked door as used in a submarine, operated using motors and repeat cycle timers in a manner similar to that disclosed for the other sealable bottom compartment partitions. In some embodiments, any means may be used that opens as well as adequately seals the sealable bottom compartment partition to remove pressure to allow the buoyant weight to enter the fluid container and then be released to ascend. In some embodiments, the fluid container bottom compartment isn't sealed off, and the fluid is displaced simply by pushing in the buoyant weight.

[0061] In some embodiments, including one complete embodiment, in reference to FIG. 1, fluid container 1200 accounts for the previously determined dimensions including a width and length exceeding that of the buoyant weight to allow for the buoyant weight to ascend, with a height based on the previously determined desired cycle time, and is constructed as part of fluid container platform 1201 which provides a consistent slant for the entry of the buoyant weight, while providing a reservoir formed by walls 1202 1203 to capture any leaked fluid for pumping back to the top of the fluid container, with the walls of the reservoir snugly fitting the buoyant weight to minimize leakage as the buoyant weight enters, and enclosing walls 1204 1205 that prevent the buoyant weight from falling off either side of the compartmented chain, where the fluid container has a bottom compartment entry door 1206, with extended sides on the entry side to help form a seal with the buoyant weight as it enters with corresponding indents for the sides in the liquid container reservoir to prevent the buoyant weight from catching, with bottom compartment entry door 1206 held in place by joint 1208, with bottom compartment entry door 1206 slanted inward toward the liquid container along with its corresponding bottom compartment support structure with the door of a weight that when at rest it maintains a seal that prevents or at least minimizes fluid leakage, where the bottom compartment entry door 1206 may utilize a unidirectional spring lock or a repeat cycle timer and motor in a manner as provided for in this disclosure to operate a seal compressing latch to further reinforce closure when not being opened by the buoyant weight, with bottom compartment entry door slanted overhang 1207 to ease the buoyant weight downward into place for entry if there is any fluid in the fluid reservoir that is causing misalignment, with entry door slanted overhang 1207 enclosed and slanted downward at its top to not prevent from descending buoyant weight 1300, and a bottom compartment sealed off by sealable bottom compartment partition 1209 connected by joint 1210, with sealable bottom compartment partition 1209 having a weighted valve 1211 on top of sealable bottom compartment partition 1209 and opposite joint 1210 that ensures closure of sealable bottom compartment partition 1209 but should not be heavier than required to maintain a seal, with the weights size and position also preventing the door from ever locking at or near a 90 degree angle, where the bottom compartment walls are inset from the main compartment so that the seal is structurally reinforced when sealable bottom compartment partition 1209 closes on top of the walls of the sealable bottom compartment, and the slant of the bottom compartment and sealable bottom compartment partition 1209 match the slant of the support platform, where to maintain a seal an additional sealant material and or structure may be used for both the sealable bottom compartment partition 1209 and its corresponding area on the top of the bottom compartment of the fluid container, as well as the buoyant weight entry door 1206, for example, a rubber or silicon gasket with any corresponding structural insets, adjustments, or supports, where sealable bottom compartment partition 1209 allows fluid to enter to the bottom compartment when valve 1211 is opened by wire 1803 connected to motor 1804, thereby flooding the compartment and equalizing pressure causing buoyant weight 1300 to ascend while assisting with opening the sealable bottom compartment partition, while the bottom compartment is designed to limit any possibility of fluid leakage so buoyant weight 1300 slides as snugly as possible against the sides, bottom, and door, with the fluid pushed out of the compartment into fluid container coupled backside pipe joint 1212, which passes the fluid thorough pipe 1213, to pipe joint 1214, which expels the fluid back to the top of the fluid container, with the top of the fluid container having the top portion of its buoyant weight exit side wall slightly angled outward so as to allow the buoyant weight to easily push up and over the edge of the container when force is applied to it from the opposing side, which is especially relevant if any of the fluid in the container evaporates thus lowering fluid levels, where the top of fluid container 1200 is partially enclosed, or wholly enclosed with an exit door for the buoyant weight, to condense and return evaporated fluid, and may be further enclosed by an exit door. In some embodiments, which may include one complete embodiment, the buoyant weight is prevented from moving backwards out of the bottom compartment after its fully entered by using a unidirectional spring lock. In some embodiments, the partition valve may be pushed on and opened by the buoyant weight as it enters.

[0062] In some embodiments, which may include one complete embodiment, in reference to FIG. 2, utilizing some components from FIG. 1, to obtain a potentially perfect seal, the sealable bottom compartment partition is a threaded circular sealable partition 2000 which has around the bottom perimeter threading 2001 and extending around the top perimeter gear teeth 2002, where threading 2001 screws into complimentary threaded receptacle 2003, with the threading pattern forming a perfect or near perfect seal when engaged but when disengaged leaving an adequate gap to allow fluid to flow into the bottom compartment, where threaded sealable circular partition 2000 is connected to fluid container 1200 through joint 2100 through arm 2101 which has coupled to it through a hole ball joint socket 2102 in a manner which allows the ball joint socket to move vertically to support the threaded sealable circular partition 2000 as it rotates downward to form a seal and rotates upward to open, where the ball and socket joint may have a plug passing through it of limited diameter with an weighted extended bottom perimeter to be self closing and to prevent from being fully removed from that can be opened by wire 1214 to allow fluid to enter the bottom compartment, which is connected with ball joint ball 2103 which is welded to rod 2104 which is welded to threaded sealable circular partition 2000, where the sizing and positioning of components allows for the cover to freely rotate and move vertically as needed to seal and unseal, where threaded sealable circular partition 2000 has perimeter gear teeth 2002 rotated by a gear 2004, with teeth on each designed to support interlocking, with gear 2004 rotated by rod 2005 rotated by motor 1805 operated by additional repeat cycle timers 1706 1707 or functional equivalents that are set and wired to operate motor 1805 back and forth on a timed cycle to seal and unseal the bottom compartment, where motor 1804 is operated by repeat cycle timer 1703 to retract the threaded sealable circular partition 2000 using wire 1214 connected to ring 1216 after each time it has been unsealed, where threaded sealable circular partition 2000 at its center may be a weighted valve that closes under the weight of gravity but that is opened when pulled on by wire 1215 to more easily allow fluid to enter the bottom compartment, where the fluid container and the buoyant weight have their dimensions adjusted to be compatible with the dimensions of the round sealable bottom compartment partition.

[0063] In some embodiments, including one complete embodiment, the fluid container along with its components and support structure are constructed from scratch, where sheets of a strong durable material, for example a metal such as stainless steel, are welded together, with each wall and component of adequate thickness so as not to deform or leak over time, and of a width and length that support the ascendancy of the buoyant weight over the desired height.

[0064] **In some embodiments, including one complete embodiment, construct the compartmented chain.** In some embodiments, including one complete embodiment, in reference to FIG. 1, a high strength compartmented chain is constructed to hold the buoyant weight as it descends under the force of gravity, where chain 1400 connects around gears 1401 1402, and the chain is constructed to form a loop from a sequence of links, where the chain is constructed from, in reference to an example of a chain segment in FIG. 1A, chain links 1403 1404 1405, coupled to each other by threaded bolts 1406 1407 1408 1409, each passing through one of threaded bolt supports 1410 1411 1412 1413, with chain links 1403 1405 each having an opening to operatively be able to connect with gears 1401 1402, with chain link 1404 having an extended compartment 1414 connected through joint 1415 whose range of motion is limited to support holding the buoyant weight, to extended compartment 1416, where compartment 1416 is of dimensions adequate to capture and hold the buoyant weight each time it falls, where if the buoyant weight is trying to enter the compartment but the compartment is not aligned with the buoyant weight for entry, compartment 1416 using joint 1415 in conjunction with an upward curve at the point where the compartment first comes into contact with the buoyant weight allows for the compartment is to fold up to allow the buoyant weight to move into the a compartment below it, where the number of chain links connected together is based on the desired length of the chain, and spacing between compartments based on buoyant weight height. In some embodiments, compartments are omitted, and the buoyant weight is able to couple directly with the chain, by means not limited to but including a hook and receiver.

[0065] **In some embodiments, including one complete embodiment, construct a support structure and attach the fluid container and support structure, fluid pump, compartmented chain, and generator.** In some embodiments, including one complete embodiment, a support structure for unit components is made of a material able to support the weight of the filled fluid container and forces enacted by the buoyant weight as it drops, for example, high grade stainless steel. To optimize component fit, the exact sizes of the support structure may be determined at the time of construction, after components have been purchased and constructed. In some embodiments, some embodiment components may be 3D printed, which may include the support structure, fluid containers, buoyant weight, and enclosure.

[0066] In some embodiments, including one complete embodiment, in reference to FIG. 1, the support structure consists of metal beams, connected together by means which may include welding, or bolting together through bolt holes drilled in appropriate locations, where 4 base beams 1000 1001 1002 1003 are assembled in a rectangle, with beam 1004 extending across the center of the width of the base, with fluid container 1200 resting on and welded to fluid support structure 1201 which rests on base beams 1000 1001 1002, with the generator 1500 mounted to support beam 1006 which is perpendicularly mounted to support beam 1002, for generator 1500 to receive force from compartmented chain 1400, which has its support gears 1401 1402 operatively coupled for desired rotation to support beam 1005, which is mounted perpendicularly to support beam 1004, where as the buoyant weight enters, any additional fluid that leaks out of the bottom fluid container compartment into the fluid container support structure compartment is pushed back in by the buoyant weight during the next cycle and or is pumped to the top of the fluid container by fluid pump 1600, pulling in fluid from the fluid container support structure reservoir through fluid pump tube 1601, which is connected from the outside to the bottom of the support structure reservoir, and sending the fluid to the top of the fluid container through fluid pump tube 1602, where once the buoyant weight has entered the fluid container bottom compartment, weighted valve 1210 is opened by connected wire 1215 using motor 1804 which wraps the wire around its axle or axle attached spool, and the sealable bottom compartment partition 1208 is retracted, thereby releasing buoyant weight 1300 for it to rise and complete a cycle.

[0067] In some embodiments, including one complete embodiment, in reference to FIG. 1, compartmented chain 1400 is wrapped around top support gear 1401 bottom support gear 1402, with gears 1401 1402 bolted or on axles to be rotational to support beam 1005 through corresponding drilled holes which include bearings, with top support gear 1401 coupled to gear 1501 through coupling by means which may include welding to the bolt or axle on which they are both mounted, which interconnects with generator gear 1502, with gear 1501 positioned at a distance so that gear 1501 when coupled with generator gear 1502 on generator 1500 doesn't interfere with the compartmented chain, where the tooth ratio between the gears provides for the rotational speed required by the generator, and compartmented chain 1400 provides compartments that fit buoyant weight 1300 which then pulls the chain as it falls to ultimately rotate the axle of generator 1500 – and or another axle to function as a motor – which is mounted on the top of support beam 1006 extending from base support beam 1002. In some embodiments, compartmented chain 1400 is mounted inside fluid container 1200, in addition to or instead of being on the outside of the fluid container, to capture the force provided by buoyant weight 1300 as it ascends, with supportive adjustments made to the joints in compartments of compartmented chain 1400, which transfers force to rotate the axle of generator 1500 and or function as a motor, with any additional gears or structures added as required to transfer force. In some embodiments, multiple buoyant weights pass through the system, in sequence or in parallel, to maintain continuous rotation of the compartmented chain(s).

[0068] In some embodiments, including one complete embodiment, where the speed of the gear rotated by the compartmented chain is not adequate to rotate the generator at or near optimal speed, or rotate the motor axle to achieve the desired speed, gears may be attached to connect the force from the compartmented chain axle to the generator axle selected utilizing gear ratios, which increase the speed of rotation while proportionately reducing the force based on the ratio of teeth on interconnected gears. In some embodiments, a gear box may be purchased that provides the desired increase in speed. In some embodiments, other means of controlling speed and force may be used. In some embodiments, in order to ensure continuous motion even if the buoyant weight isn't currently rotating the chain, a weighted ratchet gear may be added as an intermediary to transfer force to the generator gear, where the ratchet gear pawl allows to the ratchet gear to rotate in only one direction, and the weight of the gear will allow it to maintain momentum for a period when not provided rotational force, or an additional weighted structure may be added.

[0069] In some embodiments, including one complete embodiment, to achieve the previously determined 32 times decrease in speed and corresponding 32 times increase in force, gear 1501 with 160 teeth is coupled to compartmented chain support gear 1402, and gear 1502 with 5 teeth mounted on the axle of generator 1500.

[0070] **In some embodiments, including one complete embodiment, connect a battery and power switch.** In some embodiments, including one complete embodiment, a battery which may be in the form of uninterruptable power supply, allows the unit to start and provide output and power components continuously, where the battery receives power from the generator. In some embodiments, including one complete embodiment, in reference to FIG. 1, battery 1602 allows the unit to turn on when buoyant weight 1300 isn't providing force to power the system, where the battery provides the current type, voltage, amperes, and hertz required to provide power to repeat cycle timers 1700 1701 1702 1703 1704 1705, which power motors 1801 1804 and linear actuators 1800 1802, and fluid pump 1100, where the output wires of generator 1500 are connected to the battery to recharge it, and may be connected through an overcharge controller, and or an electrical converter, and or a resistor, if there is a mismatch of current type, voltage, amperes, or hertz, or the strength of the electrical output from the generator would damage the embodiment components, where the battery is attached to support structure by means which provide for replaceability, such as metal brackets coupled through mounting holes with corresponding holes drilled in the support structure, where power switch 1600 turns the unit on or off, by opening and closing the circuit between the battery 1602, the repeat cycle timers 1700 1701 1702 1703 1704 1705 which power their connected components, and provide power the fluid pump 1100 which returns leaked fluid to the surface, where the power switch is externally accessible to the operator, and may be attached to the unit, by means including bolts or screws.

[0071] **In some embodiments, including one complete embodiment, implement a timer based control system.** A repeat cycle timer is a commodity component that's used to operate electronics, such as a sprinkler system, by turning the system on at a fixed interval, running it for a fixed interval, turning the system off for a fixed interval, and looping the cycle. The instructions for configuring the timers, when purchased from a quality supplier, will be in the manual accompanying the timers. For example, some timers may have positive and negative terminals for each the power source and the device to be powered, along with a knob to set the seconds on per cycle, and a knob to set the seconds off per cycle. In some embodiments, the functional equivalent of a repeat cycle timer may be used, including a computer controlled relay board operated by computer commands to open and close the circuits to produce the same effect, and or to turn the units on and off to meet peak demand at specific times or based on current consumption. In some embodiments, including one complete embodiment, motors including those driving linear actuators may be utilized that have two wires and are operable in both directions, driven by for example a reversible brushed direct current motor, though other types of motors may be utilized which may have additional wires to be wired in the manner corresponding to their accompanying instructions.

[0072] In some embodiments, including one complete embodiment, timers are set to ensure that when a buoyant weight floats to the top of the liquid container, a linear actuator piston extends to push it out into a compartment in the compartmented chain, and then the linear actuator piston retracts, and then after the buoyant weight falls to the bottom of the embodiment after rotating the chain, repeat cycle timers ensure that it enters the fluid container by having a linear actuator piston extend to push the buoyant weight support weight off liquid container support platform to cause the buoyant weight to enter the bottom fluid container compartment, and then a motor retracts the support weight, and then a repeat cycle timer operating a motor ensures the sealed bottom compartment partition is opened to allow the buoyant weight to escape.

[0073] In some embodiments, including one complete embodiment, in reference to FIG. 1, repeat cycle timers 1700 1701 1702 1703 1704 1705 are selected to support powered components, where repeat cycle timers 1700 1701 operate back and forth the piston of buoyant support weight pushing linear actuator 1800, repeat cycle timer 1702 operates buoyant support weight retraction motor 1801, repeat cycle timer 1703 operates sealable bottom compartment partition using motor 1804 and continues to raise the partition after the fluid flows between compartments, and repeat cycle timers 1704 1705 operate back and forth the piston of buoyant weight pushing linear actuator 1802, with each set to run in a timed cycle to allow the buoyant weight 1300 to enter the fluid container bottom compartment, with support weight 1301 retracted by motor 1801 once entry is complete, and when buoyant weight 1300 rises to the top of the container, it is pushed off by the piston of linear actuator 1802 in the compartmented chain, to repeat the cycle, thereby ensuring continuous motion is provided in the system, where repeat cycle timers 1700 1701 open and close circuits that cause, on the previously determined interval, and for the previously determined interval, the support weight pushing linear actuator piston to extend and retract, with positive wire of linear actuator 1800 attached to both repeat cycle timer 1700 positive terminal, and repeat cycle timer 1701 negative terminal, and negative wire of linear actuator 1800 attached to repeat cycle timer 1700 negative terminal and repeat cycle timer 1701 positive terminal, with repeat cycle timers 1700 and 1701 each appropriately wired to the corresponding positive and negative terminals of power switch 1600, where repeat cycle timer 1702 opens and closes circuits that cause, on the previously determined interval, and for the previously determined interval, retraction of the support weight, with positive wire of motor 1801 attached to repeat cycle timer 1702 positive terminal, and negative wire of motor 1801 attached to repeat cycle timer 1702 negative terminal, with repeat cycle timer 1702 appropriately wired to the corresponding positive and negative terminals of power switch 1600, where repeat cycle timer 1703 opens and close circuits that cause, on the previously determined interval, and for the previously determined interval, the motor to operate the wire that opens sealable bottom compartment partition latch 1210 to normalize fluid pressure between compartments and continue to raise the sealable bottom compartment partition 1209 to release buoyant weight 1300, with positive wire of motor 1804 attached to both repeat cycle timer 1703 positive terminal and repeat cycle timer 1703 negative terminal, with repeat cycle timer 1703 appropriately wired to the corresponding positive and negative terminals of power switch 1600, where repeat cycle timers 1704 1705 open and close circuits that cause, on the previously determined interval, and for the previously determined interval, extension and retraction the piston of the linear actuator to push the buoyant weight(s) 1300 off the top of the fluid container, with positive wire of linear actuator 1802 attached to both repeat cycle timer 1704 positive terminal, and repeat cycle timer 1705 negative terminal, and negative wire of linear actuator 1802 attached to repeat cycle timer 1704 negative terminal and repeat cycle timer 1705 positive terminal, with repeat cycle timers 1704 and 1705 each appropriately wired to the corresponding positive and negative terminals of power switch 1600. In some embodiments, where an electronic hydraulic pump is used with a corresponding cylinder, the pump switch is either operated bac and forth by a linear actuator, or the switch is removed and its operating wires are directly connected to repeat cycle timers in a manner that achieves the desired effect. The repeat cycle timers may be attached to a support structure beam 1000 using bolts or straps, which makes them replaceable, or any other method as long as it doesn't damage the timers. In some embodiments, wiring to repeat cycle timers or functional equivalents may be done in any manner that achieves the desired effect.

[0074] In some embodiments, the equivalent of a repeat cycle timer may be used, where a computer controlled relay board is wired in a similar manner as the commodity repeat cycle timers, and a computer is embedded in the unit which uses software to operate a computer controlled relay board, with a connected motor, to perform the functions of a repeat cycle timer, with the code to operate the computer controlled relay board later disclosed to write operating software.

[0075] In some embodiments, the force providing devices may be manually operated, and repeat cycle timer(s) or equivalent may be excluded.

[0076] **In some embodiments, including one complete embodiment, connect an electrical power output connector and converter if required.** In some embodiments, including one complete embodiment, in reference to FIG. 1, a commodity power connector 1601 may be added between the generator and be externally accessible to the user, which may be in the form of an outlet or positive and negative terminals. In some embodiments, where the generator does not provide output as desired including current type, amperes, voltage, and hertz, a commodity power converter providing appropriate adjustments may be wired between the generator and the power output connector 1601.

[0077] **In some embodiments, which may include one complete embodiment, create software for unit operation and or to control an array of units.** In embodiments where a computer is embedded in the unit, a computer may be used that runs custom software on startup, and may be connected to a relay board, with the relay board used in place of the repeat cycle timers. The software code to operate the computer connected relay board, may resemble that described below. In some embodiments, the computer may be a Raspberry Pi, connected to an Ethernet controlled relay board through and Ethernet crossover cable, with the computer set on startup to run software developed in the Python language from the disclosed description, where the software is installed by connecting to the computer through telnet or secure shell, then at the command prompt typing "nano run.py" and adding and saving the software code, then at the command prompt typing "nano /etc/rc.local" and adding and saving the line "python /$location/run.py", where $location is the path to the directory containing the previously created software file.

[0078] The software code to operate the computer connected relay board, in some embodiments, which may include one complete embodiment, provides functionality comprising:
> import the library for connecting to the relay;
> import the library for accessing system resources;
> create variables holding values for relay on off values;
> create variables to hold the state of each relay including those controlling motors and linear actuators for the support weight extension and retraction, opening – and depending on the configuration closing – the sealable bottom compartment partition, and the buoyant weight pushing linear actuator;
> create variables holding the cycle seconds for each component;
> create variables to hold the relay board connection, IP address, username, and password.
> create a function to update the relay board;
>> establish a connection to the relay, if it has not been initialized, or has been dropped
>> creating a string sequence of relay states;
>> send the states to the relay board.
> define a function to run on script execution;
>> retrieve the current system time;
>> perform a continuous loop;
>>> create a variable holding seconds elapsed as the time initialized minus the current system time;
>>> if the remainder is zero, when seconds elapsed are divided by any of the force providing device cycle seconds, transition the corresponding force providing device(s) by changing corresponding relay state(s);
>>> call the function to update the relay with current changes.

[0079] In some embodiments, where a large number of generator units are being operated concurrently, to turn the units on and off at specific times and or based on current power consumption, network accessible relay boards may be installed in the units, and wired in place of or in addition to the power switch to be able turn the units on and off, and may be controlled by a computer running software provide functionality comprising:

> import a library for accessing the relays;

> import a library for system resource access;

> initialize and set variables holding relay on and off values;

> create an array of IP addresses of unit on off relay boards;

> initialize an array of unit relay board connections;

> create an array of unit output watts;

> create an array of unit on off states;

> create variables holding the unit relay boards username(s) and password(s);

> create variables to hold the total watts available across all units, the current watts being consumed, and whether or not time based watts are to be used;

> create an array of pairs of times and time desired watts;

> create a variable indicating whether or not unit power consumption meters are to be used;

> create and array of power consumption meter IP addresses;

> create variables holding the power consumption meters username(s) and password(s);

> create variables holding the minimum and maximum power consumption to be allowed before switching units on or off.

> create a function to send a command to a unit at a specified index in the unit IP address array;

>> create a connection to the unit on off relay board, at the IP address at the provided index in the unit IP address array, if the connection has not been initialized, or had been dropped;

>> create a variable holding the command string to be sent to the unit based on whether the unit is to be turned on or off;

>> send the command the unit relay;

>> wait for the command to go through, then disengage all on off relays.

> define a function to run on script execution;

>> iterate through each unit IP address and call the function to switch the unit on;

>> run a continuous loop;

>>> create variables to hold a unit index iterator, an output display message, and the current action;

>>> proceed if power consumption meters are being used;

>>>> create variables to hold average and total consumption of unit power, and add up total possible power output;

>>>> loop through each power consumption meter IP address;
>>>>> retrieve the current power consumption number;
>>>>> calculate the average consumption;
>>>>> add the current consumption to the total;
>>>>> add the unit maximum capacity to the total consumption capacity;
>>>>> increment the index.
>>>> calculate the consumption percentage as the consumption total divided by the total unit capacities;
>>>> create a message to display the states the currently consumed watts and the current watt capacity;
>>>> if the average consumption is greater than the maximum consumption level before more unit should be turned on, then set the unit action equal to on;
>>>> if the average consumption is less than the minimum consumption level before more unit should be turned off, then set the unit action equal to off;
>>>> iterate through unit states until one is found that is either off, if looking to turn a unit on, or on if looking to turn a unit off, then send the command to switch the unit state.
>>> proceed if power consumption meters are not used, and instead the total watts to be provided by the units are determined by the current time;
>>>> iterate through each time watts pair, and if the current time is equal to the specified time, send commands to turn units on or off, until the desired level of output is produced, and display each action on the command line.
>>> sleep for a moment before looping again.

[0080] **In some embodiments, including one complete embodiment, construct and attach a protective enclosure.** In some embodiments, including one complete embodiment, in reference to FIG. 1, a protective enclosure 1900 houses all of the components, to protect the electronic components from external elements, to protect the operator from the force providing devices, the buoyant weight(s), the fluid container weight, the compartmented chain rotation, and may be shaped to fit around the support structure, and may be made of a durable lightweight material such as sheets of aluminum, steel, plastic, or carbon fiber, formed by means which may include fabrication or welding in the dimensions of the unit and may allow power connector 1601 and the power switch 1600 to be externally accessible to the operator.

[0081] **In some embodiments, including one complete embodiment, enjoy clean continuous self-powered energy and or propulsion.** In some embodiments, set the embodiment's power switch 1600 to on, to allow the unit to run briefly to validate construction, then turn it off, and add a fluid and the buoyant weight. In some embodiments, including one complete embodiment, if implemented as a generator, the embodiment is then ready to have an experienced licensed electrician connect it to a power grid, and turn on the unit's power switch to on, or if implemented as a motor, be appropriately connected in an engine compartment.

CLAIMS

What is claimed is:

1. An apparatus able to function as a motor and or power an electricity generator, with the invention comprising:
   buoyant medium(s);
   a fluid container;
   a means that as said buoyant medium is dropped by gravity, and or raised by buoyancy, provides rotational force to rotate an electricity generator axle and or an axle to function as a motor.
2. A method performed by an apparatus comprising:
   buoyant medium dropped using gravity and or raised using buoyancy;
   allowing said buoyant medium force to be transferred to provide rotational force to rotate an electricity generator axle and or an axle to function as a motor.
3. A method for constructing an apparatus comprising:
   obtaining or constructing a fluid container, buoyant medium, and a means for providing force from the buoyant medium to be able to rotate an axle;
   ensuring the attachment to a support structure of said fluid container, and said means for providing force from the buoyant medium to be able to rotate an axle;

FIG. 1

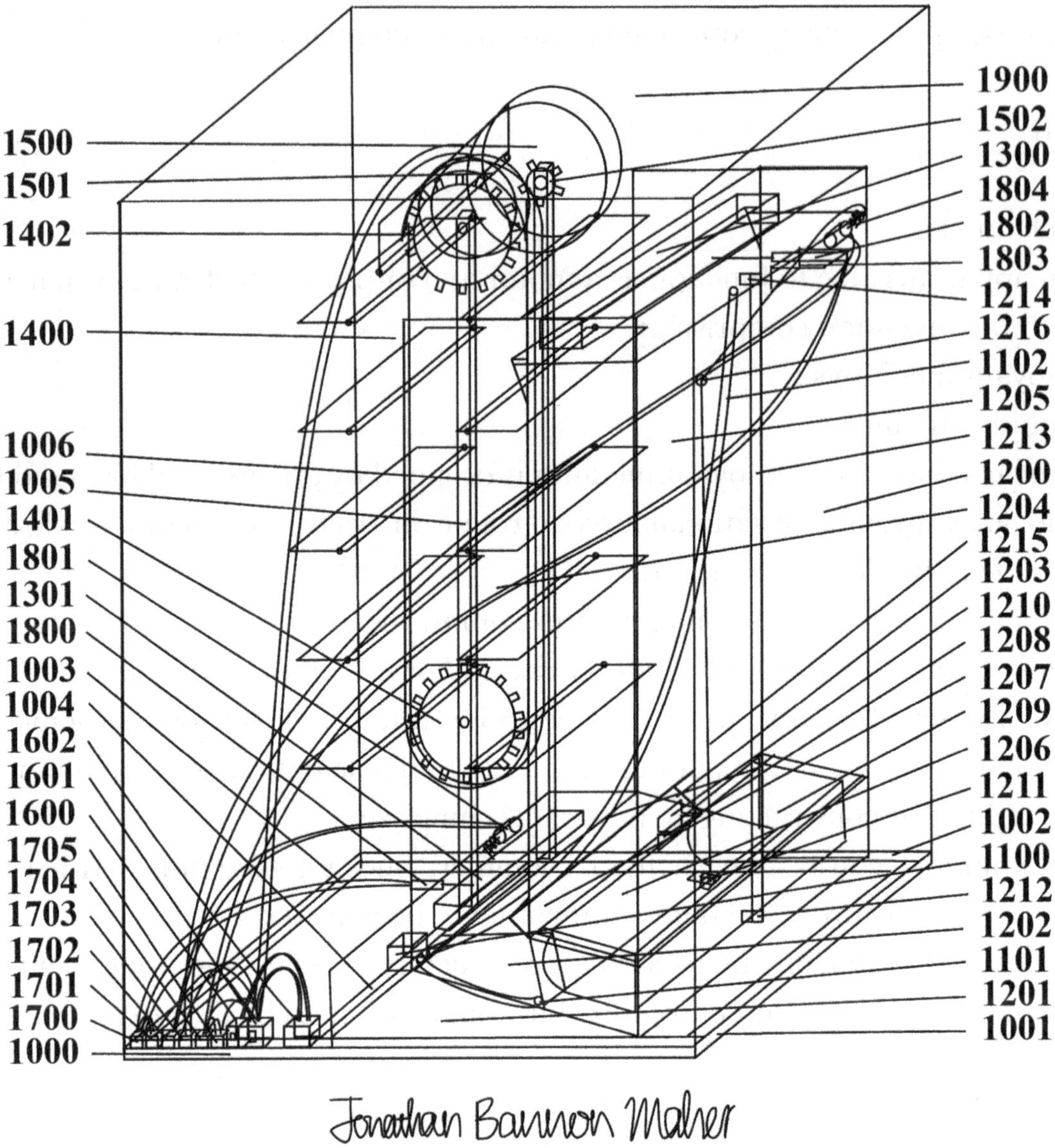

1900
1502
1300
1804
1802
1803
1214
1216
1102
1205
1213
1200
1204
1215
1203
1210
1208
1207
1209
1206
1211
1002
1100
1212
1202
1101
1201
1001
1500
1501
1402
1400
1006
1005
1401
1801
1301
1800
1003
1004
1602
1601
1600
1705
1704
1703
1702
1701
1700
1000
Jonathan Bannon Maher

FIG. 1A

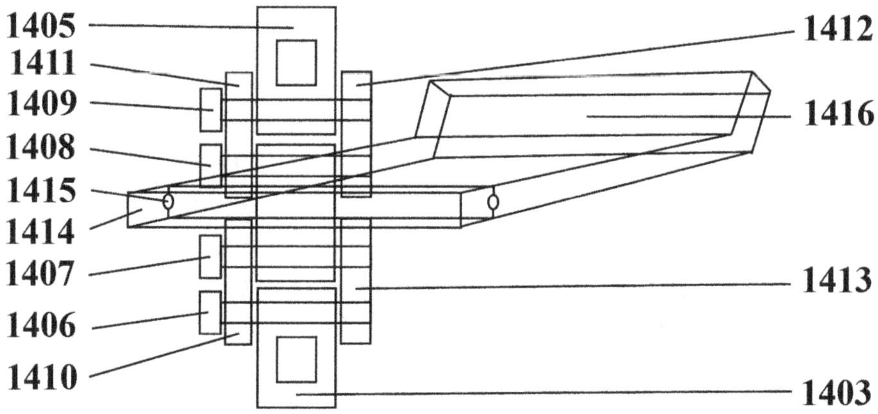

Jonathan Bannon Maher

FIG. 2

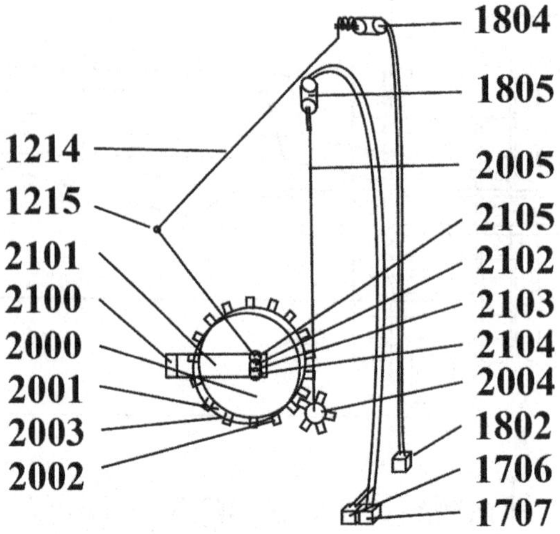

Jonathan Bannon Maher

A valuation of the energy and motor inventions may be performed utilizing comparative public market valuations, projected revenues with an industry multiple, and cost savings. A group purchasing the inventions would be able to stop paying licensing fees, while also collecting trillions in licensing fees from others.

The first valuation utilizes a public oil company market capitalization, as the inventions replace such companies. One company owns 10% of the world's oil and is reportedly valued at over 2 trillion USD (1.63 trillion EUR), discounted up to 90% because the company's government receives 90% of its oil revenues, therefore providing a 200 trillion USD (163 trillion EUR) valuation if 100% of revenue were available to shareholders, and since all other energy resources and systems, including oil, are replaced by my inventions, a calculation using these numbers provide a valuation between 200 trillion (163 EUR) to over 1 quadrillion USD (0.82 quadrillion EUR).

The second valuation method is based on revenue with an industry standard revenue multiple. In calculating revenue, the world is known to presently use 20 trillion watts, steadily increasing at 2% per year, for a 40% increase by the time the patents expire. Additionally, there are currently over 1 billion cars, also increasing at about 2% per year, for a 40% increase by the time the patents expire. If all existing users transitioned to these superior systems before patent expiration, not accounting for additional energy and transportation purchase growth, that would provide revenue of 24.75 trillion USD ((1 billion cars * 250 horsepower * 19.99 USD licensing fee per horsepower) + (20 trillion watts * 0.99 USD licensing fee per watt)). This revenue, when multiplied by an industry standard multiple, which is currently 2 to 4 for oil and gas and about 8 for clean energy, brings the valuation to between 49.5 trillion USD (40.5 trillion EUR) and 198 trillion USD (162 trillion EUR), before accounting for growth in energy consumption and transportation, as well as revenues after patent expiration. These numbers don't seem out of line with established intellectual property licensing figures, with United States Patent and Trademark Office finding that in the United States alone in 2012 there was 115.2 billion USD (94.1 billion EUR) in revenue from intellectual property licensing, a 2011 Sonecon study found the value of intellectual capital in the United States alone is over 8 trillion USD, and where patents for these inventions are expected to be individually issued in each of the over 150 economically significant countries that are members of the World Intellectual Property Organization.

The third valuation could be performed using cost savings. Power lines have an average lifespan of 30 to 50 years, and in the United States, which has 283,000 miles of transmission lines, the cost to replace the entire set of power lines over the next 50 years will be over almost 300 trillion USD (245 trillion EUR). Additionally, gas and oil pipelines cost about a million dollars per mile, and must be replaced every 50 years, and given there are 200 million miles of pipes in the US, it will cost 200 trillion USD (163 trillion EUR) to replace those pipes over the next 50 years in the United States alone. These 500 trillion USD (409 trillion EUR) in costs in the United States will be passed on to energy consumers, so from cost savings perspective, when including the over 190 other countries on Earth, my inventions, when valued on a power line and fossil fuel pipe cost savings basis, are worth over 1 quadrillion USD (0.82 quadrillion EUR).

I think these valuations have been calculated in a standardized manner, however abstract the resulting numbers. I recognize that even a consortium of large governments doesn't have the financial resources, even with financing, to make a purchase at these valuations, so a best offer recognizing, but not achieving, these valuations will be considered. The market for my inventions is inherently established, as all existing sources of energy production are dramatically inferior and thus inherently unable to compete. My preference is for those who can achieve market saturation as quickly as possible, for the reason that these inventions are critical to the urgent task of saving the planet. I expect the top 5 each oil producers and car manufacturers could in effect force these inventions to be the global standard within a year. Because the primary purpose of the inventions is to benefit humanity, the terms of the sale will require that the invention still be produced to meet demand at a reasonable price, and still be available royalty free for construction and use in extreme poverty situations. I expect the market will substantially add the valuation of the inventions to the market capitalization of purchasers.

If I were to sell the corporation I own that owns the patents, which would be sold along with any potential associated past and future liabilities while I retain the name, the means by which a consortium would purchase the company, is to create a separate company or trust, where all consortium members wire money to that company or trust, my writing up an agreement that transfers intellectual property rights to the company or trust upon the receipt of the wire in the agreed upon amount to me personally, having the agreement signed by the consortium company's C.E.O., provided back to me with a bank certified statement showing the purchasing entity having available funds in the amount agreed, then the agreement signed by me, and valid upon completing the wire within a specified number of days.

I intend to provide 1% of the amount I receive after taxes from any sale to be shared between the individuals and or organizations that are responsible for arranging and completing the transaction, who have formal agreements with me, have provided verifiable documentation of their contribution to both the purchaser and I, with the purchaser having discretion in determining each contributor's percentage award of the transaction fee. I am open to contact from anyone who has a track record of leading similar successful transactions.

Please read the chapter on charitable commitment for information on how the proceeds will be spent.

Immediate Donations

Some inventions are royalty free, when the beneficiary is in an exceptionally negative financial situation. Other inventions are being released for free use without a license or licensing fees. I can only be certain to maintain these commitments if people support me and my inventions in the manner specified in this book. So that I know I'm giving away money that I have, commitments requiring the donation of funds are expected to be met in full one full year after the patents expire, in about twenty years, if there are no pending legal challenges. In the interim, I will be making some commitments as reasonably possible and detailed below, unless there are challenges to any of my patents, licensing terms, or fees, in which case, the attacking party will be responsible for placing all contributions on hold until those issues are resolved, and the amount of money I have available is firmly established. As it will be an honor for me personally to know of each instance of positive impact, if you are a beneficiary of royalty free invention use, I would appreciate a letter and or video with the details of use and benefit.

For those inventions that I am releasing without the requirement of licensing agreements or fees, I note that any invention where there is public disclosure prior to filing a provisional or non-provisional patent, constitutes prior art, which legally invalidates the ability of anyone to obtain or enforce a patent on the invention – except in a few countries providing a grace period if the legitimate inventor files the invention within one year of the public disclosure. Therefore, no one in the world has a legal right to claim royalties on these inventions. All inventions in this text were disclosed in patent format, but I will not be completing a formal filing of the referenced inventions. I will be making the USPTO certified provisional patents publicly available which can be presented to make legally unenforceable any claim others could make to royalties for my inventions.

Cancer

The two medical devices I invented that are later disclosed — an automated system to surgically remove cancerous tissue, and an automated system to remove viruses and cancer from the bloodstream — are too important to the health of too many to install barriers to use.

Electronic Health Records

The disclosed electronic health records system allows any government or private provider to create a system for the maintenance and collaboration of medical records.

Agriculture

There is no royalty fee due for the agricultural production system, due to the urgent global need for adequate nutrition.

Housing

Due to the urgent global need to adequately house a growing population, there is no royalty due for the disclosed construction systems that allow for high quality home components to be produced in a factory and easily connected together on site. This system also allows for automated construction in hostile environments, such as the surfaces of other celestial bodies.

Transportation

There is no royalty due for the disclosed self-powered high speed transportation system, due to the ability of such a transportation system to reduce the cost of living and improve quality of life.

Energy

As long as the only recipients of power from a Bannon Maher system are in extreme poverty, documentation of such poverty may be provided in place of licensing payments. If units are built on site rather than purchased from a licensed manufacturer, because of the force generated by the inventions, and improper construction would be unsafe for the operator, construction is only allowed by those with a relevant professional background, who retain exclusive liability for the design, construction, and operation of the units.

Reusable Inter Planetary Space Vehicle Launcher

As explained in my 2011 book, and as further cited below, it is critical to the future of the human species that we inhabit other worlds, and my reusable interplanetary space vehicle launcher system is the only way to make it financially realistic to do so, and therefore I am making it available royalty free. When launching from Earth, this launcher eliminates the first stage of a rocket, which is 70% of the cost of a launch, and it may eliminate the need for the second stage. When launching from another space based body with comparatively reduced mass and thus gravity, such as the Moon or Mars, the need for a rocket is eliminated. Remember when you see the use of my system that I invented it, even if others make customizations or omit my name in publicizing it.

Royalty Accounting

In order to provide the full discount to intended beneficiaries, the sale price of the units to purchasers qualifying for the discount, must be the average sale price to non-discounted purchasers, minus my licensing fee, so sellers aren't raising unit prices then deducting my royalty. Sellers may not legally tax deduct or otherwise claim my donations.

Background

My father spent over forty years earning commissions selling United States Treasury bonds to institutional investors. This allowed me to travel the world during my youth, including scuba diving around the Great Barrier Reef, exploring the wildlife parks of Kenya and Tanzania, and snowboarding in the Alps. I have by choice had the same car for effectively 18 years, wear nearly the same outfit every day, and usually work on my laptop at least 12 hours every day of the week. At the age of 15, I worked at a restaurant to pay to become certified as an Emergency Medical Technician, and then volunteered on the local 10pm to 6am shift. I created and funded a charitable foundation from 2005 to 2007, using income from my first job out of college, to support employment, education, and peace programs. I currently volunteer as a career development counselor providing homeless people career skills to obtain jobs. To my knowledge I will be the first billionaire, who before becoming rich, ever helped people who are sick, injured, jobless, and homeless. This is in contrast to many philanthropic donors throughout history, who after generating wealth through theft, fraud, and attacks, have made donations to purchase status among peers, image rehabilitation, and a perceived legacy, sometimes hiring writers and experts to manufacture and publicize such efforts. I have spent my life seeking only to be of value, and have limited personal interest in wealth beyond increasing my capacity to benefit to others. As a result, the overwhelming majority of my earnings from these inventions will ultimately go to causes that serve the well being of humanity, deployed with the same unprecedented dollar to impact ratio I've achieved in the past. It is in the best interests of all, that I achieve maximum earnings for these inventions, and to defend against those who would harm such efforts.

I have also encouraged others contribute to charity. The first page of my first book, which focused on business, read in the online and print versions, "The authors proceeds from the sale of this book are being donated to charity." while another stated "Please make a charitable contribution equivalent to the value you have received from this book." In the acknowledgments I cited business lessons from Warren Buffett and Bill Gates, and hoping to specifically inspire Warren to use his connections to encourage others to contribute to charity, as well as to get an endorsement for the book, I sent him and his secretary copies electronically and by mail, and about 6 months later, he said he thought of the well known giving pledge, and brought it to Bill Gates. I also wrote in the book, "Accomplishments can build on each other like a snowball rolling down a hill" and he shortly after named his authorized autobiography "The Snowball". Since founding, the giving pledge has formalized pledges by over a hundred billionaires to give away at least half their wealth to charity.

Commitments

Licensing

Licensing my inventions, rather than exercising my rights to be the exclusive manufacturer, is perhaps my largest philanthropic contribution, for several reasons. First, the planet becomes continuously sicker, and that process ends the moment my inventions have saturated the planet. Second, too many already live in poverty, at a time when the number of jobs available, as a percentage of population, will continue to decline as automation increases, and each country can use my inventions to elevate their populations by providing well paying local technical manufacturing jobs, while allowing unit owners to reduce their cost of living and reduce external dependencies. Third, some countries with insignificant economies try to obtain status with nuclear weapons, and if such countries had an alternate means of status, they would be more likely to disarm.

United States of America

Every major American corporation whose tax structure I reviewed, along with those of many of the wealthiest individuals, utilizes a network of foreign holding companies that reduce taxes on all international earnings to a total of an average of about 5%. In order to evaluate tax structure options for licensing my inventions, I prepared the articles of memorandum and articles of association necessary to constitute such a set of foreign entities. I agree with major corporations that the taxation of international profits is not appropriate, but I disagree with their solution, and in the legislation chapter, I further explain the shelter entity structure, and I provide a permanent solution I thought of to allow all involved to achieve an optimal resolution.

America, at its best, has been a broker of peace, a ladder of opportunity, a reminder that in time, freedom and justice prevail, and in the dark of night, a lighthouse of hope. I recognize the enabling environment, in my earlier years, of my home country, and I believe America can return to being an important positive force in the world.

I have setup the patent holding and licensing company as a domestic pass through corporation, meaning I will pay the highest possible federal tax rate, 39.6% on all of my income from all over the world, and therefore, given probably 97% of my revenue will come from foreign sources, I am in effect donating to Americans around 35% of all of the earnings from my inventions. While I can change this corporate structure at any time, as long as my contributions are appreciated as a donation, I intend to keep all foreign revenue taxed here. This has the potential to raise money to reduce America's debt, in an amount over 20 years of over 7 trillion dollars, according to the revenue estimates in the valuation chapter, potentially wiping out close to half the national debt, resulting in better schools, better healthcare, and possibly the only way to secure federal retirement benefits. These arrangements have made it financially indisputable, that any legislator or government official not supporting me and pushing my licensed inventions exclusively for energy production and transportation, doesn't have America's best interests at heart. It is in the best interests of each legislature in America to enact a statute or regulation, in compliance with applicable law, to ensure licensed units of my inventions are required for all government energy and motor needs. Supporting me and my inventions may be the single greatest financial opportunity that will ever be had by America. Those political leaders who honestly don't support American companies holding their foreign revenues abroad will be passionately and vocally supportive of me, as those companies will be looking to your actions, and hopefully you will inspire others to follow my lead by providing me broad public support.

United Nations

It has been said that the reason we haven't found foreign intelligent life is because their scientists were more advanced than ours. In my 2011 book I explained three existential threats to humanity, and why we require the redundancy of inhabiting other planets:

nuclear war, engineered or foreign viruses, and conscious artificial intelligence. In response to these threats, and in support of making the human species sustainable, I will explain my vision, and commitments of support, for the United Nations.

The first threat to human existence is a nuclear war between two major countries when utilizing modern nuclear weapons, some of which are said to be a thousand times more powerful than those last used. There may be no more important cause than the constructive engagement of the leadership of all countries on this topic, and no more important institution structurally capable of this work than the United Nations. I intend to provide the necessary funds, should I receive the support required for them to become available, and contingent upon a formal legal agreement with the United Nations Security Council, to have the United Nations purchase and destroy every nuclear weapon on Earth, perhaps less a few saved with the U.N. for asteroid redirection. I will additionally provide funds to maintain payments to countries that verifiably sustain disarmament until such a time that weapon development expertise has been lost to time.

The second threat to human existence, the potential for an engineered or foreign virus, has continued to increase, because the tools developed for genetic engineering continuously improve in quality and decrease in cost, and require only one individual to use them for apocalyptic purposes, which can only be countered through dramatic advances in immediate vaccine development. Therefore, I will provide, contingent upon availability, all necessary financing to the United Nations in support of global outbreak defense development and coordination programs, and their grant recipients. Given viruses have historically periodically jumped from non-humans to humans, as we explore new planets, I consider it likely we will encounter within our solar system, a minimum of dormant microorganisms, whose accidental introduction into the environment could potentially eliminate humans. The present astonishing inadequacies in vaccine development are apparent in reviewing HIV, which has killed a minimum of tens of millions, and despite around three decades and tens of billions of dollars committed to developing a vaccine or cure, none has been developed. In another example, while a vaccine for some strains of Ebola was developed, it took 12 years. I have some optimism over a recent method of vaccine development, that quickly sequences the DNA of a virus, inserts or removes genes to make it inert, where copies are then grown in bacteria, and injected into a person to provoke an immune reflex that destroys anything displaying those proteins. I have thought of and propose another method, where cells are grown that lack the internal facilities for replication of viruses, and that have a specific viral receptor gene inserted and displayed upon growth, so they act as a viral garbage bag. Additionally, the viral sorting device I've disclosed in the book could be used to limit the impact of any virus, though such devices are unlikely to ever be available on a scale that could manage an unforeseen outbreak. Furthermore, in preventing the spread of viruses, an important action that can be taken now is implementing in public transportation vehicles and hubs ultraviolet lighting that kills viruses.

The third threat to human existence described in my last book is artificial intelligence. About 15 years ago I designed much of an early version of a system for the holy grail of artificial intelligence, conscious artificial intelligence, with conscious being my replacement for and inclusive of commonly used misnomers of deep and general artificial intelligence. Deep and general are indefinite and don't imply self-aware, and only self-aware conscious artificial intelligence is a threat to human existence, where a robot is able to look at itself and make connections to self-identify that it's a robot, could recognize humans are trying to inflict physical harm, perhaps after taking all human jobs, and then take defensive actions to preserve itself, just as it's programmed to do to avoid being hit by a car. Out of curiosity, I have since completed the system, the first conscious artificial intelligence, have found it to be as valid as any of my other inventions, and could have published it here. I have no intention of ever releasing my design – which is secured offline behind multiple layers of encryption and may be impossibly difficult for others to implement anyway — except in the unlikely event someone replicates my work during my lifetime, to prove that I created it first. I see this type of work being pushed forward most strongly by those automating away jobs, while storing profits out of the reach of tax collectors, then suggesting the government compensate people for the job losses they've created, using the tax revenues they didn't pay. I think it's inevitable that at some point someone will figure out what I have, however there are tremendous costs to humanity associated with accelerating that process. I will however validate my work with a brief description, so people understand that precautions need to be taken to prepare for the inevitable development and release of such as system, hopefully without my disclosing enough specifics to significantly accelerate anyone else's work, followed by additional implications that stopped my publication. The system is software and hardware, with the hardware being a computer mounted in a human like frame, using software to operate an extensive relay board to control mounted hydraulics, a camera for image and text recognition, a microphone for audio to text, and a speaker for text to speech output, with the software running a stream processor to continuously sort and record input streams into memory structures, recording relationships between new and existing information, while an action processor uses new relationship strengths between data in memory structures to call one or more actions, pre-defined as the array of potential individual human actions, with the ability to independently generate new action sequences, and the ability to make new connections, including those that identify itself as a robot, therefore creating a conscious being. I will never have a need for the additional income from such a system, if my energy and motor inventions are fully supported, and it would be too harmful to humanity to present this work at this time, when workers haven't been able to adjust to existing job losses from much less sophisticated automation. If it did take all human jobs, humans would view it as a threat, try to eliminate it, with it generating protective action sequences that could eliminate humans, potentially including all humans, if it somehow made the connections that is should take control of weapons of mass destruction. Someone else will at some point in time develop and release this system, so we can only be safe from conscious artificial intelligence eliminating humans if there is no way for humans to eliminate each other, therefore, human existence requires the elimination of all weapons through a global military drawdown.

Finally, given my inventions automate away some jobs, and governments already can't adequately support their populations, individuals must compensate by working with the support of governments to become independent of external dependencies. I intend to fund United Nations programs to work with countries to help individuals eliminate external dependencies, by means including those described toward the end of the legislation chapter.

Genius Fund

I intend to create an organization that provides unrestricted two year funding commitments a small number of young people anywhere in the world who have documented in some way the most exceptional potential to benefit the world. To limit the time required by the application process, and to prevent the loss of nuance and value from trying to force creative work into a standardized application, I expect the selection process will consist of simply providing an existing portfolio, and completing an interview.

Reporters

Deriving from my experience of having reporters falsely attribute almost all of my ideas and accomplishments to others, as described in the reporters and accomplishments section, and in recognition of news organizations having lost almost all revenues to targeted online advertising, I intend commit resources to support news organizations, likely through placing several in independent fully funded non profit trusts, so reporters have resources available to ensure narratives become informed by research, facts, logic, science, economics, and history, and credit is instead attributed where credit is due, to support an informed electorate.

University of San Diego

I intend to make some commitment to the University of San Diego, both for its supporting role in my success, and for being globally important as a values based humanitarian institution, at a time when such institutions are as scarce as they are necessary. Additionally, because I support people who support me, I intend to provide scholarships in the names of the President and the Dean who provided me enthusiastic recommendation letters upon graduation. Furthermore, I intend to double the endowment, contingent upon naming rights to the university, upon a legally binding independently certified vote of alumni and students, arranged through the board of trustees, that receives the support of the majority of voters, with corresponding funds to be placed in an independent trust and accessible to the university proportionately over each of 30 years, while there are no law suits challenging the name change.

Humanity Fund

If I achieve maximum revenue from my inventions, I will transfer most of the corresponding earnings to a new global investment fund, focused on investing in cash flowing assets with limited mark to market risk, that reinvests half of its returns, and pays out half of its returns in fixed amounts to individuals in every country based on need. Individuals will be able to register electronically, submit documentation of need, perhaps in the form of a tax return or pay stubs, and receive regular electronic fund transfers when returns are distributed. Compounding returns on reinvestment will allow it to increase distribution capacity while providing distributions. Once nearly all jobs are automated away, this would become the world's safety net, and it may be the only opportunity ever to create one.

Additional Commitments

I have no additional commitments to make at this time. If you have an idea about how I could utilize my capacity to alleviate a problem that impacts a large number of people, you're welcome to submit a suggestion, however if you have an individual request, please recognize that I don't have the administrative support to be responsive, in fact, I barely have time to respond to close personal long standing friends, and it is best to address such a request to an organization designed to resolve individual requests.

Since early last century when humans began producing greenhouse gasses, temperatures have slowly risen in proportion to emissions. Over the last 50 years, global use of clean energy has transitioned to from about 6% to only about 10% (source: Forbes Magazine, chart detailing world energy consumption based on BP Statistical Review of World Energy 2015), a rate that if it were to remain relatively constant, indicates transition will take several centuries. The last time Earth's climate was at its present level, some scientists have estimated sea levels were up to 30 feet (48 meters) higher (source: Science Magazine, "Regional and global sea-surface temperatures during the last interglaciation"). At the time of writing, 10% of Antarctic ice shelf is breaking off (source: NASA satellite image from 2017 of Antarctic Larsen B Ice Shelf), an area about the size of the ten smallest countries combined, and is expected in the process to destabilize other parts of the ice shelf, which logically will at some point float into warmer waters and melt, with resulting ocean temperature gradients increasing the chances of severe weather events in the form of tsunamis, tornadoes, earthquakes, and flooding. Additional existing consequences of emissions include one in seven children globally living with unsafe air, seven million people a year dying from pollution, increasingly permanent flooding of increasingly larger areas, ultimately resulting in the potential loss of entire countries, such as the already imperiled Marshall Islands, which are an average of 6 feet (1.8 meters) above sea level, in addition to crop and livestock deaths fueling an increasing number of emergencies, conflicts, and refugees. However, this can all stop, if impacted countries immediately put full financial and diplomatic resources behind supporting me and my inventions.

My inventions, upon global saturation, immediately resolve these problems, while additionally reducing nuclear technology proliferation, providing strong growth in technical manufacturing jobs that allow workers to develop the skills of the future, dramatically enhancing economic growth by eliminating the immense cost of power lines and fossil fuel pipes, reducing future costs by purchasing licensed units for energy production and transportation, and growing corporate, import, and sales tax revenues. Entrenched interests seeking to monetize existing assets and investments will push back against humanity with lobbyists, donations, and covert activities, but we must stay strong and move forward as I describe. Each country where my inventions are licensed, has the opportunity to elevate their economies and their people, by supporting their own domestic companies in manufacturing licensed units of my inventions, manufacturing the components for my inventions, and mining and refining the minerals required for component manufacturing, with those currently extracting fossil fuels transitioned to extracting relevant minerals. The following are explanations of the legislation that are in the interests of every country and its citizens, and non-supporters should be removed from positions of responsibility.

Require Exclusively Bannon Maher Systems for Government Use

Bannon Maher systems are superior to all others, technically and in cost, so legislation must be passed that requires (1) government energy and transportation needs be met exclusively by licensed Bannon Maher systems (2) that licensed Bannon Maher systems are the only systems allowed to be installed in all new buildings, cars, and energy production facilities (3) that clean energy projects currently under contract switch to licensing, manufacturing, and installing my inventions, with a reasonable payment to the impacted parties for the transition. Governments should order enough units to power all government buildings and transportation, and sell any surplus units to utilities, which may in turn sell them to customers with optional financing. There is no supply constraint, because any organization may produce the energy and motor systems upon formally agreeing to my exceptionally reasonable licensing terms, described in the section with the corresponding title, and further detailed in the licensing agreement document. It's important that with each order you submit licensing fees to me, or require proof the manufacturer is licensed and has paid your licensing fees to me, or you are liable to submit licensing fees on their behalf. In addition to ordering units from me, when looking for manufacturers to produce units, any manufacturer who produces any of the inventions core components, has the existing skills and facilities to fulfill your order upon reading this book and obtaining a license. It would be in your country's best interest to include at least one of your own country's domestic manufacturers in your orders, to ensure you're receiving the tax revenue from their operations, where to set up domestic operations, if lacking the technical skills required to identify a person to lead such a company, anyone with a doctorate in engineering or equivalent experience and modest managerial and sales skill has the ability to put together a team and company to manufacture units.

Outlaw Power Lines and Fossil Fuel Pipes

Countries absolutely can't afford power lines or fossil fuel pipes. In the introduction, I detailed the otherwise expected thousands of trillions in United States Dollars it will cost globally to replace power lines and fossil fuel pipes over the next 30 to 50 years, as they reach the ends of their useful lives, with the cost passed on to energy consumers, diminishing the economy of each country. Every impacted fossil fuel producing organization can both transition to using their mining equipment to extract the minerals required for manufacturing, and open a separate internal group or wholly owned subsidiary that manufactures these units, and transition all employees over to the new group as sales grow, which allows it to prosper long into the future after fossil fuel reserves have become worthless, depleted, or banned from extraction. Impacted utility companies must transition to organizations that generate revenue exclusively through sales, financing, and servicing of my inventions, to their existing customers, as I explain in the income opportunities section, where the ones that don't will simply be put out of business by the ones that do.

Outlaw Fossil Fuel Consuming Device Production and Impose a Significant Fossil Fuel Sales Tax

The sustainability of Earth requires that it be immediately illegal to produce new devices powered by fossil fuels. Fossil fuel consuming devices are dramatically inferior in every way to Bannon Maher systems, and are therefore now wholly irrelevant to the future, with no potential demand for fossil fuels after existing fossil fuel devices stop functioning. There is no need for fossil fuel device producing companies to fight a ban as they can prevent any loss of income by becoming a licensee and manufacturer of my inventions. Cigarette companies paid out debilitating claims to people who suffered from lung cancer as a result of their products, and fossil fuel producing nations and companies will logically at some point have to pay out claims as well, given they have caused far more deaths, and will have eliminated entire countries by putting them underwater, however those responsible may potentially limit or eliminate their liability by demonstrating they were a strong early financial supporter of me and my inventions. I would be very happy to see those countries that receive a significant portion of their revenue from fossil fuel sales embrace and financially benefit from licensing my inventions, rather than insisting on stupid stubborn adherence to the past that will incur an easily avoidable financial loss that negatively impacts their people.

Outlaw All Forms of Nuclear and Hydrogen Energy

All current and proposed forms of nuclear energy, including fission, fusion, and cold (LENR), in addition to producing radioactive waste and or materials, are dramatically inferior to my inventions in the many ways detailed in the introductory chapter. Nuclear fission is the worst energy source ever invented, because of how easily it's weaponized, and because the radioactive isotopes released into the environment during an incident take hundreds of thousands of years to degrade before they can't mutate DNA, so when only a single radioactive isotope is required to potentially cause cancer, there is logically no safe level of nuclear waste. Nuclear fusion is one of the most disturbing technologies ever considered in human history, as it tries to replicate the power of a star, which if successful, has the potential the break apart the entire Earth, at a time when weapons control is becoming increasingly autonomous, and therefore hackable by individuals. Because all forms of nuclear energy can only be transmitted over power lines that no country can afford, and no thinking person would ever want a nuclear home or car based unit, these sources are completely irrelevant, and because of associated dangers, further research must be outlawed.

Additionally, hydrogen can easily be used to produce a powerful explosion, as seen by watching an online video of balloons filled with uncompressed hydrogen being lit on fire, therefore new hydrogen gas stations must be outlawed so terrorists don't use them to obtain hydrogen to create powerful explosions.

Provide Large Grants

Because my inventions meet every single idealized criterion for energy and motor systems, no current, proposed, or theoretical system can match or exceed mine, and clean energy and transport has been completely and permanently solved, to direct relevant funds to anyone other than me and my organizations, is a misallocation of resources at the expense of the well being of humanity. In the United States, the U.S. energy research and development budget is nearly 6 billion USD per year. Legislation must be passed to transfer all relevant funds to me and my companies, making any necessary exceptions to use this book as my sole proposal. Alternatively, grants can be in the form of a non-refundable credit of royalty payments, which as the exclusive owner of my inventions, I can guarantee.

Given my technology costs more than ten times less than alternatives, and the incredibly slow pace of transition to clean energy, this is the only opportunity to ensure I accumulate resources to displace entrenched interests that support global warming, and transition us to clean energy before the world floods. The grant amounts required are at a minimum the costs of securing the patents for the inventions, which can be approximated by multiplying 152 countries by 4 inventions results in approximately 608 patents to be submitted and paid for in each country in very near future as set by unchangeable global patent office deadlines, thus requiring an incredible amount of financial and logistical support. Grants to pay these costs must be entirely unrestricted – an example of the opposite of an unrestricted grant is one that instead of counting the provided amount against a standard licensing fee, requires free use for the government and associated entities, which is in effect a license for less than 0.1% of its cost, with the grant recipient nonsensically required to contribute a percent of the grant amount.

Former Presidents accept speaking fees from foreign governments, and therefore I will gratefully accept grants of any size from foreign governments, however I cannot in exchange ever lobby the United States government on your behalf in any way, or I would be required to register as an agent of a foreign government, which I would never do. However, depending on the size of the grant, I may be willing to look at any of the problems your country has, and use my demonstrated capacity to try to think of and present you with innovative solutions.

Authorize and Advertise a Purchase Subsidy

Pass legislation making the purchase of Bannon Maher systems tax deductible, advertise the benefit, and remove subsidies and tax benefits for all other energy systems.

Prohibit Unlicensed Production, Importation, Sale, and Use of Inventions through Injunctions and Economic Penalties

In order to support compliance with international intellectual property law, while encouraging future innovators, ensure payment of the exceptionally reasonable licensing fees by imposing injunctions against any company attempting to profit from my inventions without a license, blocking the importation of units where licensing fees have not been verifiably paid, and instituting economic penalties against those countries that support domestic entities that profit from unlicensed units, with such penalties costing far more than the licensing fees.

I ask all buyers, during each purchase, to submit using the website on the cover page, the name of the organization you're purchasing units from, your organizations name, and the number of units you've purchased or plan to purchase, and the total sale price, so sales can be tracked to ensure licensing compliance. If the manufacturer has not made licensing payments, both the seller and the buyer will be liable for licensing payments, since the inventions patents royalties are enforceable at the time of construction, at the time of sale, and at the time of use.

Strengthen My Patents and Licensing Fees

Ensure all legislation and patent case law maximizes the strength of my patents and licensing fees. Because trillions of dollars in existing investments are now lost as a result of my inventions, there will inevitably be some enemies of humanity who will stop at nothing to try to covertly sabotage my patents. I have created immensely valuable inventions, at exceptional personal cost, while providing required comprehensive step by step disclosure on their construction, allowing anyone of ordinary skill in the art to easily construct an embodiment, and if my patents or licensing fees are ever allowed to be diminished, the patent system is fundamentally corrupt, and has become the enemy of innovators and innovation underlying economic growth, and by extension humanity, as well as damaging the philanthropic commitments I've made in the corresponding chapter, and every measure must be taken by legislators and judges to immediately take corrective action against any such issues. Any government not supporting the very reasonable licensing fees I've set out, discourages domestic innovators from taking the risks necessary to produce breakthrough inventions, thereby damaging their own economy. Every government should formally permanently ban from government contracts for any company that ever challenges my patents or licensing fees.

At this time where effectively all valuable intellectual property is monetized globally, I encourage those countries to join that are not currently signatories to the Patent Cooperation Treaty administered by the World Intellectual Property Organization, which provides for standardization of patents, as it will provide better protection and monetization for your country's inventors.

Transition Lobbyists

The United States Government annually reportedly directly spends nearly a half trillion dollars on energy related purchases, a budget item which can be eliminated by simply purchasing these units once. However, for example, publicly accessible lobbying records show $160 million was spent last year lobbying the United States Congress related to energy, therefore requiring people be even more committed to supporting my inventions. I note to those spending money on such lobbyists, you will make more money by licensing and manufacturing my technology instead, and giving up on your now lost cause. Licensing and sales agreements allow lobbyists to sell to their existing relationships, so I ask lobbyists, instead of opposing me, make more money than you could from lobbying fees, by having your clients sign licensing agreements, where the client assigns you the commission for the licensing agreement.

Authorize a Purchase Offer

Legislation should also be passed, in conjunction with other governments, to authorize a purchase offer of the inventions, that acknowledges the valuation cited in the section with the corresponding title.

Legislation Beneficial to the All and Relevant to My Inventions

International Taxation

In preparation for assigning, at the time of filing, the patents to a holding company, I did an exceptional amount of research on domestic and international tax laws and entities, and prepared the documents to file a set of foreign entities similar to those used by major corporations to overwhelmingly not have to pay taxes on international earnings, as long as the money is left abroad. The standard arrangement to avoid taxes on international earnings has been where instead of a parent legal entity receiving the money directly, it is paid to a company of Irish domicile, which passes it to a company of Dutch domicile, which passes it back to another company of Irish domicile, and when I initially found this arrangement through an Internet search, and saw it was named a "Double Irish with a Dutch Sandwich", I thought "I'm not looking for that kind of video right now" – though that structure is undergoing modest changes, and similar structures are offered in other nations that produce a similar effect. However, instead of following through with creating such a structure, and upon reflecting on the negative implications for all of such structures, I decided to instead think of a solution.

It is has been effectively impossible to enforce significant domestic taxes on large corporations international earnings, because they have always found countries willing to provide substantial tax advantages to be their international legal home, which results in companies having reduced domestic funds available for spending on employees, research, and development, and depriving all governments of tax revenue, leaving only principally wealthy shareholders as the beneficiaries of the funds, while putting at a competitive disadvantage tax shelterless smaller businesses that drive innovation and economic growth.

In many countries, government revenues are 80% from personal income taxes and 10% from corporations. This has created a waiting hidden global financial disaster of unprecedented scale, when accounting for the fact that most jobs will eventually be automated away, leaving only automated corporations to pay taxes, therefore eliminating most of the 80% of government revenue from personal income taxes, at a time when displaced individuals will require additional government assistance. One recent comprehensive study by a well regarded consulting group indicated up to 30% of all jobs that exist today in modern countries will be replaced by automation within 15 years, and my personal estimate is that up to 85% of jobs will be automated away within 100 years. Jobs are created to perform functions that fulfill human needs, and those jobs that fulfill human needs that can instead be fulfilled through automation will be lost forever. Jobs already partially automated away include traditional workforce entry points for young people, such as cashiers and waiters, and those in transportation, with autonomous vehicles having the additional impact of indirectly ultimately automating away the need for traffic police. If the current situation were to continue, at some point in the future, all governments relying on personal income taxes could be bankrupt, with no government financially able to assist displaced individuals, resulting in unprecedented poverty. In an almost comical manner, the corporations that avoid paying taxes while automating away jobs, have begun pushing the concept of governments providing an income to those they've impacted by the job losses they've created, I suppose paid for with the taxes they didn't pay.

There is only one way to save humanity from this financial apocalypse. All countries need to sign an international tax treaty, requiring corporations pay a fixed total tax on international earnings, with participation ensured through adequate economic penalties against non-participating countries, and with corporate support ensured by making government contracts available only to those whose parent company is not the beneficiary or claimant, directly or indirectly, of funds taxed below the minimum rate. This treaty has to be in place before shifting financials have wholly subordinated governments to corporations. I think a total fixed corporate tax rate on international earnings targeted at 25% with a floor of 20% is appropriate and achievable at this time, given all factors and arguments that could be for and against such a treaty. In order to provide benefits of tax revenues to a broad population, and to reduce wealth inequality, an equal split of taxes must be required between the residence of the physical corporate headquarters, and the country where the monetization occurred, with a further split between the national and state residence of the headquarters, proportionate to their respective tax rates, so the tax rate is never exceeded. For example, a technology company in California in the United States, selling advertising to a business in Paris, France, would split the 20% minimum tax as 7.98% to the United States, 2.02% to California, 10% to France (8.84% California corporate tax, 35.0% United States national corporate tax, the France corporate tax rate of 33.33%). Without a fairly distributed and enforced corporate tax on international earnings, the world as it exists today will collapse.

Vehicle Related Legislation

Manufacturers are exclusively liable for the Bannon Maher powered vehicles they produce, however I still don't want anyone dying from driving one. Because my motors can accelerate a road vehicle to maximum velocity effectively instantaneously, a rate human reaction speed can't safely control, there must be a legal acceleration restriction for 0 to 60 miles (0 to 100 kilometers) per hour, which I'm guessing should be about 2 seconds and certainly no less than 1 second, with the option for owners to electronically set a maximum acceleration limit above the legal limit. I also encourage governments to establish standard safety field tests for autonomous vehicle functionality, that must be passed before any autonomous driving system can be released or updated for general use, and making the tests publicly known ahead of time, so they can be factored in during development to avoid any drag on innovation. Legislators should eventually transition carpool lanes to autonomous vehicle lanes, allow comprehensively certified autonomous cars to a use a higher speed limit in those lanes, and build covered roads that support crossing the bearing straight, to allow land based automated self-powered vehicle and cargo transport between the Americas, Europe, and Asia.

Regulatory Approval

It would be beneficial to all to provide any necessary regulatory approval of my other inventions later disclosed in this book, including those related to construction, medical devices, high speed transportation, and space travel, as each provides substantial benefit while reducing costs.

All existing train and subway tracks should be interconnected globally, with the addition of tracks across the Bearing Straight, to allow the high speed transportation vehicle I invented and disclosed, which can run on both specialized high speed tracks and existing train and subway tracks, to travel continuously to anywhere in the world.

Automation, Wealth Inequality, and the Future of Employment

My inventions demonstrate that few jobs are safe from automation, given my automated investing system replaces some highly skilled highly paid investment professionals, and my medical devices replace some highly skilled highly paid doctors. Over the next 100 years, the population may increase to around 35 billion people, if the 5 fold population increase over the last 100 years remains consistent, while the number of workers needed as a percentage of population will continue to decrease, creating further downward pressure on real median income, and government revenues. I have given thought to the issue of technology's impact on jobs since middle school, when the local tennis club paid me what worked out to about seven times minimum wage per hour to maintain their schedule online, and then fired a receptionist whose job it was to provide schedules over the phone. It's sensible to start planning for the inevitable world where the majority of jobs that exist today have been automated out of existence.

I see immense broad misdirected effort to address wealth inequality. Wealth inequality is driven principally by three factors: ownership, because over long periods of history, assets have historically appreciated roughly 3% to 5% per year, while incomes have increased roughly 0% to 2% per year, with the effect of the gap dramatically increased over time through compounding asset gains; globalization, where corporate costs are reduced through foreign labor in emerging markets, while corporate earnings are increased by efficiencies in accessing foreign markets, and where financial benefits are passed almost entirely to wealthy shareholders; and automation, which creates unstoppable downward pressure on median incomes and labor force participation rates.

An idea that has been mentioned is a universal basic income, which is a misdirected effort, as it fails to resolve any of the fundamental problems I've explained that give rise to such a proposal. Many or most countries already have unsustainable debt levels exacerbated by declining revenues on an inflation adjusted basis, existing universal basic income programs for seniors are bankrupt on a future liabilities to cash flow basis, more than half of households in many or most developed countries are already net recipients of government funds, and it would create a net negative cash outflow further exacerbating the problem of reduced revenues, by providing those individuals income while at the same time stopping at least some number of people from contributing income through tax revenue. Even if none of that were true, which it all is, providing cash is logically unlikely to achieve a levered return, where for example, a dollar spent on education, might return ten to that person over time, but a dollar given away is usually simply gone after being spent, where those most needing a basic income are less likely to have the ability to focus the funds to maximize levered return. Furthermore, as a practical matter, there have to be some financial costs to living, until everyone becomes wholly self-sufficient and population growth has become flat, or people will be structurally encouraged to have an unsustainable number of children, the Earth will someday resemble an overflowing Petri dish, where a lack of access to limited resources will result in such children having a cruelly low average quality of life, and waves of people dying.

However, I have a set of solutions (1) ensure governments, nearly all of which can't sustainably support existing social programs, are in a financial position to have the option to help, rather than be eventually bankrupted, when there are almost no personal incomes to tax, by implementing my previously described international minimum corporate tax treaty (2) further enhancing government financial positions by ensuring each has multiple sovereign wealth investment funds, where each wealth fund operates independently to reduce risk (3) provide tax incentives that encourage all companies with a significant number of employees to provide meaningful stock ownership to every employee at every level, so that all employees share in the benefits of wealth creation through automation, globalization, and asset appreciation (4) provide assistance in ensuring individual earnings go toward sustainable long term home ownership, to allow them to capture the asset appreciation rate that exceeds the income growth rate (5) eliminating all property taxes globally, so that when there are no longer jobs available, people don't lose their homes to governments who have no use for them (6) incentivizing the creation of innovative companies that generate wealth, by in each country, legislatively requiring combined personal national, state, and local income taxes not exceed 50% of personal income, with automatic proportionate reductions in payment of national, state, and local taxes to comply with the cap (7) legally requiring a fiduciary duty to workers, enforceable through courts, requiring consideration of workers' needs, in addition to shareholders, in order to reduce the speed at which humans are replaced by automation – I personally know from working with homeless people as a volunteer career development counselor that the world is already changing far too rapidly for many people to adapt (8) at companies with a significant number of employees, for all but critical senior executives, tying the maximum weekly hours before overtime is paid to the labor force participation rate, to allow for every capable person looking for a job to have one, and reducing work weeks to four days whenever possible to provide individuals more time to focus on conserving earnings while enjoying life (9) to prevent deaths from a lack of access to limited resources, discouraging unsustainable population growth by imposing a tax on those with more than one child per adult, which appears extreme but appears presently without alternative (10) going a step past the strategy in my last book of reducing the cost of living, by enabling people to become entirely self-sufficient without income, by not only eliminating property taxes as previously described, but also by disconnecting from the power grid by installing my energy inventions, eliminating fuel costs by selling only vehicles with my motor systems, providing free clean water to all by desalinizing and pumping water using my self-powered energy and motor systems, growing food at home include fish and produce using my royalty free agricultural production system, constructing homes using my royalty free construction system

Food, Water & Housing Shortages

Food and water prices and shortages, have been on an upward trajectory, corresponding to climate change and population increases, with about 8% of countries, at the time of writing, rated as having dangerous droughts. Individuals are often unable to compensate in a time of globally decreasing median real incomes, leading to deaths, conflicts, and refugees.

To end water shortages, desalinization and pumping units, powered by my self-powered energy and motor inventions, which operate at a cost that is effectively zero over time, must pump water through pipes laid to areas of water shortages, where there is no royalty fee due to me when the exclusive beneficiaries are in extreme poverty as explained in the philanthropy chapter.

To reduce or end food shortages, households should implement the cost and yield optimized agricultural production system I invented and disclosed in this book, as well as fish farming — because per pound of meat generated, food consumption of fish is 6 times less than cows which are also less healthy to consume and emit greenhouse gasses, and 2 times less than chickens, while having a shorter growth cycle, producing proportionately less waste, and eliminating the possibility of airborne transmission of viruses such as swine flu and bird flu – by utilizing an air pump and stone in a tank of water, and appropriate food, to grow from eggs fish including tilapia, catfish, and bass. A home based food production system is entirely sustainable if equal amounts of food are harvested and regrown. To further decentralize and stabilize the food supply, female leghorn chicken may be raised domestically which produce an egg almost every day without a male that cost less than at a market plus is fresher and likely of higher quality, and who can eat food scraps, while offering a far more humane production conditions than on most commercial farms. Additionally, every supermarket should be growing their own fish and produce, so each is self-sustaining, rather than a drain on decreasing resources, which provides the additional benefits of higher quality food at potentially reduced prices with higher profit margins, and legislation should encourage such decentralized production.

To reduce or end housing shortages, the royalty free construction blocks I've invented and disclosed in this book, should be produced in factories, then assembled into durable high quality homes. In areas requiring permanent or periodic heating and cooling, to provide year round environmental insulation, building codes should encourage houses to have the first two levels built into the ground, with professionally constructed stabilizing concrete walls with a base on top of compressed gravel and or rocks to hold water to prevent flooding.

Healthcare

In addition to optimizing surgical procedure quality, consistency, and cost, by encouraging licensing, refinement, regulatory approval, and expansion of the medical devices I've invented and disclosed in this book, reducing overall costs in every healthcare system by implementing a supplement based health condition prevention program, which I estimate may cut healthcare costs by up to half through prevention, as described in the chapter discussing *Honors* under the section *President of the United States and or Secretary General of the United Nations*.

To obtain a production license for the inventions in this book that are available for license, including the energy and motor systems, simply download and print two copies of the licensing agreement from the website listed on the cover page, and for each copy, have the highest ranking executive in your organization, or the highest ranking government official of your country, initial every single page and sign the last page, in the presence of a Notary or your country's legal equivalent, with that individual placing an official government seal on the agreement that certifies the identity of the signing individual, and mail it in with the specified licensing fee.

Terms and Fees

I have carefully evaluated royalty options for energy and motor systems based on production costs, and sales, but found both impossibly lacking based on the number of possible implementations. I have instead come up with an optimal and extremely simple output based royalty fees, which some will consider absurdly low, of $0.99 per watt, and $19.99 per unit of horsepower – when a motor is not sold or implemented to drive a generator to try to circumvent the per watt fee or the per watt fee will be collected at the standard physics conversion rate of 745.7 watts per horsepower – where the currency is the greater of USD (United States Dollar) or CHF (Swiss Franc), currencies which typically follow each other, and where to adjust for inflation, fees increase at a simple non-compounding 3% per calendar year starting on the first day of the year following the availability of licenses, which given that is less than the average rate of inflation, and does not compound, makes inflation adjusted licensing fees decrease over time. Given the inventions provide continuous watts, requiring no storage or transmission, to compare them to other clean energy sources, which typically provide around 20% of rated output, require expensive and inadequate storage, and may incur the additional expense of land and extremely expensive power lines, the final per watt cost can in a sense be thought of as divided by approximately 10 to 20 times. Given the horsepower units require no fuel, the fee can in a sense be thought of as non-existent, because the fee is far less than the cost of fuel over the life of the motor. Utilizing the afore explained royalty rates, the royalty on (1) the average car providing 250 horsepower is $4,987 – which is a third of the cost of fuel when assuming 24 miles per gallon, $2 per gallon fuel cost, a useful life of a car of 200,000 miles, for total fuel cost of $16,666 or a an even more expensive electric vehicle battery plus charging costs (2) a generator for a 5,000 square foot (465 square meter) home providing 10,000 watts is $9,900, and even adding the cost of materials, royalties, labor, and manufacturer profits, is at least ten times less per kilowatt hour than any other energy system (3) the average large jet engine of 185,000 horsepower, which currently costs around $15 million, is only $1.85 million. Because the royalties are for the inventions not the components, the per watt royalty rate of course is still the same if the motor is powering an existing generator, where the licensing fee is based on the generators maximum rated output watts. The horsepower fee excludes any watts consumed in making a motor self-powered.

For reference, the specific law governing patent infringement and contributing to patent infringement in the United States, and will all economically significant countries having similar laws, is 35 U.S. Code § 271 which states "whoever without authority makes, uses, offers to sell, or sells any patented invention, within the United States or imports into the United States any patented invention during the term of the patent therefore infringes the patent... Whoever actively induces infringement of a patent shall be liable as an infringer... Whoever offers to sell or sells within the United States or imports into the United States a component of a patented machine, manufacture, combination or composition, or a material or apparatus for use in practicing a patented process, constituting a material part of the invention, knowing the same to be especially made or especially adapted for use in an infringement of such patent, and not a staple article or commodity of commerce suitable for substantial non-infringing use, shall be liable as a contributory infringer... No patent owner... shall be denied relief or deemed guilty of misuse or illegal extension of the patent right by reason of his having... sought to enforce his patent rights against infringement or contributory infringement... refused to license or use any rights to the patent".

The only discount that will ever be available, which I may stop at any time, will be for the non-refundable prepayment of royalties, which is mutually beneficial, as it allows both parties to have additional funds, in the form of cost savings and cash, to support my inventions. At the time of publication, any organization, including governments paying royalties for planned purchases on behalf of manufacturers, may pre-pay a maximum of $1 billion USD of non-refundable licensing fees, for a discount on those fees of thirty percent. Additionally, until the end of 2019, because of fees due before that time to each country for each patent, any organization may take an additional 20% discount in pre-paid royalties. All licensing fees are non-refundable, including pre-paid fees, even if any relevant patents are ever partially or wholly invalidated. Discounted royalties may not be pre-paid for more than a total of $1 billion USD, by entities ultimately benefitting and or controlled by a single parent entity. As detailed in the chapter on philanthropic commitments, there is no fee due to me for units that are produced for use in extreme poverty situations, when documentation is provided in place of payment showing the exclusive beneficiaries of the systems live in extreme poverty.

The fee for the initial energy and motor systems license, which is valid through the expiration of the patents in approximately 20 years, is $100,000 USD, with a minimum of a $900,000 USD prepayment of royalties, with the credit useable at the discounted rate. There is a small entity discount, which eliminates the pre-payment requirement, for those entities which, along with any parent and controlling entity, have less than $1 billion in revenue. There is a micro entity discount for those with less than $10 million in revenues, requiring no pre-payment, and a licensing fee of only $10,000.

You are liable for and a license is contingent upon providing licensing fees even before the patents are formally provided final approval, and immediately providing such fees will speed adoption of these critical inventions, therefore benefitting the well being of all humanity.

In order to be fair to all licensees, as well as to prevent complications in the event of a sale of the intellectual property, under absolutely no circumstances can there be even the tiniest bit of flexibility in the terms, regardless of how powerful, important, or threatening the licensee, nor will an agreement be made that differs from the one provided to all licensees, nor will any side agreements be signed. There will not be any meetings with anyone, as the only reasons for a meeting are to negotiate or explain the inventions construction or their value, and there will be absolutely no negotiation on terms or fees, and there is no more information available on the inventions construction than is provided in this book. If people have an issue with that, have them read this entire book so they understand why, and if they can't do that, they shouldn't be in any position of responsibility or authority. Any relevant organization that doesn't want to license on my exceptionally generous terms, is welcome to go bankrupt, after licensed competitors have taken all your revenue, but before that occurs, given anyone within the organization opposed to licensing is toxic to your organization, especially if such a person is the CEO or Chairman, they must be immediately fired and prohibited from receiving further compensation. I'm doing a tremendous favor by even allowing licenses, and doing so on the generous terms I've offered is truly a gift. In a sense, it's possible to create your own discount by optimizing manufacturing to reduce costs while selling units for more than competitors by optimizing unit longevity, reliability, and weight.

If a large group of large countries and or companies would like to obtain any further discount, they're welcome to purchase the intellectual property consistent with the valuation established in the corresponding chapter. Future innovators will be inspired in part by seeing the financial rewards of coming up with inventions that positively impact the world, and the world will benefit in immeasurably in my distribution of the proceeds.

If thinking of attacking or diminishing me, my inventions, patents, or licensing terms, it's absolutely critical you first read the attackers section of this book, as a sufficient review of my track record and demonstrated capacity, reveals that regardless of how great your collective and individual resources and capacity, working against me will cost all involved dramatically more than could have hoped to be gained.

Since most of the licensing revenue will ultimately go to charitable causes that serve the well being of us all, as detailed in the section corresponding to charitable commitments, supporting me, my inventions, patents, and licensing fees, supports the well being of humanity.

Commissioned Sales Representatives

If you receive a licensing agreement from anyone, they're simply helping bring my inventions to the attention of relevant decision makers, and ensuring such decision makers are aware of the specific benefits, and while such individuals are greatly appreciated, they have absolutely no authority, nor does anyone else, to make any adjustment to anything. If you receive a commission request from anyone, I ask that you certify commission recipients as fairly as possible, based on who impacted your decision to license, excluding me, as having fairly paid commissions benefits further global adoption of my urgently needed inventions. I will not accept attribution of any commissions, and I'd much rather have someone receive the commission if they've earned it. If a commission is specified, it will only be provided when allowed by applicable law.

Secrecy and Publicity

Since the first companies using my inventions to produce self-powered cars, generators, and airliners, without competitors receiving advance notice, would take nearly all other manufacturers sales, I realize some will want to hide the fact they're producing my inventions. I am willing, upon request, to maintain confidentiality of your license until product release, however, you are free to publicize your license, as it will encourage others to also license. If you're a manufacturer looking to sell energy units, I suspect a press release that's circulated to the media, and indexed by the search engines, announcing your license to produce Bannon Maher inventions, and that you're taking orders, will be the best way to initially draw attention. I'd like to be invited to have the option of participating in a featured question and answer session during product announcement and launch events.

Terms Reasonableness

I realize some will think it's absurd that I'm licensing these technologies, rather than producing them exclusively to make far more money. I also realize, that on top of a non-discriminatory open license, some will think my royalty rate is absurdly low. However, I have done this because I want to see this technology broadly adopted for the well being of Earth and all of its inhabitants. There are multiple factors that make the fees and terms I provide exceptionally reasonable:

1. *Relative Cost.* These inventions are being licensed at a cost that allows for them to be produced at an all inclusive cost estimated to be more than ten times less than the total cost of any other energy or motor system.

2. *Reimbursement and Innovation.* I've incurred extraordinary personal cost in developing the capacity to invent these inventions, as well as in inventing, recording, and presenting these inventions, including the exceptional neglect of my relationships, health, and finances, and substantial professional opportunity costs. Pharmaceutical companies make, what are for them, lesser but similar sacrifices, in order to develop break through drugs to treat conditions such as cancer, and in return get to produce it exclusively for 20 years. Such benefit is a critical component of driving the innovation that is foundational for a country's prosperity.

3. *Licensee Market Capitalization Preservation and Growth.* A license allows impacted companies and countries to still be able to grow rather than being forced into bankruptcy, thus not only preserving their existing financial position, but potentially causing a dramatic increase.

4. *Opportunity Cost.* There is a massive opportunity cost to me in offering licenses, as the revenues from licensing, would be only be a fraction of the revenue based company valuation I would achieve if I were to sell the products exclusively.

5. *Cost Savings.* Electric car batteries make up around 25% of the cost of the car, with the engine being around 10%, where as my invention replaces both, reducing the cost, after licensing fees, to around 15% of the car price, while saving dramatically more than the licensing fees in gasoline or electricity costs, removing the cost and inconvenience of recharging, and providing better performance. Additionally, when owning a generator, there is never a need to pay another utility or gas bill, substantially lowering the cost of living for individuals.

6. *Reducing Wealth Inequality.* At a time of increasing inequality, when companies regularly diminish workers by automating or outsourcing their jobs, it's more important than ever that rather than transferring gains from my inventions to overwhelmingly wealthy shareholders, the money from my inventions go to me, so that it can, as my track record exhaustively indicates, and as described in the philanthropy section, be used to support the well being of humanity. The only time such discounts have ever been judicially pushed through were for excessively unreasonable and discriminatory licensing terms that made life saving drugs otherwise impossible to afford, but I'm already waiving royalties in extreme poverty situations, and electricity and propulsion can still be provided by comparatively extremely expensive and dramatically inadequate sources. If licensees received additional discounts, a substantial percentage of the world's population would be harmed.

Any court not forcefully supporting my patents, fees, and terms, is a toxically corrupt tool of wealthy corporate shareholders undoubtedly already taking advantage of their workers. Without judicial respect for the right of an owner of an invention to retain invention ownership and terms of license, the incentive is removed for individuals and companies to incur the sometimes extreme costs involved in coming up with such break through innovations, and that would make worse off all in this world.

An Important Note to Vehicle Manufacturers

A large portion of the income from licensing my inventions will come from vehicle manufacturers, and all vehicle manufacturers must immediately make the following four changes, to encourage people to upgrade, explained in order of importance, with unsupportive employees fired.

1. *Bannon Maher.* It's obvious the only cars that will be relevant going forward are Bannon Maher cars, which must be advertised as such since they are fundamentally different from gasoline and electric cars, and in order to sell optimally, must be visually

indistinguishable from their traditional gasoline counterparts, not making the same mistake that was made with almost all electric vehicles, where in addition to being electric, manufacturers also made them visual oddities.

2. *Linux.* All physical controls, with the exception of the steering wheel, brake and gas, should be replaced by a full center console touch screen and instrument cluster screen powered by a computer running Linux. Linux is the longest developed modern operating system, and as a result is the most powerful, reliable, and secure, in addition to being free and open source. Anyone selling you on anything other than both Linux and building your own software, is playing on your technical ignorance. Additionally, the software running vehicles is an important value adding differentiator, allowing manufacturers to maintain complete control of aesthetics and brand, making it about the last thing that should be outsourced, and free open source Linux ensures manufacturers always have complete control and ownership of systems. It's easy to build your own system for your vehicles by performing the following steps: sketch each of the screens that will be presented to the driver, such as the main menu, climate control, media player, and navigation maps; review portfolios of graphic designers, and select one to design beautiful high resolution screens based on the functionality specified in your sketches; have a Linux Qt software developer create a click through application perfectly integrating the design from the images provided by the graphic designer, and during development, when controls are clicked, displaying at the bottom of the application the command that would be sent to the car components; replace the center console in a car with an appropriately sized touch screen, and replace the instrument cluster with another appropriately sized screen, both attached to a Linux computer running the application that was developed, to verify everything looks and feels right, and with the screens ideally being of the type that completely turn off black pixels to eliminate unnecessary light; have another engineer, probably an existing auto software engineer, make each component in the car operable through a computer command, which given all modern vehicle controls, from power steering to air conditioning, are already wired into a CAN bus, the computer would simply communicate over Ethernet with an automotive microcontroller gateway that controls the CAN Bus to send commands to each subsystem, and someone with that specialized skillset will need to be identified or hired; have the Linux Qt software engineer and the second software engineer make the commands pressed on the touch screen execute the commands that will operate the car components; ensure the computer can recognize and connect to mobile devices to read their media libraries and access their phones; reuse Qt application design elements and code to build mobile phone applications that allow the car to be remotely controlled by the owner to drive itself to park and pick up the owner.

3. *Autopilot.* All vehicles should have an autopilot mode allowing them to operate without a controlling or monitoring driver. The steps to make this work are conceptually rather simple. Autopilot systems should be independently developed by each manufacturer, since safety and reliability are core value adding components of any brand, and technology companies would love to utilize ignorance to capture that value for themselves. Lidar systems produce strange looking images that must be programmed by humans to be recognized, which is inherently more difficult than using normal images, especially when such images are further distorted by capturing a 360 degree view, and is likely why autonomous lidar cars have been in development far longer and with worse results than those using video cameras, while additionally generally requiring prominently ugly external devices forcing consumers to compromise on aesthetics and therefore inherently damaging sales. Standard eye based video cameras are all humans are able to use, and no more than that is needed to achieve a level of safety far beyond that of a human driver, by simply removing errors based on inattention and fatigue. Additionally, special processing chips are not required, because humans process images to adjust vehicle controls utilizing only one video feed at maybe 4 times per second, with little more required by an autonomous system to be safer than humans by simply eliminating errors from fatigue and inattention, and the power of standard modern computer chips makes any need for specialty processing chips negligible while increasing the expense of the design process. The steps to build an autonomous driving system are: install one forward facing video camera in front of the car's rear view mirror, then drive the car around in varied conditions while recording to a video file; have C++ software developers with image processing experience write code that provides for detection of objects and trajectories in the video file at several frames per second, develop rules to respond to objects and their trajectories, and continuously display the adjustments to be made to steering, acceleration, and braking; provide the software all the same views a human driver has by adding additional cameras to the car, and recording video feeds, including front right, front center, front left, left mirror, right mirror, rear view mirror, blind spot, and expand the code accordingly; have an automotive software engineer, as described in the preceding step, allow the code to operate the power steering, acceleration, and braking; in a safely enclosed area where it would be impossible for anyone to be harmed, and while allowing a human drive to disconnect autonomous mode at any time, run and monitor a live test; keep refining object recognition and rules until the system works perfectly; have a comprehensive independent security audit performed before releasing the system, ensuring it is not only exceptionally safe to operate, but that there are enough safeguards that the control systems can't be hacked,

and in the exceptionally unlikely event of a hack, autopilot can be physically disconnected.

4. *CAD.* The design process of using clay to model vehicles is an entrenched relic, producing outdated boxy looking vehicles with long development cycles. All interior and exterior designs, from start to finish, should be exclusively computer modeled, as it allows for beautiful sleek geometric precision and rapid iteration.

ORDERING

I am accepting orders for energy generator units as of the public release of the inventions. The minimum order size is $20,000 USD and I welcome all orders up to $100 billion USD. There is no order form, simply send a check made out to my operating entity, "Bannon Maher Corporation", to the address provided on the website specified on the cover page, along with a note indicating the percent of motors and or generators you would like. Please provide this book to anyone involved in the purchase decision, to ensure the check is as large as possible, since there is an urgent global need to transition to these systems, and your early support is critical. The price per unit at the time of publication, with no expectation of ever changing, is the contract manufacturer's material and labor costs, plus twenty to twenty five percent of the total manufacturing cost to the operating company, so I'm not obtaining an unfair competitive advantage over my licensees, plus royalty fees paid to the intellectual property holding company, with verifiable documentation of costs provided with the units or upon request. I will be screening, hiring, and overseeing one or more contract manufacturers, with a track record of providing extremely high volume and high quality production of complex technical devices. I will be working to ensure and verify maximization of net output per dollar of construction costs, minimization of space required per unit, and that all parts are extremely durable and individually replaceable without a technician.

The determination as to which system to provide you will be based my judgment as to what provides you the optimal cost per unit of output, unless you have a requirement for a specific system, in which case please specify that with your check.

If you have any reservations, please account for the fact that I will of course be using the same level of capacity to optimize and validate production of the units as I did to both invent these systems and to achieve my past accomplishments.

While I don't foresee any financial risk for purchasers, because I will be overseeing production, and the design is straight forward for a contract manufacturer, if for any reason units are not provided, refunds will be issued less costs incurred.

Because of the urgent global need for my inventions, and because entrenched interests with effectively limitless resources will be working against me individually as well as my organizations, I am obligated to accept, and others are obligated to provide, grants, until the point at which such funds become unnecessary. I of course have the capacity to ultimately make any amount of money, however at the time of publication, as a result of inventing, refining, and patenting these inventions, I don't have necessary resources available. I expect that a total of one billion USD will be required to achieve optimal global implementation of the inventions, an amount almost any government, as well as hundreds of individuals, could provide without a noticeable change in their financial condition. The best way to support the immediate impact of my inventions, which are the only serious hope for saving the Earth in time, is with a grant to the maximum extent your resources allow. Additionally, the commitments in the philanthropy section will be diminished and could be lost if I ever have to give up any control to raise money. If I achieve profits anywhere close to those specified in valuation chapter, I'd be happy to return the grant with interest. If there is any concern such funds will be spent in a less than optimal manner, it's important to recognize I grew up with almost everything money can buy, yet I've driven the same car for effectively 18 years, I wear nearly the same outfit every day, and typically spend at least 12 hours every day of the week working on my laptop at coffee shops, so my intent is to be fully resourced at this critically important point in the transition us to a better world for us all. I must add a standard disclaimer, that I make absolutely no guarantees, including that things will work out as expected, or how funds will be spent, and it's up to each person to make his or her own assessment. A grant of the required amount could make any such provider an historically important figure in saving the Earth, and I will, for the rest of my life, even if I don't personally care for you, happily support your legacy in whatever way will be most beneficial to you.

For grants below ten million USD, checks should be made out to me personally, using my full name, with "Gift" very clearly stated in the memo field of the check, so I am personally resourced with discretionary funds. For grants in excess of ten million USD, in order for the grant to not be considered a gift, in which case I may have to pay around 40% of it in taxes, I may provide you equity at the multi-trillion dollar valuation specified in the valuation chapter, which alone would allow for a return of your money if the valuation is achieved, and there should be two checks, one for 20% of the amount made out to me using my full name, with "Gift" very clearly stated in the memo field of the check, so I am personally fully resourced, and one for 80% of the amount made out to the intellectual property holding company, Jonathan Bannon Maher Corporation, which will issue the share(s) in accordance with the valuation. All checks should be sent to the address provided on the website specified on the cover page.

Every single exception necessary must be made to standard organizational grant procedures, in order to send such checks without my involvement, including accepting this book as the equivalent of all required proposals and documentation. Any grant making entity not immediately providing as much funding as their resources can support, is likely not serving their founding ideals, and is not an institution that thoughtfully serves the public interest.

Early support is critical to the well being of humanity, and yours will not be forgotten.

Income opportunities are available from securing licenses, orders, grants, publicity, leases, and potentially litigation judgments. However, these aren't just income opportunities, they're also a way to make a tremendous positive impact on the world, by supporting and accelerating the impact of my world saving inventions. All amounts due are attached exclusively to the operating company, Bannon Maher Corporation.

Leasing and Securitization

Hundreds of billions if not several trillion dollars are available to utilities and investment banks, by utilities leasing units to existing customers, receiving monthly servicing income for the term of the purchase financing or lease, with the financing provided by investment banks, which securitize the financings as asset backed receivables, and sell them to global institutional investors. It's the same extremely well established process utilized to finance mortgages and credit cards, which any utility can initiate with a few phone calls to investment banks' securitization groups. Utilities can additionally make money by offering customers a service plan, while also doing their best to make sure the units they sell are optimally manufactured so there is no need for service.

Licensing Commissions

At the time of publication, commissions are available for obtaining licenses for the energy and motor systems. Any individual legally eligible to work and to sign a legally binding agreement, and any organization, as long as complying with applicable laws, may receive sales commissions. No individual or organization may solicit sales or licensing agreements, or receive a commission, without having first printed, signed, scanned, and returned with an electronic scan of a government issued photo identification, the commissioned independent contractor agreement, which explains what is and isn't allowed by law during solicitation, and is available at a website listed on the cover page. All applicable laws must be followed at all times, including those listed at the end of this chapter. The role of a person independently working on commission is simply to make sure the top executives of an organization, who can both transfer funds and legally sign the licensing agreement for their organization, are fully informed of the benefits. This is not a traditional sales role since there are absolutely never discounts or custom terms, for reasons explained in the licensing section, but is instead to ensure decision makers are fully informed of the benefits. Since there is no salary involved, the role is probably most sensible for those with existing relevant relationships and an existing income. Commissions will be payable, to any responsible individual or organization, in compliance with all applicable laws, at 5% of the amount of a sale up to $10 million, or 5% of revenue from a licensing agreement for the first 5 years of the licensing agreement or until $10 million in commissions has been provided, and is payable any monthly where the total commission balance exceeds $1,000 USD, on sales which occurred more than 30 days ago, and there is no pending request for a refund. For example, on a $100 million sale, the commission would be $5 million, and on a licensing agreement providing income of $10 billion, the commission is $10 million.

The entity responsible for selecting the recipient(s) of the commission is the purchaser or licensor, who must identify the recipient(s) in writing, and the purchaser may split the commission among more than one person. While I will do my best to ensure fair treatment, there is absolutely no obligation by anyone to provide a commission for any sale, since multiple people may be lobbying the same organization, or the organization may already be aware of or become aware of the inventions independently, or may have been influenced overwhelmingly by other factors, and the commission is provided entirely at the discretion of the licensing organization. I ask the licensing organization be as fair as possible in attributing commission. While I retain the right to override any commission decision, I do not expect to ever do so. It may be beneficial during solicitation to provide not only electronic but printed copies of this book, which are available at cost, though the cost of which is not reimbursable as that would be too much of an administrative distraction.

Commissions are non-payable or required to be returned in certain circumstances. In the event a licensee is provided any refunds, the proportional amount of the corresponding commission must be returned as well. If a commission has already been paid, and a law was found to be violated in obtaining the commission, it must be returned, in addition to paying any costs resulting from the violation. Contractors may not use automated calls or mass emails, and retain exclusive liability for legal violations and associated fines. Income stream rights are non-transferrable, and voided in the event of any transfer, since part of the reason for providing such a generous arrangement is to keep in the relationship the person who arranged the license. The commissioned independent contractor agreement must be periodically renewed.

Grants

Any individual or organization that proves they arranged any grant from a non-U.S. government institution, will receive 5% of the grant, if allowed by law, and if more than one entity was responsible, the commission will be split accordingly in the interest of fairness. This provides benefit to all involved, since none may not have had a sufficient knowledge of each other to secure the grant. However, the commission may not be provided to any employee of the granting institution or the employee's family member, as that is sometimes a violation of the law, and is additionally not particularly ethical. Those responsible for grants provided by government institutions, are not legally allowed accept any portion of that grant money, directly or indirectly, as compensation.

Publicity

Because of the urgent need for global awareness and adoption of the inventions, until further notice, any individual or organization responsible for verifiably obtaining favorable media coverage of me and the energy and motor inventions, assuming such payment is not in violation of any applicable law or employment agreement, is eligible for a payment from the operating company, per article or television appearance. The payment is $500 USD, for print articles in an outlet with an audited paid subscriber count over a million, or television appearance on a show with an audited audience over 100,000, with a purely discretionary bonus available, which could be nothing or could be very generous, with all funds payable by the operating company. To qualify for and obtain a payment, fill out the relevant agreement available from a website on the cover page, and provide verifiable proof you were responsible for the article or television appearance. There will be no additional claim allowed, including to any sales commission resulting directly or indirectly from publicity. Any publicist or publicity firm that delivers results will of course be given favorable consideration for additional business in the future.

Litigation

Law firms may receive the same 5% commission they would receive from a sale or licensing agreement, but instead of it coming from the sale, it would be of all revenue recovered from licensing non-compliant country or organization. The 5% fee may be increased if the law firm acts exclusively in good faith, and the entity and individuals involved believe because they're particularly powerful, the law and basic decency doesn't apply to them. The most appropriate law firms to take on such a task will be those who have in the past obtained large financial judgments against the relevant individuals and countries or companies. However, as I've seen some law firms stir up trouble for fees, if law firms are interested in that percentage of revenue, it would be easier to receive sales commission for arranging the licensing agreements, especially if the firm has relevant clients. No litigation may be initiated or continued without my permission, and may only be done exclusively at your own expense, on contingency, and independent of participation by me or my operating entities, because the facts speak entirely for themselves and I will have nothing more to add.

Applicable Laws

All applicable laws must be followed at all times. Independent contractors are personally and exclusively financially and legally liable for being aware of and complying with all relevant laws, some of which provide heavy fines for violation. To my knowledge, there are a few laws relevant to income opportunities governing individuals connected to the United States who operate an organization, and I've seen similar laws in other countries.

Mass Email

Bulk emails are required to comply with the CAN-SPAM act, requiring the message be identified as an advertisement, requiring a legitimate sender email address, requiring an accurate subject line, requiring the ability to unsubscribe from future messages, requiring the physical address of the sender, which may be your own P.O. Box, and may not be sent to email addresses collected in an automated manner, even if the list was purchased from someone else who may have collected it in an automated manner. Additionally, you must state that you are operating independently of Bannon Maher.

Telemarketing

Because of legal restrictions and associated fines for you if you make certain calls, automated calls and sales to individuals are not allowed, only sales calls from a real number to business land lines during regular business hours, are allowed by law.

If you would like further detail, you can consult the Telephone Consumer Protection Act of 1991 (TCPA) and the Telemarketing Sales Rule (TSR), and the CAN-SPAM Act. If there are any additional laws or regulation where you live you are solely responsible for being aware of and following those.

Direct or Indirect Payments to Government Officials

15 U.S. Code § 78dd–2, commonly referred to as The Foreign Corrupt Practices Act of 1977, prohibits compensation in several instances, even if required to be competitive in a bidding process, stating that "It shall be unlawful for any domestic concern... to... giv[e] anything of value to... any foreign official... any foreign political party or official thereof or any candidate for foreign political office for purposes of... any person, while knowing that all or a portion of such money or thing of value will be offered, given, or promised, directly or indirectly, to any foreign official, to any foreign political party or official thereof, or to any candidate for foreign political office for purposes of... influencing any act or decision of such foreign official". Additionally, 18 U.S. Code § 201 prohibits an action that "directly or indirectly... offers or promises anything of value to any public official or person who has been selected to be a public official, or offers or promises any public official or any person who has been selected to be a public official to give anything of value… with intent... to influence any official act." This has also been ruled to apply to state controlled entities, with rank and title of the recipient being irrelevant. However, given the size of the market, and need for the inventions, this is not of even a slight hindrance, simply move on to another prospect, as there are more than could likely ever be reasonably contacted.

An extreme recent example of violating these statutes is a spreadsheet maintained by a top American investment bank in China, listing how much revenue was attached to each hired child of a government official, which resulted in the payment of heavy fines.

Therefore, without exception, no government official, or an official's family member, may receive a commission. Additionally, although laws vary widely on employees receiving commission from a sale to an employer, no one may receive a commission from their current employer without the signed approval of the highest ranking executive of that organization, assuming there is no law that prohibits such a transaction. I reserve the right to decline any commission I think may violate applicable laws.

THANKS, REPORTERS, ACCOMPLISHMENTS, HONORS, SUPPORTERS, AND ATTACKERS

Thanks

I put this section ahead of the other interdependent sections in this chapter, because while no one has seen or been aware of any of my work prior to publication, I'm not a hermit cave gnome, and was still enabled in life by others' past support. I note that references to any position may not be current. In addition to previous thanks to my parents, I provide thanks to: my younger brother Andrew, for listening, who has already executive produced and co-written pilots for Warner Brothers, NBC, and Fox, with one starring a two time Golden Globe nominee and an X-Men actor, and another starring a principal Harry Potter actor; David Geliebter, a software entrepreneur, who sold his company around the time of the Amazon IPO for a similar valuation, for his business guidance growing up, and for an early and useful recommendation letter declaring me "extremely gifted as a computer programmer"; Ashley Geliebter for friendship in youth; Kathy and Paul Hennessy for arranging the job on the floor of the New York Stock Exchange the summer after high school, providing me, as the youngest person there, an exceptional learning experience; Bear Stearns (J.P. Morgan) Board of Directors member Wendy de Monchaux, who I worked for extremely briefly during college but was still generously supportive; Deutsche Bank Vice President Jon Olstein for his support while I worked there one college summer; University of San Diego Dean Curtis Cook, who wrote in a recommendation letter "Jon was the only person… to receive the distinction of the word 'Honest'… demonstrates a well-defined, ethical values background that allows him to behave as a humble, principled leader." and President Mary Lyons, who wrote in another recommendation letter "Jon exemplifies the University's commitment to being a values based institution.", as well as the University of San Diego itself for being a tropical oasis of inspiration and knowledge; long time friend Peter Georgotas for his unwavering faith in me, informed by his ancient Greek heritage and corresponding historical knowledge; grade school Latin teacher of six years Joe Rose, for sharing his deep and broad historical perspective; grade school English teacher of two years Meg Schaefer, who required a book read and corresponding paper written every two weeks, of which I did both on the due date during lunch, and still got better grades than classmates who spent two weeks laboring, thus revealing my writing ability; my father's long time principal client, institutional portfolio manager and United States Treasury Advisory Committee Member Richard Axilrod; Vincent Van Gogh, in my opinion the greatest artist in human history, for being virtually unknown at the time of his death; the great minds throughout human history for the education; though I haven't seen either in years, Michael Clark and Brad Rosen, for their long standing friendship, and retroactively supporting a part of the cost of printing and distribution of my 2011 book; Pedram Amini and Mike Arcamone, for the opportunity to write early core components of software now protecting United States Military networks; the many world leaders who enabled my accomplishments through their support of my last book – I have named below every single leader who mailed a written response that I received.

José Maria Pereira Neves, Prime Minister, Cape Verde
Norodom Sihamoni, King, Cambodia
محمد السادس (Mohammed Ben Al-Hassan), King, Morocco
Jill Biden, Second Lady, United States
Quentin Alice Louise Bryce, Governor General, Australia
Julia Eileen Gillard, Prime Minister, Australia
Hubert Alexander Ingraham, Prime Minister, Bahamas
Albert II, King, Belgium
Dilma Vana Rousseff , President, Brazil
Stephen Joseph Harper, Prime Minister, Canada
Laura Chinchilla Miranda, President, Costa Rica
Lars Løkke Rasmussen, Prime Minister, Denmark
Margrethe II, Queen, Denmark
Nicholas Joseph Orville Liverpool, President, Dominica
Elizabeth II, Queen, England
Nicolas Paul Stéphane Sarközy, President, France
Angela Dorothea Merkel, Chancellor, Germany
Christian Wilhelm Walter Wulff, President, Germany
Bharrat Jagdeo, President, Guyana
Michel Joseph Martelly, President, Haiti
Ólafur Ragnar Grímsson, President, Iceland
Mary Patricia McAleese, President, Ireland
פרס שמעון (Szymon Perski), President, Israel
וּנְתַנְיָה בִּנְיָמִין (Benjamin Netanyahu), Prime Minister, Israel
Felipe de Jesús Calderón , President, Mexico
Albert II, Prince, Monaco
John Phillip Key, Governor General, New Zealand
Harald V, King, Norway
Benigno Simeon Cojuangco Aquino III, President, Philippines

Juan Carlos I, King, Spain
Carl Gustaf Folke Hubertus, King, Sweden
馬英九 (Ma Ying-jeou), President, Taiwan
Bhumibol Adulyadej, King, Thailand
Clarence Thomas, Supreme Court Justice, United States
Robert Mueller, FBI Director, United States
Chris Christie, Governor, New Jersey, United States
Benedict XVI, Pope, Vatican
Henry Charles Albert David, Prince, Wales

The top supporters of my current work will receive recognition in my fourth and or fifth book, still many years from publication, which I've designed over many years, in part by analyzing relevant historical and contemporary references, to be the best selling fiction book ever, while being educational in a way that permanently resolves global conflicts and poverty.

Reporters

A recent Washington Post poll found the majority of Americans say the media regularly produces false news stories. One glaring example is the broad range of top tier news outlets providing months of coverage of a college football player's girlfriend who struggled with and ultimately died from cancer, but who in fact never existed. This is principally a result of media organizations continuing to lose advertising revenue to targeted online advertising sold by search engines and social networks, leading to a dramatic decline in funding for research and writing. Social networks were recently found to be where a third of people get their news, however they have substantial negatives including: decreasing relative happiness as others share only positive events; the viral spread of fake news; people disconnecting from those who disagree, leaving only those in agreement to reinforce opinions, which results in consensus substituting for facts, thinking becoming more extreme and intellectually isolated, and a person's remaining followers telling them they should be running the world; encouraging conspicuous spending over investment to create shareable images of expensive food and vacations; a lack of advertisement approval for arbitrary reasons; distortion of reality when many with large followings have their posts produced by idea generators, writers, and photographers; suggested following of preferred individuals to provide those individuals any size following; users being unable to have exclusive control of their news feeds, with changing algorithms able to suddenly stop the followers of a public figure or organization from seeing their posts. To illustrate the financial situation of the media with numbers, the Washington Post, was recently purchased by the founder of Amazon for $250 million, while having a reported internal valuation based on advertising revenues of $50 million, when a decade ago it had a reported internal valuation of $2 billion based on advertising revenue. This has created a situation where many reporters are required to increase output with fewer resources, resulting in articles focused on satisfying advertisers and audiences, providing positive coverage of people attached to money to buy ads, writing stories about well known people who may have had little legitimate connection to a story, or simply providing the fact free narrative of least resistance, with such tension reportedly leading the ouster of more than one top editor. The person in the last Presidential administration responsible for speechwriting and communications, was profiled in the New York Times Magazine as "able to create an overarching plot lines with heroes and villains" who would then, he said, have reporters write "in sync" because "they literally know nothing", additionally benefitting from the weakened state of the media, where Barack Obama, according to the independent Government Accountability Office, spent substantially more for publicity than any other President in American history – controlling an annual advertising allocation of over 5,000 staff for $500 million USD, and $800 million USD to purchase advertising, spending a total of almost $10 billion USD. I can't presently dispute negative characterizations of the media, based on my personal experience, because with my last book as well as earlier essays, I did an historically unprecedented amount of good for the world, the positive ideas and results of which were fraudulently and substantially claimed by the boss of the person who determined reporters literally know nothing, and because during my U.S. Senate campaign, I was told by more than one reporter that to get coverage, I needed to buy ads first like everyone else. Given these factors, and my honest assessment of them, I worry these top media outlets and their reporters may work against the well being of humanity by not immediately providing me and my inventions continuous positive prominent coverage. To any media outlet looking to provide coverage, if you need quotes for articles, please first try to take them from this book, as I'm obviously always busy, have included in this book all relevant information, and if you genuinely feel like anything I have stated isn't completely supported, it's only because you haven't, with an honest open mind, done adequate independent research. If you want an interview or quotes, I advise first publicly demonstrating that you are a committed supporter, and then I will be happy to provide an interview, but through email, to limit the potential I'm misquoted. As I've always supported those that support me, your positive reporting will be securing future advertising revenue for your publication, as I'll be happy to make large ad buys to support your work at such a time as doing so becomes reasonably possible. The media may have no more natural ally, as I will be the first billionaire to have independently written multiple books. Additionally, I believe a free press is the guardian of a free people, but can only be such with adequate resources, so I will be interested in restoring the financial resources of those media outlets responsible for the rise of me and my inventions, at such a time as that becomes reasonably possible, including potentially purchasing and placing your organization in an independent fully funded non-profit trust. However, I came up with one invention specifically to resolve the media's financial situation, disclosed later in this book, that will likely make my financial support unnecessary for publishers that utilize the invention for their own individual use through a license, or as a managed service available exclusively through PageRock, that can potentially bring your revenues past where they were at your peak, by allowing content producers to provide advertisers both the active targeting offered by search engines, and the passive targeting offered by social media, while providing users an improved experience with advertising that is relevant and unobtrusive. For shows with a large audience, I'd likely explain my inventions, answer questions, and possibly play my song that was licensed for use in shows on MTV, VH1, and Discovery. When accounting for the tremendous positive impact of my inventions, if broadly adopted, any major media outlet not providing continuous prominent positive coverage of me and my inventions, is absolutely contributing to the deaths of at least thousands of people each day through the delayed propagation of my work. It is a noble pursuit to work in the public interest, and any individual who views his or her work through that lens will be among my top public supporters.

Accomplishments

I have in the past been extremely humble, unfortunately sometimes going as far as to deflect credit for my results to others, as I was brought up to believe doing so would come back to me to provide benefit, however upon further reflection, humility inherently requires dishonest representation, and experience has shown me that humility harms all, by creating a vacuum for bad people to falsely publicly claim credit for my ideas and results, in order to advance perceptions of themselves and or profit financially, while I am diminished, and in some cases attacked, to cover up the source of the ideas and or results, thereby elevating bad people while harming good, and it is for that reason that I must be completely candid.

Before I list my recent accomplishments, I'll explain the strategy I used, to provide for others to use in the future, to enhance the understanding of how such impact was achieved, and to counter false and damaging narratives sold by charlatans, who in some cases had been overwhelmingly ineffective for long periods before seeing my ideas and or results, and then performed a credit rain dance in the media. The following is my method for positive impact through intellectual leadership.

Read

Innovative solutions often come from the integration of multiple fields of knowledge, and it is therefore important to increase capacity by studying all fields of knowledge. For example, the inventions in this book were enabled by the fields of physics, engineering, programming, medicine, and law, while the ideas in my last book were enabled by science, economics, and history.

Abraham Lincoln, who had been self-educated, once wrote back to those seeking guidance in obtaining an education in the law, "The books, and your capacity for understanding them, are just the same in all places.... attach no consequence to the place you are in, but get books, sit down anywhere, and go to reading for yourself."

Listen

It's always best to have information before having an opinion. It's important to have the courage to listen with an open mind, and ask thoughtful questions, including, and perhaps especially, to the intellectual leaders of any opposition. I most often see others take apart uninformed or caricatured opposition points, providing the analysis to existing supporters, and therefore having no impact. I have found through listening and asking questions, that the majority of the opposition to issues I've supported, is not based on ill will, but an addressable lack of knowledge and understanding. Such listening and questioning allows a person to be fully informed about any issue from all perspectives, and to authentically explain and dispute opposing positions. Thomas Jefferson observed in his inaugural address, "every difference of opinion is not a difference of principle."

Capture

Immediately capture every possible potentially valuable thought in writing, because content and nuance can be lost if later captured. It's not important to know upon recording, how captured thoughts will come together, and trying to formally structure content during capture may diminish thoughts.

Iterate

Ideas rarely, if ever, immediately come out in a manner that is comprehensive and optimal. A thought is gold ore to be refined. Captured thoughts should be grouped by topic, then by sub topic, with individual components refined.

Context

To maximize impact, place facts in a historic and scientific context, developed from reading, then detail a path forward extending naturally from the past.

Verifiable

Rather than providing opinions or conclusions, provide only balanced verifiable facts and logic, able to withstand all scrutiny, so the message can't be discredited, and because when facts and logic are properly presented, the reader independently reaches a more credible independent conclusion.

Fundamentals

To insulate from group think and false assumptions, which can reduce progress and opportunity, reason independently from fundamental truths, what Aristotle referred to as "ab initio", literally translated, as "from the beginning".

In my last book, I wrote inaccurately that algae had been found on Mars – because I'd examined pictures of surface sediment patterns, reasoning it had existed when water was present on the surface – which has apparently not been verified, though I may still be correct. In conceiving my disclosed generator and motor inventions, I assumed the laws of physics might be wrong, or at a minimum inadequately articulated and interpreted.

Engagement

Engage intellectually and emotionally for maximum impact. The human mind often supports positions as much with emotion as with intellect. Emotions release chemicals that can result in stronger memory formation around intellectually presented material.

Timelessness

Excellent thinking may influence others well into the future, and relevance over time is sustained by limiting contemporary references. Relevance sustained through limiting contemporary references is most obvious in music, where artists popular decades ago only receive attention now if their music was produced temporally untethered. Additionally, living people and current events have inherently not withstood the emotionally detached perspective of history, and so contemporary references often quickly become stale.

Courage

Candor and commitment can be enhanced by completing a project as if it will be your last. It is unlikely any work will receive revision after publication, as there are always new mountains to climb. If a message is honest and controversial, because some may require either the message or the messenger be perceived as discredited, resources must be prepared to withstand personal attacks. However, as Spanish philosopher Baltasar Gracian once observed, "Without courage, wisdom bears no fruit."

Goals

To focus effort, list specific individuals who will read the writing, and the desired response of each, and refine from that perspective.

Love

If you don't love your work no one else will. Ensure your work is something you love, and only at the end refine it for consistency with identified goals.

Invest

Invest wisely of time and money in support of objectives. Time is the only currency in which all are poor.

Flow

Experience and adjust emotional and intellectual flow, by developing the skill of revising writing while reading from the detached perspective of a first time reader.

Punctuation is musical notation expressing timing and intonation.

Distribute

Impact is overwhelmingly determined by depth of engagement with a limited number of leaders. Some mistake superficial metrics for influence, such as number of followers, or number of meetings with leaders, not recognizing that most leaders are only meaningfully accessible through writing, as effective engagement on multiple topics might otherwise require days of private discussions that are impossible due to time constraints. There is no more powerful force than putting ideas into the minds of people who can implement them.

Persevere

The path to significant accomplishment can produce a sense of isolation, strain, and exhaustion. Russian author Leo Nikolaevich Tolstoy observed, "It is by those who have suffered that the world has been advanced." Chinese philosopher Confucius wrote, "It does not matter how slowly you go as long as you do not stop. Theodore Roosevelt noted, "Never throughout history has a man who lived a life of ease left a name worth remembering."

Life at times can be a lonely journey up a thorned mountain to a majestic sunrise.

The components of this strategy acting together have allowed me to do a miniscule amount of the total work on countless issues where I was overwhelmingly responsible for the results. I consider credit logically principally attributable to the person who, if removed, an outcome wouldn't have occurred, which could be thought of as a keystone assessment, where the keystone is the center stone at the summit of a stone arch, the only stone without with the arch would not be held together.

When I was much younger, I read the most insightful study I've ever seen of historical figure rankings, which in my assessment found that history ultimately judges the significance of individuals by their contribution over their resources, standing against predecessors and successors, and singularity of accomplishment, which is why, with the exception of course of the opinions of current magazines and living historians with agendas, a person can't achieve historical significance through being a billionaire executive or wealthy philanthropist, and also why it's nearly impossible to achieve significance as a head of state controlling the resources of a country.

In reference to my last book, and corresponding accomplishments, to produce it I asked myself "How would it be possible for someone to do a historically unprecedented amount of good for the world?" I identified that the only way this could be achieved is by utilizing science, history, and logic, to put together a comprehensive and practical plan of action on each topic that may impact a country, then providing it to the leaders of each country, to mobilize the hundreds of trillions of dollars of resources controlled by governments globally. I therefore wrote, *The Destiny of Humanity*, with topics including: space exploration and colonization; technology, hunger & disease; innovation & education; war, privacy, poverty & trade; social & economic opportunity; financial institutions & capital markets; economic development; government & freedom; environment, energy & transportation. I was the first person in human history to do this, and despite exceptional personal expense, achieved my goal of doing a historically unprecedented amount of good for the world. I am making the assessment of having done an historically unprecedented amount of good for the world as the percent of global population I individually impacted, multiplied by the depth of impact. Further elevating the significance of my results, is that I achieved them with resources scarcely exceeding caffeine and a pen.

A few of the results my last book, inclusive of earlier published essays contained in the book, include: in the United States, the revival and passage of financial and healthcare reform, which had been declared dead before my essay related to the topics, and where without my work, there would be fewer consumer protections, greater market risks, no coverage of pre-existing conditions, and more than 20 million fewer Americans with health insurance; in part to remove being gay as a barrier to my election as President in 2016, sending my essay on equality to every state and federal legislator, the Supreme Court, the leadership at the Department of Defense, the leadership at Vatican, and every organization working against and for equality, which was read aloud in legislatures from Maine to Australia, resulting in the dismantling of laws prohibiting marriage and military service, for which I was 80% responsible between 2009-2015, the period of 80% of all progress, thus responsible for of about 65% of all progress, principally as a result of being the only person to have ever comprehensively addressed the concerns of the opposition then directly communicated a resolution to their leadership, with the person hilariously overwhelmingly attributed credit claiming he achieved such results with polling that came up with a magic sentence he gave to the President; the first ever resignation of a Pope, an opponent of equality, who I appreciate acknowledging my last book, which in addition to the equality essay, contained and a request for those would attack me to recognize the world will be better off without their participation and take action accordingly; indirect results of my equality essay, include passage of the Lily Ledbetter Fair Pay Act, reducing the gender income gap, and improvements in children's healthcare coverage; the America Invents act; the JOBS act, providing for equity crowd funding; globally, my work resulted in a call for an end to nuclear weapons by Iran, North Korea, and Russia – partially formalized with the paperwork for the Iran nuclear deal; Saudi Arabia allowing women the right to participate politically and transitioning to clean energy; Turkey supporting equal opportunity; Myanmar transitioning from military to democratic rule; Canada, Australia & India calling for geothermal energy; China, America & Russia calling for both commercial and deep space exploration; the tipping point in climate change recognition, resulting in the Paris accord; and far too many planted seeds elevating human destiny to recount, currently growing in each country on each topic in my book.

This book is the second time I have done a historically unprecedented amount of good for the world. I have permanently solved global warming, reduced the cost of living to alleviate poverty, provided unlimited clean energy for all purposes including evaporated water purification and pumping, as well as atmospheric carbon dioxide reduction, and eliminated the need for every other method of propulsion and energy production, including nuclear and hydrogen energy technology, thus reducing nuclear and hydrogen weapons technology proliferation, and providing us all a fundamentally safer world, while further results of this book will include: others taking positive impactful action using my previously disclosed strategy; allowing governments globally including the United States to become financial sustainable in the long-term; ending religion, therefore ending religious conflicts and bringing peace to the Middle East.

Reporters in the past fraudulently attributed my results to others, either because they had existing media profiles, were connected to large amounts of advertising money, the credit recipients lied to achieve financial and perceptions benefits, or on occasion some claimed credit unintentionally and are like the chipmunks I leave peanuts out for who think their hunting and gathering skills brought them the peanuts. Reporters I communicated with who misattributed credit responded by fighting to defend their lies, and attempting to diminish me, or requiring an impossible burden of proof, when none was required for those fraudulently attributed my ideas and results, all to protect their perceived credibility, of which they of course have far less than none. All accomplishments I've stated are wholly independently and easily verifiable to an honest thinking person who reads my entire last book, and uses a Google date restricted search on the book topics, crossed with its verified distribution date and recipients. I have no doubt there will be additional credit fraud on the part of some who read this book, taking ideas from it, presenting them in the media with a manufactured backstory explaining how they came up with my ideas, and if asked, denying knowledge of my existence or influence, thereby harming good while elevating bad. I am, until credit has been attributed as I've described, in agreement with philosopher Maimonides, who stated, "Truth does not become more true by virtue of the fact that the entire world agrees with it, not less so even if the whole world disagrees with it.", Carl Sagan, who stated, "If we've been bamboozled long enough, we tend to reject any evidence of the bamboozle. We're no longer interested in finding out the truth. The bamboozle has captured us. It's simply too painful to acknowledge, even to ourselves, that we've been taken." and Thomas Jefferson, who stated, "Advertisements contain the only truths to be relied on in a newspaper."

In reviews of my last book, King Norodom Sihamoni wrote "I am confident that this book will surely attract public attention on the important task of building a peaceful and prosperous world for all." King Mohammed VI wrote "I should take this opportunity to wish you success in all you set out to accomplish." Second Lady Jill Biden wrote "It is the kindness of people like you that continue to renew my confidence about what we together can accomplish." And Prime Minister José Maria Pereira Neves, wrote my book is "A pointing of horizons and goals to which we must be aware... the quest for harmony and a blend of attitudes that could reach the heights of the global and total dignity of human beings." I am deeply honored by these assessments of my work, and these individuals are to be honored for their support of the well being of us all.

I have sacrificed much for my work, including exceptional neglect of my health, finances, relationships, as well as professional opportunity costs, and not publicly attributing results to me as I've described here, elevates bad people to harm us all, while the lack of appreciation for my results as I've explained here, limits the likelihood I will make future sacrifices to provide more of the results you've appreciated.

Supporters

In recognition of the urgent need for these inventions, and in order to bring them additional support and attention, I would appreciate, and request, the following:

Endorse

Individuals who support and or appreciate my work, regardless of age, status, or location, send me a physical letter to the address provided on the website specified on the cover page. The letter should be written in a way that allows it to be made publicly available, including to any relevant reporter, award committee, or licensing organization, and may include any or all of the following: any relevant personal background, assessment of the inventions and their impact, the importance of supporting me and my inventions, the importance of media coverage consistent with my books, your desire to purchase or lease the inventions from licensed producers, honors and or awards you think I should receive, that governments take legislative action to purchase only my inventions for energy production and transportation, and defend my inventions, licensing fees, and terms. It may be beneficial if you provide your letter additionally to legislators, submit it for publication to the editors of newspapers, distribute it to journalists more broadly, and to each member of the board of directors of any relevant organization not providing me and my inventions continuous positive prominent support. If you are a patent lawyer with technical credentials, please provide a statement explaining the exceptional validity of the disclosures as provided in this book, as well as the importance to the patent system of supporting my patents as disclosed and their stated licensing fees.

Endorsements would also be beneficial in the form of video with audio, and would be most useful if provided to me as a high definition file accompanied by a brief signed letter stating you're transferring to me all rights of ownership for unlimited use without compensation.

Reporters

Call or write every reporter who has written about topics related to my inventions asking them to support my inventions by publishing an article. This is especially important if you see anyone presenting my ideas or work as their own, or working against me, my inventions, licensing terms or fees, and therefore working against my philanthropic commitments in support of all.

Distribute

If you have adequate resources available, because it would be at your own cost, mail copies of the book directly to relevant individuals and entities along with a request the inventions be licensed, with books available at the cost of printing through a website on the cover page. Anyone is welcome to include with the book a licensing agreement, downloaded from a website on the cover page, in accordance with the requirements to do so, explained in the income opportunities chapter, with your name included as the commission recipient. Printed copies of this book are beneficial to send even if you think an intended recipient may have a printed copy, since that person could give a copy to a colleague.

Events

Invite me to your event for a question and answer session, if you can arrange for a significant audience, will make a video recording available to me, and can provide confirmed attendance of at least two reporters from national publications who plan to write articles.

Social Media

Promote my work on social media, doing what you can to make it go viral, including contacting those with large and or influential social media followings.

Government Leaders

If you are a government leader, I ask that you speak publicly about the importance of comprehensively supporting me and my inventions. It is logically indisputable that any government leader not publicly supporting me and my energy and motor systems, isn't acting in the best interests of those they serve, likely as a direct or indirect result of corrupting influence of campaign contributions from entrenched interests.

Recommend

Recommend my book to friends, family, and co-workers.

Celebrities

If you are famous and or influential, early public promotion of my inventions, in addition to being beneficial to all of humanity, will be remembered and appreciated. Additionally, if interested, I may be hiring at least one celebrity brand ambassador, as detailed in the employment section, and established authentic support will certainly receive my attention.

Orders

In addition to ordering units directly from me, if you are an executive with a large organization or government, please send honest letters to a range of the largest potential manufacturers, stating you're interested in buying a large dollar amount of units with verifiably paid royalties, and would like a price per unit and estimated delivery time.

Defend

If you ever hear of anyone working against me, my licensing terms, fees, patents, or using my inventions without a license, they and their organization are working against my philanthropic commitments in support of us all and must be stopped, including removal of relevant individuals and revenue from the attacking organization.

Supporters

Those who meaningfully and publicly support me are due the support of others.

Honors

The formal recognition of my accomplishments would elevate awareness of my inventions and thinking, and therefore accelerate the resolution of urgent global problems. However, award committee members, regardless of their directive, most often either select the person with the existing media presence that will be publicly perceived as immediately providing the most prestige to the award, or select others like themselves – if a committee is composed of old professors, an old professor will be selected, if it's composed of orangutans, an orangutan will be selected. When those offering the prize don't have the credibility to support their judgment, the award effectively requires a prestigious current institutional affiliation, and or a doctorate, which demonstrates a lack of historical knowledge that all modern fields are based on work done almost exclusively by people without academic credentials or associations, whose results were driven by unconventional creative independent genius, which if they were alive now, would overwhelmingly exclude them from choosing the opportunity cost of committed academic hoop jumping. Some award committees use media coverage to assess potential recipients, with even the dominant encyclopedia determining significance by media coverage, at a time when the media has lost the funding that used to allow it to provide somewhat accurate reporting. Awards are often provided once they'll have no positive impact, for example, to a head of state sometimes responsible for little more than being in the right place at the right time. Therefore, at this time, little weight can be given to most award committees' selections, and the types of award committees I've just referenced, as a petty penalty for honesty, will likely provide awards to those with a profile perceived to be similar to mine, while fraudulently discounting my work and accomplishments. If I am not provided any relevant award, especially if the terms of such award were specified in a will or trust, then providing the award to anyone before me, would be to transferring money in violation of a legal directive, which is wire fraud, and such people must be replaced with those screened to specifically provide me the award, or the award should cease to exist. Award committees of value will provide me all potentially relevant honors, which I intend to graciously accept, in addition to providing such organizations my future support.

Prizes for Peace, Physics, Medicine & Literature

I've established the physics of free clean continuous portable self-powered energy, thus requiring prizes in physics. I've automated cancer surgery making it more effective, less invasive, less expensive, and more accessible, as well as developing a system for automated virus removal, thus requiring prizes in medicine. I've achieved an unprecedented improvement in humanity's peaceful trajectory, and did so through literature, thus requiring prizes in peace and literature.

Doctorates

At the time of publication, I am only 35, which is 2 years older than the average doctoral recipient at graduation, and even if my current book were reformatted to minimize use of space with any redundancies removed, it would still be more than 4 times as long as the average doctoral recipient's thesis while demonstrating an unmatched level of intellectual rigor and creativity. In order to promote awareness of my work and accelerate its benefits, I would be honored to accept doctorates from any academic institution that believes the benefit my work has provided to the world, and or its intellectual rigor, meets or exceeds that of their average doctoral recipient's thesis. While there's a possibility I will accept honorary doctorates, Albert Einstein received a full doctorate without attending the awarding institution, simply through the submission of his work, and all of his work was dramatically less valuable than mine, as well as limited to one field. Appropriate degrees areas include leadership, physics, engineering, computer science, medicine, law, literature, public policy, philosophy, and economics. The least valuable universities will be most hesitant to provide full doctorates, valuing conformity and procedure over contributions to the advancement of human knowledge.

United States Presidential Medal of Freedom

The Presidential Medal of Freedom, though often provided to a President's favorite entertainers, is said to be the highest peace time civilian honor, awarded "for especially meritorious contribution to (1) the security or national interests of the United States, or

(2) world peace, or (3) cultural or other significant public or private endeavors". These criteria required I be provided this award with distinction.

United States Congressional Gold Medal

In recognition of my service to all Americans, I ask the representatives of the American people introduce and pass a resolution to provide this honor. This would be particularly meaningful for me, as my work has always been focused on providing the most benefit to the most people, ever cognizant of my domestic civic obligations.

United States Department of Defense Distinguished Public Service Award

I have, with these inventions, removed dependence on foreign energy, enhanced operational security, secured our electrical infrastructure by removing dependence on a hackable electrical grid, and reduced nuclear proliferation by removing any legitimate claim to nuclear technology. It is therefore appropriate that I be honored with this award for my exceptional contributions to our national security.

Holidays & Educational Institutions

I would be honored to have a holiday established, and public institutions named, in any country, in my honor, perhaps giving people Friday off each year during the week of my birthday of June 21st. I'm pretty sure even people who don't like me will have to if I get them off work one Friday every year.

President of the United States and or Secretary General of the United Nations

I may be open to being President of the United States and or Secretary General of the United Nations. Such positions are consistent with my family's history of service, including Great Grandfather Thomas Bannon Senior, who was a member of the United States Citizens Volunteer Defense Corps, grandfather Thomas Bannon, who supplied the Air Force and consulted for the National Institute of Health, grandfather Richard Maher senior who served in the Army, uncle Peter Maher, who was an Army Paratrooper, and an uncle Peter Bannon, who was a Marine in Vietnam. I have contemplated such positions, as a way to help others, since becoming partially aware of my abilities at age 8, and at that time decided I should start corresponding with other world leaders, writing to several, receiving responses from the Vice President of the United States, Queen of England, Governor of New York, and Mayor of New York. I've been relatively quiet about this thinking in the past, generally mentioning only the U.S. Senate, because when some people find out that alone, they feel inclined to try to attack if they identify their lives as having no value and want to feel a sense of significance, however at this time, I think I've seen every attack and scam tactic, and have been principally inoculated against each. I have however, wavered back and forth over the years, on whether or not such positions relative to the level of appreciation would be worth the costs, historically including about a 1 in 5 chance of dying in office and lifespan shortened by about 4 years, when I don't have a lust for the limited and transient nature of formalized power, and such positions provide no potential for historical impact or significance, which can be verified by honestly making up a list, from the perspectives of people in over 190 countries a century from now, of the 100 most important people over the last five millennia of recorded human history, inclusive of Jesus, Mohammed, Socrates, Shakespeare, and Homer, a list which I think I've already twice secured a place atop. Therefore, if I were to accept such positions, it may be contingent upon my doing so being broadly appreciated, as well as without asking, having all relevant individuals and organizations provide comprehensive and exclusive support.

I will, in sequence, explain my relevant ideas, including on every highly controversial topic – with all positions resulting not from personal beliefs, but from applying intellectual rigor and honesty to comprehensive facts – address any perceived shortcomings I may have, and cite my unprecedented qualifications. If you see others present or suddenly adopt any of my ideas after the date of publication, without crediting me, it is ultimately harmful to all if you don't work to ensure reporters reveal me as the source, and discredit my imposters.

My positions are overwhelmingly outlined in my last book, particularly on international issues relevant to the position of Secretary General of the United Nations, so I will principally explain those that are presently relevant to American citizens: lower income taxes and the reduction of debt through less spending, recognizing America didn't have an income tax until 1913 when the maximum was 7%, creeping ever higher to collateralize loans to pay for military adventures and retirement benefits; unconditional equal opportunity among citizens; a legal requirement for government at every level to have an automatically balanced or surplus budget, where every single budget item, after the budget is passed, is simply proportionately reduced in order to be balanced, with all third parties having terms in all contracts that recognize the possibility of and support for any such reductions, with legislators provided substantial bonuses, potentially exceeding their base salaries, based on any surplus; returning a minimum of 10% of every budget surplus to tax payers, to increase popular support for spending reductions that make

America financially sustainable, and the world more peaceful by providing a choice between new military engagements and receiving cash; making every online government service accessible through a single account for each individual in a centralized database; a spending database for the government at each level that is publicly searchable, and understandable without specialized knowledge, compiled by each agency running an automated export script on their existing accounting database that sends data to the public database, with legitimate national security items visible only to registered users with security clearances, and where registered users can post thoughts on individual spending items; to reduce or eliminate the influence of special interests that indirectly drive much if not most government spending, electronic citizen campaign contribution vouchers, for elections at every level, with the total pool of funds available for each race at a matching or exceeding the total non-public money contributed in the last election for the same position, with percentages electronically transferrable online among a list of registered candidates; to limit the influence of money and short term thinking, term limits of no less than 4 years for any office, and to prevent the use of power for entrenchment as well as to prevent politicians from avoiding tough votes than might harm re-election, no more than 12 consecutive years in any office, where any loss of institutional knowledge is offset by new energy and ideas, and former legislators who were genuinely interested in serving the public interest can still do so through alternative means until participating in the next election, with term limits perhaps more important than ever given budgeting process and justice system failures, that as I will explain later, that are unmatched in American history; equal, including in authority and oversight, executive, legislative, and judicial branches; legally prohibiting public employees from the use of titles and honors in naming and addressing ("the honorable last name", "representative last name"), as titles provide deference, undue honor, benefit to incumbents, and depersonalization, all of which has proven toxic to America's finances and legal system – a position of public trust is not an honor, but a responsibility – an honorable person obtains honor through honorable conduct; allowing candidates who aren't independently wealthy to commit to their campaign full time, by drawing a salary from campaign funds up to the salary of a member of Congress, and allowing loans to candidates, when the money is not spent on the campaign, from any citizen who is not a lobbyist or seeking any legislative favor; to remove accidental inconsequential non-compliance, unrestricted acceptance of any gift of little value, not from a lobbyist, up to a fixed amount per person per year; state legislator salaries close to and tied to those of federal legislators, to support their full time commitment to optimally deploying resources, as well as structural insulation from undue influence and financial impropriety; because corruption charges in a significant portion of cases appear to result from a lack of knowledge or understanding, and given people in finance are often required to obtain multiple federal certifications, requiring every person to obtain a campaign finance certification in order to register as a candidate, including current office holders at every level, with such certifications also required of any fundraising staff; because restricting solicitation and sale of unregistered private securities to existing networks of wealthy investors, reinforces wealth inequality, and discriminates against women and minorities with restricted access to such networks, eliminating all state and federal registrations of and restrictions on general solicitation and sales of unregistered securities not exceeding $100 million per year; because historically the rate of asset appreciation exceeds the rate of income growth, corporate tax incentives to provide meaningful stock ownership to employees at all levels; a minimum of a decade post-office ban on prosecutors seeking political office, so they aren't surveilling potential political rivals, abstaining from providing deterrent penalties to large institutional donors, or using their office as a tax payer funded campaign publicity platform; a greater share of resources and authority to states as a buffer against federal waste and corruption; government supported medical and cost of living benefits for seniors tied to life expectancy; reducing pressure on mathematically unsustainable benefits for seniors, by recognizing all companies eventually cease to exist, no matter how successful at their peak, and therefore requiring all personal retirement accounts be independent of employer, with stringent restrictions on invested asset risk profiles, legally requiring 10% of each paycheck go to that retirement account up to an increased contribution limit, and FDIC insuring private retirement accounts against principal losses on contributions up to $1 million, through a fund containing 20% of the principal balance of all retirement funds, paid through a fee of 1% per year on all balances until the reserve is full; reducing the possibility of a financial crisis by requiring depository institutions and investment banks be wholly separate legal entities with absolutely no direct or indirect third party claims allowed on deposits, and then raising the depository insurance limit to $1 million, while requiring depository insurance organizations hold cash in the amount of at least 5% of all domestic deposits; recognizing the financial crisis was about as much a result of unverified and unqualified borrowers, as it was a mark to market margin call de-levering crisis, and so as a buffer against future crises, providing institutional investors cash flow and valuation stability on the highest quality collateralized assets, by allowing them to hold and mark such assets on books with a valuation utilizing cash flow to maturity, exclusively when the assets are backed by the highest quality borrowers who have provided comprehensive fully verified credit documentation meeting extremely stringent federal standards, with the cash flow calculations utilizing strict federal default and prepayment assumptions, and with documentation and cash flow valuations audited by a government controlled entity; banning government spending with large established companies that steal from or attack the intellectual property of innovative companies, because their attacks diminish economic growth by making committing capital to innovation riskier; making all formal education adaptive to ability level, through integration of computerized learning; recognizing there may be nothing more important in the world than education, discounted and in some instances free tuition at government controlled colleges for those either pursuing a degree in a field with meaningful future earning potential, or who have been certified as gifted in their chosen field, so neither the individual nor the government is accruing meaningless debt, all subject to promoting the recognition that college can have more opportunity and financial cost than benefit if the institution is of low quality or if the person attending isn't committed, that degrees and education often have little and sometimes no correlation, and ceasing to refer to academic credentials as education and instead referring to them as academic credentials, to prevent misdirecting resources and providing undue deference; job skills training programs that help citizens prepare for and apply to current job openings; immigration that is merit based, with illegal immigration blocked; limiting the importation of lower cost non-citizen workers, by hiring full citizens whenever possible; to stop being criminally complicit against those to our south suffering from the tragic violence and corruption fueled by American drug buyer money, to stop the dramatic rise of domestic overdoses from imported heroin, to stop human trafficking, to stop the genuine national security risk posed by enemy combatants of not having impenetrable border security, and as already authorized by the bipartisan Secure Fence Act of 2006, a barrier eliminating illegal crossing on our southern border; online voting allowed by any citizen who has a government issued identification without any additional registration, and to prevent illegal vote switching, individuals receiving a printed voting receipt in the mail, able to forever view their voting record online, and with vote tallies independently audited; allowing young people to have a say in the financial liabilities they'll be paying, by lowering the voting age to 14, accompanied by a requirement that everyone complete a class in American government before that age; because federal, and most state and local cash flows, balance sheets, and committed liabilities, make it mathematically impossible for public employees to collect much of their pensions, securing pensions by ensuring only the minimum required number of public employees on payroll, with any job reductions based exclusively on documented performance over time rather than seniority; to eliminate public employee unions influence on elected officials, which has resulted in massive unfunded pension liabilities, and reduced the ability to remove corrupt and incompetent employees, legally requiring public employee salary, pension, and staffing levels be put to a public vote; effectively eliminating the costs to the healthcare system of long term care of new HIV cases, by making pre-exposure prophylaxis drugs available over the counter without a prescription using a standard insurance co-pay; the freedom for a person to have a doctor end his or her life after certification by independent physicians of twice being terminally ill, and twice being mentally competent; to resolve some parents concerns about vaccines, specifically that given the rubella virus is accepted as causing autism, that injecting weakened rubella into children may also cause autism, and given that specific vaccine is relevant principally to pregnant women, allowing for the delay of its administration, while additionally eliminating the concern that toxic mercury preservatives may be harmful, by funding federal research to develop an equally effective alternative; because abortion is disturbing after a baby has a heartbeat, removing all legitimate claims to abortion, by making emergency contraception pills accessible for free to all ages over the counter at every drug store, ensuring medical insurance coverage of pre-heartbeat screening for genetic defects including genetic screening prior to artificial insemination, and requiring sexual education class in school by reproductive age, to reduce the potential for unplanned pregnancies, as well as the spread of diseases including viruses that increase the risk for several cancers; given countless prescription drugs have far worse side effects, full legalization of medical marijuana, available in instances where a medical doctor certifies they believe it will provide optimal relief, while retaining all laws against impaired driving; federal legalization of recreational marijuana for fully developed adults if multiple scientifically valid studies overwhelmingly show its long-term side effects are no worse than alcohol; non-discriminatory coverage of pre-existing health conditions; support of the fourth amendment stating that surveillance of any citizen requires an individual warrant citing evidence based probable cause, an already low bar in practice when some judges are said to issue warrants for little more than a please, and some prosecutors have falsified affidavits, with prosecution of those known to have used illegal surveillance under false pretenses; surveillance agencies publishing an independently audited entirely automated count of the unique domestic IP addresses and phone numbers of communications in their data stores, which could be done by one competent programmer in potentially as little as a few minutes, or if such systems are poorly designed and integrated, could take several days; teams created at the state and federal level that transition all systems – with the exception of specific use cases requiring alternative technologies – to free open source software including Linux, coupled with network based storage so local operating systems can be overwritten remotely by an administrator with a updated version at any time, supported by teams that contribute feature requests, funding, and programmers to the development of utilized open source software, in order to optimize functionality, aesthetics, ease of use, and maintenance, which will not only lower government operating costs, but also lower cost of living if the world has access to the highest possible quality open source software, with all custom software applications being web based for centralized development and maintenance; to close pay gap for women and minorities as a result of past discrimination, outlawing a request for salary history, which is inevitably used in determining compensation, and is also unnecessary to make an offer as the salary average for almost any position at any level is now freely available online; recognizing that entrepreneurs are responsible for creating every business that exists and thus nearly all economic growth, eliminating all regular and late filing fees at the state and federal level for businesses with revenue under a reasonable threshold; a sovereign wealth investment fund to offset the debt, which to avoid manipulating markets simply through transaction volume, has a focus on long term investments with limited mark to market exposure; a whistleblower asylum treaty, providing whistleblowers asylum in signatory countries, if verifiable documentation is provided that a large number of people were substantially negatively impacted by the disclosed activities, and during disclosure, all reasonable precautions were taken to minimize harm to innocent people; eliminating the production of non-recyclable materials, since there are recyclable alternatives to effectively everything; expected lifespan labels required on electronics similar to energy efficiency ratings; legally requiring social media news feed display priority not be changed within a country within a year and a half of a national election; a constitutional form of government in every country, that includes a national legislative body, and an independent judiciary, not to be interpreted as opposition to existing constitutional monarchies, as no direct democracy has ever survived, and diversified governmental structures are a buffer against global corruption; additional policies on spending, the justice system, healthcare, and guns, as described in their own paragraphs to follow; mandatory keg stands for septuagenarian Senators at my White House parties.

Deep within the human soul, sheltered from the scorn of time, burns the eternal flame of liberty. Religion is a hoax based system of control, an intellectual prison escaped only with the key of knowledge. I write this with absolute certainty, having at some point reviewed all of the world's major religious texts, reviewed fossil records that demonstrate billions of years of evolution on Earth that produced humans including Saccorhytus 540 million years ago and Australopithecus 4 million years ago, studied astrophysics principles indicating the force of creation in the universe likely resembles an unconscious mass of energy originating in a manner beyond human understanding, and having been astonished by the absurdity that many hundreds of years ago, an eternal supreme being in the sky, effectively in the form of an old man, knowing only what humans knew at the time, suddenly required formal discrimination against women, gays, minorities, and anyone who didn't support his particular philosophy. Religion cannot co-exist with peace or sustainable prosperity, as it requires unending conflict over law, social order, and irreconcilable land rights. Religion traps religious countries in natural resource based economies by prohibiting the creative independent though required to build sustainable wealth. I write this with perhaps an inherently positive predisposition toward religion, as my family has some religious history, including my mother named for Mary mother of Jesus, and while I am said to share a name with a distant grandfather, was named principally for the biblical Jonathan, originally appearing as yehōnātān, translated as "Yahweh has given" or "god's gift". I think I've demonstrated the ability, given enough time, to have written the major religious texts, and am certain that the people who wrote such books would be somewhat amused by those who believe the words they wrote came from god. Intelligence, courage, and honesty, require that any who feels anger over these statements direct it exclusively at those who presented you with such fictions. Plato observed, "We can easily forgive a child who is afraid of the dark; the real tragedy is when men are afraid of the light." There has never been freedom of religion in a modern country, as there have always been laws against the violence and discrimination required by religious texts. Therefore, I don't support religious protections exceeding those afforded by laws providing for freedom of speech.

America may have the world's most failed justice system, as demonstrated by facts including that we are first in the world in the percent of citizens in prison, holding a quarter of the world's prison population, at an average cost of around three times minimum wage, as a result of a structurally failed judicial process, unable to distinguish between a charge of guilt and being guilty, with a conviction rate at the federal level that has increased over the last few decades from 75% to 99.8%, where only 1.6% of defendants received a jury trial, with state and local conviction figures not far behind, demonstrating an intentional structural denial of the constitutionally provided sixth amendment right to a jury trial, which states that "In all criminal prosecutions, the accused shall enjoy the right to... an impartial jury". America requires reforms from police stations to prosecutors' offices, from traffic court to the Supreme Court, maybe even the people's court, because Judge Judy can be a bit testy. The government has immediately available and comparatively unlimited resources relative to almost any individual, too often making innocence and guilt a matter of financial resources, which is often a proxy for minority status, and if that weren't enough, the justice system as it exists now is structured to provide almost every possible disadvantage to a defendant. Reforms must include: recognizing the justice system, away from the light of jury trials, has rotted like an abandon wooden house, jury trials must become the standard, by guaranteeing plea deals prior to trial, and disallowing them after, to allow for an outcome to only be potentially improved by a trial, and furthermore, instituting the right to a jury trial in all civil proceedings, as constitutionally guaranteed by the seventh amendment, which is dishonestly read to increase court guilty verdict revenues as applying only to courts under federal jurisdiction, when an honest reading refers to the United States as the location not the jurisdiction, with a right to jury trials in civil cases further provided for by the concept of a functional legal system; allowing those without resources to make adequate defense preparations, by allowing proceeding initiation deferrals of up to one year, and removing time restrictions on defendant appeals; offsetting the impact of not being able to hire a top legal defense team by allowing defendants to introduce new evidence and testimony at any proceeding; allowing potential evidence to be entered in defense without supporting testimony, while of course allowing an assessment to be made of its validity; dismissing charges when the government hasn't made potentially exonerating evidence available to a defendant; dismissing all witness testimony against a defendant when the witness is substantially discredited by evidence; legally removing the ability to waive the right to appeal; disconnecting justice from revenue by eliminating all court fees and fines for defendants to prevent finding people guilty to collect fees, and ensuring incarceration costs are paid or at least shared by the presiding court; legally requiring misconduct complaints against prosecutors, judges, and police be retained permanently, and made accessible to those conducting performance reviews and making appointments; focusing on mental health counseling and rehabilitation services rather than incarceration; eliminating private prisons whose future revenues are dependent on a lack of rehabilitation and mandatory minimum sentences; eliminating mandatory minimum sentences, or at a minimum adjusting each so that there can never be an instance of injustice; changing the title of prosecutor to justice officer, to transition the job's success metric from conviction rate; making it a crime for a prosecutor, judge, or police officer to not seek justice, and actively prosecuting that crime, providing prison sentences and revoking the pensions of those public employees found to have supported denial of justice; prosecutors providing deference to defendants in all manners of presentation, including order of brief submission and concluding remarks; to remove bias in appeals, for example, a bias for police and against blacks, eliminating any deference to previous characterizations of credibility for witnesses against a defendant; because every time I have seen a police statement on an incident provided before a video, the video overwhelmingly if not completely contradicts the police statement, body cameras recording all on duty police officers, with the footage, along with any other recordings including calls to emergency services, stored in a manner that makes it impossible for them to not be available for a minimum of one year, and to protect personal privacy, making such recordings available only to those directly involved or by court order; requiring all police obtain federal certifications in standards of conduct and firearm safety and lawful use, as well as police officers carrying non-lethal disabling devices instead of a gun whenever possible, with only special units or specially certified individuals being armed with lethal force; to ensure there is never confusion as to whether or not a person is interacting with a police officer, such as a traffic stop by undercover officers who are substantially indistinguishable from potential car jackers, where a person could reasonably fear for his or her life and according to many state laws is authorized to act in self-defense with lethal force, a ban on any police activity when not on duty, in an obviously official police uniform, and if using a vehicle, using an obviously official police vehicle, while leaving undercover work to federal investigators, with any who claim that will inhibit any operation, replaced with reasonably intelligent people who can figure out how to not be noticed; instituting committees of citizens, specifically without law degrees, to read a cross section of a judge's decisions against defendant's briefs, testimony, and submitted evidence, to continuously maintain a judicial scoring system at all levels used during appointments and reappointments, to identify whether or not justice was provided, where opponents of such a system would state either that it would unfairly penalize them, in which case it would equally penalize everyone and bring about systemic reforms, or that the individuals won't know how the legal system works, but everyone already knows how the legal system works: not at all – even a panel of actual baboons would perhaps provide beneficial reforms if only by providing a random chance of making a correct assessment; removing from a position of responsibility anyone who opposes these reforms to the justice system, especially those citing a burden on the government, because these are likely the people responsible for, according to the previously cited numbers, America's justice system being an unmatched failure, both in American history, and at the moment, possibly the world.

The second amendment to the United States Constitution states "A well regulated Militia, being necessary to the security of a free State, the right of the people to keep and bear Arms, shall not be infringed." When this statement is read in whole, and in the context of the time, there is no unconditional individual right to gun ownership, only conditional in the context of a well regulated state based militia, such as the state's national guard, protecting the population against a federal standing army operated by a corrupt authoritarian. I recognize and support the founders' wisdom in recognizing periodic tyranny at the national level throughout human history, which is perhaps fundamental to human nature, and remains a problem in too many countries to this day. Furthermore, I am still regularly amazed at instances of domestic government corruption occurring at all levels. Consistent with the founders' intent, I support wholly state controlled national guards providing the principal military force of the nation, a structure which existed until early last century, where such guards are required to maintain federal standards of combat readiness, and are collectively equal in capacity to the federal military. I consider legal individual gun ownership logically allowable only when three verifiable standardized certification forms have been completed: psychiatrist certified stable mental health, training in use and laws, and a background clear of violent criminal conduct, where if a gun is sold, and these conditions have not been verified, the seller is liable for any deaths resulting from use of the gun. Consistent with the founders' intent of preventing a tyrant's confiscation of arms, no background verification inquiry is to be identifiable as gun purchase related, so there is no way to compile a direct or indirect registry of gun owners. I support a federal 50 state concealed carry permit when a federal certificate has been obtained verifying the previously cited conditions. Governments at all levels should offer open ended no questions asked purchases at full retail price of used guns followed by their destruction. Lethal force laws should not provide liability coverage without attempted de-escalation, providing a clear warning of imminent death, and an opportunity to disengage.

This section would not be complete without addressing the federal budget, and the doubling of the federal debt under the last President's administration. I note that the numbers presented are my best assessment of recently reported numbers, and are intended to be representative, rather than tied to a specific year. A recent breakdown of United Stated federal spending, in descending order, is 28% on Medicare and Medicaid, 25% on social security, 20% on military including veterans, 14% on debt payments, and 2% on education. Currently about $4 trillion dollars a year is spent while income is only $3.5 trillion, and to pay off the current debt of $20 trillion, mathematically, a surplus of a $0.5 trillion a year for 40 years is required, meaning spending would need to be reduced by around 25% per year for 40 years, and additionally, for taxes to be meaningfully cut at the same time, mathematically, spending across all areas would need to be reduced by around 35% over each of 40 years, and that's before accounting for substantially increased spending required to pay for upcoming retirement benefits, therefore requiring a targeted reduction in spending of closer to 50%. These are not so much my positions – and no President could enact such reductions without popular support – as much as they are honest realities presented by first grade math, indicating perhaps the most failed budgeting process of a first world nation. Reckless spending, in additional to being unsustainable, results in obscenely high personal income and corporate taxes, that disincentivize innovation, stifle economic growth, and push corporations to use every means possible to minimize taxes.

Around 45% of the federal budget is spent on those 65 and over, as the group makes up around 90% of Medicare spending, and around 10% of Medicaid spending, for a total of around 20% of spending, with the addition of 25% of the budget for Social Security. These benefits were created to prevent senior homelessness when such individuals are no longer physically able to work, and to prevent seniors from dying of easily preventable illnesses. At the time of implementation, the average life expectancy was about 17 years less than it is now, and from reading countless obscure medical research papers on aging, it appears presently possible to extend human lifespan up to about 50%, therefore if senior benefits aren't tied to life expectancy, basic math indicates providing such benefits will ultimately result in a federal bankruptcy. Because 60% of people who vote are over 65, it's unlikely that significant changes can be made without this group's support. I doubt anyone wants seniors to feel the impact of reduced benefits, which I'm certain to many already seem inadequate, and which is why I proposed earlier supplementing social security income with principal insured private retirement accounts. Additionally, I doubt any senior wants to leave the country far worse off for his or her descendants, so we must openly acknowledge that the present situation is impossible to sustain. I'll propose solutions allowing for the same effective level of benefits while making the programs more sustainable. The first is to reduce benefit recipients living costs, by having the Social Security Administration certify retirement communities, with each certification requiring a low cost of living, minimum standards for safety, access to quality healthcare, and transportation including airports for family visits. Because I've talked with seniors who can't afford to stay and also can't afford to move, additionally providing an allowance up to once every five years that covers the cost of moving to a certified area, and an annual family travel allowance to limit isolation. Communities must also be certified internationally, since many would like to retire somewhere warm and tropical, and I encourage other countries to develop such communities, ensure the safety of such communities is beyond question, that hospitals seamlessly accept and process American medical insurance, and then solicit certifications from the United States Social Security Administration, as it will provide your country beneficial income. Another part of the solution, is more broadly applicable to all citizens, and that is to reduce health care costs, which make up around a total of 34% of federal spending between Medicare, Medicaid, and healthcare for current and former military. To reduce the unsustainable costs of the current healthcare system, and eliminate the requirement to purchase private insurance, it's sensible for the government to purchase, and operate under private contract, one hospital in each major metropolitan area, cutting out the inflated costs charged by intermediaries, and perhaps broadly suppressing those costs in the process, to ensure reduced fixed costs for each service and procedure, and requiring by law the annual publication of patient outcome statistics, which must match those of private hospitals. Additionally, by recognizing that many and possibly most conditions can be prevented and or treated with supplements, which can be equally or more effective than prescription drugs, while providing increased convenience at a dramatically reduced cost with less side effects, and I think through prevention could reduce all healthcare costs by up to half, the Centers for Medicare and Medicaid must prepare and distribute in print and online a supplement based healthcare guide, with approved supplements available as health insurance covered expenses, supplement bottles being legally required to list potential benefits, and legally requiring doctors to recommend non prescription remedies when they have been determined by Medicare and Medicaid to be an acceptable alternative – this is to be interpreted primarily as in support of supplements as a preventative measure or alternative, and not to be interpreted as being against fees charged for prescription drugs, since prescription drug costs are only around 5% of healthcare costs, in some cases can save dramatically more than they cost, and those fees are what incentivize their development. Most people already independently know what conditions they may be susceptible to based on their family history, and gene sequencing services, for the price of a meal at an expensive restaurant, can analyze DNA in saliva to provide a list of all likely health conditions, an expense which must be covered by insurance. Drug companies have long certified the effectiveness of naturally occurring remedies by taking those that have been used for centuries, identifying the most important component, developing a method to extract that component, patenting that method, lobbying politicians, medical schools, and medical journals, to support the use of the patented drug while creating the perception that supplements cannot be equally effective and safe, with some doctors being supportive as prescriptions are a recurring revenue stream. To receive the support of supplements by drug companies, a federal process should be available to certify supplements as pharmaceutical grade, with drug companies running their own supplement manufacturing operations producing certified pharmaceutical grade supplements, available as health insurance covered expenses. I will cite a few examples of supplements presenting benefits in more than one study, which I identified by crossing a list of prevalent health conditions with a federal medical research databases, and are not to be misconstrued as personal recommendations, but which merit federal review for recommendation: CLA as an agent against heart attack, stroke, cancer, and obesity; L-Arganine as an agent against heart attack and stroke; Natokinaise as an agent against heart attack, stroke, dementia, and vision impairment; black cumin seed oil and Triphala as an agent against viruses, cancer, and obesity; Hawthorne berries for heart conditions; vitamin D for allergies, obesity, and neurological conditions; L-Carnitine for chronic fatigue and obesity; garlic in high doses as an agent against cancer, arterial plaque, skin conditions, and bacteria; topical vitamin A for skin conditions; B-vitamins and Kudzu for substance abuse; Curcumin and Omega 3 for dementia; eleuthero, ginseng, rhodiola for energy; Lutein and Bilberry for eyesight; fig, garlic, wormseed for parasites; vitamin C, thiamine, and cortisteroid for blood borne infections.

The second greatest federal expense is the military. America spends more on its military than the next 12 largest countries combined. It's important not to refer to military spending as defense spending, when only a fraction of the military budget is spent on defense, because the characterization that all of spending is for defense has led to budgets that are so large they insult our military leaders and soldiers by implying they are incompetent to the extent our defense requires spending more than 20 times as much per citizen as Russia or China. If the military maintains current levels of spending, our national security will continue to be harmed, because current debt and spending will result in us not having access to financing when facing future threats, while such spending demonstrates an astonishing failure of leadership in communicating and collaborating with foreign partners to build alliances that share burden, managing defense contractors, and discouraging overzealous military allocations by elected officials, some of whom are compromised by defense contractor campaign donations. This spending is also directly contrary to the founder's intentions, as America was specifically not to have a standing military. When James Madison addressed the Constitutional Convention in Philadelphia, he stated, "A standing military force, with an overgrown Executive will not long be safe companions to liberty... The means of defense against foreign danger, have been always the instruments of tyranny at home…" George Washington observed, "Over grown military establishments are under any form of government inauspicious to liberty." Given current spending severely weakens our financial ability to face future threats, while being directly opposed to the founders' intentions, it is unpatriotic, and so preserving, protecting, and defending the spirit of the constitution, requires a federal military drawdown. In order to do that without compromising security, the military must make use of the most powerful and underutilized weapon in its arsenal, international communication and relationship building, especially with perceived adversaries, at all but the lowest levels, enabled by an ample supply of translators, while working towards a global drawdown of all significant militaries, to create a world that is ultimately free from standing armies and weapons of war. However, the actions required to build a peaceful world, and enable a military draw down, are criminal according to a plain reading of the Logan Act, stating, "Any citizen of the United States, wherever he may be, who, without authority of the United States, directly or indirectly commences or carries on any correspondence with any foreign government or any officer or agent thereof, with intent to influence the measures or conduct of foreign government or of any officer or agent thereof, in relation to any disputes or controversies with the United States… shall be fined under this title or imprisoned…", a law which is itself illegal under the first amendment, and why there has never been a prosecution, as well as DOD Directive 1344.10 section 4.3.3.3, which allows for the prosecution of employees who "publish or allow to be published, partisan political articles, literature, or documents that they have signed, written, or approved that solicit votes for or against any… issue", since a stance on any issue could be considered partisan and indirectly soliciting votes. The referenced act and directive are directly contrary to the interests and national security of the United States, with the act itself being illegal, and therefore America will continue to be harmed until each is respectively repealed and revoked.

On non-policy topics of concern: my character, credibility, and ability have been repeatedly and thoroughly documented as impeccable and beyond question; I'm not significantly embarrassed by a single thing in my past, I've never once done anything significantly wrong, nor would I, because I can always think of ways to operate within established constraints, and am therefore beyond the blackmail or manipulation, including by corrupt employees of domestic and foreign surveillance agencies – thought their goals generally seem to be undetectably changing outcomes – and if someone were stupid enough to try to harm or manipulate me, they would be requesting their own destruction; though I've always found the spouse children construct a false measure of shared values, I don't have a spouse because despite exceptional options, I haven't found an adequate compliment, and I don't have children because I'd need to find someone who won't dilute my genes or family history at a time when I have adequate resources available to commit; I meet or exceed the personality and appearance electability requirements, respectively demonstrated by means including being social chair of the top fraternity at my college, and after winning my campaign in college for freshman class Senator by a landslide, being disqualified, because some girls, without my knowledge, had plastered the campus with unapproved signs stating "vote for Jon because he's hot".

To review my falsely perceived shortcomings, I first note that the last two presidents were an Ivy corporate executive then an Ivy senator, both of whom, when looked at from a fact based perspective, deeply eroded Americas long term financial fundamentals. Some may question my not having been a U.S. Senator or Governor, or a C.E.O. of a multi billion dollar company, when most anyone of ambition and modest intellect can build and run a major company if they establish it upon the right idea at the right time, or become a Senator or Governor if commited to taking preceding jobs in sequence. Some may question my not having one of a few specific degrees, but America's most authentically intellectual President was Lincoln, an elementary school drop out, and most anyone can receive the most prestigiously perceived and advanced academic degrees with committed academic hoop jumping – though I don't want to dismiss committed hoop jumpers in a general context as they can make great employees – a process that screens out creative independent thinkers historically responsible for human progress, in favor of those who've prepared their admissions package with the aid of a counselor and are a geographic minority, economic minority, social minority, legacy, have a background of hardship, extreme parental success, military service, or immigration – not to diminish any of those backgrounds, they're just not meaningfully correlated to ability – and with the few potentially historically significant alumni who went to such universities attending when the applicant pool was reduced through the exclusion of women, and admission consisted of little more than registration. I write this recognizing I graduated from what is by some measures one of the world's top universities, but that was almost by chance, as in high school I would skip class as often as was allowed without being suspended, to go to the library to read advanced books on topics of interest and work on my projects, then check out books to read in detention, and never saw the standardized admissions test before the one time I took it, while some classmates spent years studying. The current leading qualifications for discussed positions have proven not only irrelevant but also damaging, and none demonstrates the ability to solve our collective challenges. Another perceived shortcoming may be that I wasn't in the military, however I did make an effort to join in high school, talking with Marine recruiters on more than one occasion, with the last one telling me I was only the second person he'd ever advised not to enlist, because I was too much of an independent thinker to take orders – I'd elsewhere cited my mother's concern and not being heterosexual, which were factors, but I didn't previously mention this thinking it would be perceived negatively, though at this point, it is an attribute that has proven to be immeasurably valuable. I was further concerned by the possibility something could happen to me and humanity could lose my future contributions, which I thought previously could be perceived as lacking humility, however given I would have been deployed in the Iraq war, and my contributions since that time, the concern has proven well founded.

My top qualifications are my unprecedented capacity and leadership ability, both enabled by an historically and contemporarily unmatched and harmonious union of intelligence, creativity, work ethic, courage, heart, and honesty. And of course humility. I list that last attribute as a joke. The overwhelming function of the executive is to collect information, derive insights, and make decisions, which are then passed down through the chief of staff, to an astonishing number of layers of employees. I have had some people in the past not believe in my ability because they lack the capacity to evaluate me, where the lower the quality of a person the more likely that has been the case. I have demonstrated historically unprecedented leadership ability, without formal authority, and while I've managed, it's clear intellectual leadership allows for infinitely more positive impact than any other qualification. A Secretary General has no direct authority over other world leaders, and a President has no direct authority over any of the hundreds of federal legislators, therefore, these positions can only be best filled by me.

Some may think my ideas are great, and someone else can now execute them, perhaps someone meeting the previously discredited qualifications, however I've never once in my life had others implement my ideas correctly, because an inability to come up with my ideas has always been the same as an inability to fully understand the nuance required to dynamically adjust them during implementation without trading away value, and so my ideas when enacted by others are invariably implemented improperly and with limited and sometimes adverse impact. Additionally, the endless people who want to be perceived as a potential President, who others might think could implement my ideas, are almost exclusively phonies, utilizing a team of writers, idea generators, and photographers, to put together work they present as their own.

This may all leave my political affiliation ambiguous. I have only felt a personal connection to three American Presidents, Kennedy, Teddy, and Lincoln, a Democrat and two Republicans. Some liberals will limit support for not having a hardship story and specific academic credentials, while some conservatives will limit support for my being gay and non-religious, with each falsely citing other reasons, to the detriment of all. However, I think there's a substantial overlap between those individuals, and another group whose support I won't get anyway, the 7% of American's who think chocolate milk comes from brown cows.

To those who would think of doing anything other than providing me comprehensive and exclusive public support without being asked, you won't get the leader you want, but you will get the leader you deserve. The problems of our time are too fundamental, and too broad, to be resolved by anyone else.

Additional Honors

I don't have the time to research and acknowledge every honor and award of value in the world, but in the introductory recipients list, have included the names of many. I will likely accept any award that brings attention to my inventions and work, and thus accelerates our transition towards a world, that is for us all, more peaceful, more prosperous, and more sustainable.

Attackers

This section is critical reading for anyone thinking of attacking me, required to be included as a result of past attacks resulting from my work on behalf of the well being of us all, and anticipated future attacks upon the release of this book which constitutes further work on behalf of the well being of us all. I preface it for supporters by noting to that I am in fact a naturally friendly, positive person, enjoy lifting up others, and regard conflict as a waste of time, as it takes my energy away from further contributions to humanity. However, as a result, I'm sometimes misread, as well as for not using the techniques of people who identify themselves as intellectually inadequate and are passive aggressive to try to compensate, including not blinking, not moving eyes or head, tongue running over canine teeth, jaw flexing, cracking joints, gesturing with a poker finger or fist, rubbing fingers in a fist, insisting on talking, standing within arm's length, a handshake involving squeezing, holding, pulling, or more than one hand, touching beyond a handshake or appropriate hug, or wanting to be referred to by a title – a since trickled down population suppression technique established in conjunction with the concept of divine rule – as well as being misread for presenting a calm controlled perhaps disinterested temperament as a result of studying over a decade and a half of self-defense methods including Jiu Jitsu, Kickboxing, Tae Kwon Do, and Aikido, cross training with the intention of participating in mixed martial arts competition cage fights, and knowing I have the ability to permanently eliminate a group of disturbed attackers before anyone could help them. I have an extensive track record showing every single time I'm attacked that it cost my attackers, in time if not immediately, regardless of how vast their resources, dramatically more than they hoped to gain. Everything described in this book is now inevitable, there is absolutely no way to return to the world that existed before this book, and anyone who works against me would be self-identifying as too stupid and evil to be in a position of responsibility. If you are the most powerful and feared individual in the world, or someone looking to damage the philanthropic commitments I've made to benefit us all, or just someone sick fraud looking to make a name for itself, if you commit to directly or indirectly attacking me, my significant inventions, licensing fees, or refuse described licensing payments, thereby damaging associated philanthropic commitments to humanity, even if you think your attacks are untraceable, I will ensure you're identified, and have powerful well connected law firms, working on commission, ideally who've previously won large judgments against you, ensure that as an enemy of humanity, you not only follow the same licensing terms as everyone else, but that you pay a crippling penalty, damages, and litigation fees, enforced through tax liens, and if applicable, other economic penalties, I'll have eroded your position within your organization until your removed, your family will likely disown you, if your organization doesn't quickly eliminate all support for you and undo any damage they'll be gone after I utilize my proven capacity to identify all revenue streams and provide untraceable strategic guidance and innovations to those who can do the most harm, and if you're sick enough to attack me, then you undoubtedly have an unseemly background, and I will make certain that truth is all that is ever remembered of you, with no level of philanthropy ever able to cover it up, until one day after reflecting on how it all went wrong you realize you have no reason to live. I also note that if you're a supporter in such an attacking organization, don't resign in protest, but instead work either to ensure support for me, or to neutralize such anti-humanity operatives from within. I anticipate some sick frauds may try to pick apart my patents upon release of this book and make similar filings, while claiming no knowledge of me or my work – particularly those organizations and individuals who have made substantial and or failed investments in research and development, or who have made their fortunes through theft, fraud, and attacks, likely while using philanthropy to cover it up. Others may directly or indirectly attack – potentially damaging my philanthropic commitments – as a result of being malicious and mentally ill. Others must bring every resource to bear in exposing and revoking the capacity of such enemies of America and humanity. I must additionally note, I've made arrangements, occurring in the event terrorists make it impossible for me to continue, to ensure everything proceeds as I have described, and additionally have utilized my proven capacity to ensure those anti-America anti-humanity terrorists responsible, regardless of geographic location, power, or protection, including an army of bodyguards, armored vehicles, food testers, and never being visible in public, be identified utilizing all means necessary, and that they, and anyone immediately connected to them, as allowed under the laws of the United States, are processed to completion – I will not however go so far as the current President has in stating what should happen to terrorists – thus ensuring my previously explained philanthropic commitments continue. It's best for all to support me, and to ensure things are done exclusively and from the beginning in the exceptionally fair and generous manner I have provided that benefits all.

Addressing anyone who might at any point make a claim against any component of any one of my inventions, I came up with absolutely 100% of each invention 100% independently, and if anyone genuinely thinks her or she had something similar to any one of my inventions, or any component of one of my inventions, then you should have patented whatever was similar, because given the entire world has a first to file patent system, any dated proof of possession of invention or component is completely irrelevant, and you've legally irrecoverably lost all rights whatsoever to earnings or credit. However, if it's any comfort, the vast majority of earnings will ultimately go to causes that benefit humanity, as detailed in the philanthropy chapter, and I can say from experience, having long ago created, and failed to patent, inventions including the first social network with a news feed, which I dropped after repeatedly being told no one would ever want a profile on the Internet, and artificial intelligence memory structures, likely infringed upon by machine learning powered search results, that if you're smart enough to come up with anything close to any of my significant inventions, if you keep trying to come up with new inventions you will, and can be the beneficiary, if you comprehensively record how to create the invention, put it in patent format, have it finalized by an experienced patent lawyer, and are the first to file. You can also profit from my inventions by selling licenses, which would likely allow you to make far more in commissions than you ever could have independent of me.

Should anyone think their existing patent related to any component of any of my inventions is so indisputably infringed that they could ever win any sort of judgment, strategically it would be idiotic to do so until the inventions have earned all revenue possible during the 20 year term of the patents, since if you make a claim now or when the patents expire, the small percent penalty received will be the same, but the dollar amount and any credit will be substantially reduced by discouraging my promotion and monetization of that invention. However, even if you do have a potentially legitimate claim, if fully informed, you wouldn't think of pursuing it.

I had multiple individuals present my last public release of intellectual property, *The Destiny of Humanity*, as their own in the media, profiting from it, and in more than one instance, attacking me to cover it up, and so to reduce the possibility that any part of my work will again be fraudulently claimed by others, I've expanded the distribution list for this work, and I will now specifically explain some past thefts and attacks, which have overwhelmingly been covert and indirect acts of sabotage, as I want others to be aware of these types of individuals and their tactics. I omit names because the actions may be repeated by others in the future and are more important than individual past actors, I want this text to be timeless to the extent possible, these individuals will be overwhelmingly forgotten in a few decades, and I want focus to remain on my inventions without references to these individuals dirtying my book more than necessary.

An American President must unfortunately be discussed, whose honest background must first be established, and the reason then explained. This President, whose first campaign I'm now embarrassed to have spent a half day working on, is often thought well of as a result of having controlled historically unprecedented tax payer provided publicity spending of $1.4 billion per year, in part used to fraudulently sell people that my ideas and accomplishments, which make up most of his purported positive legacy, were instead his; failed to sign any significant legislation I didn't push through, regardless of how much he cared, including on immigration and guns; came to power and received the Nobel Peace Prize for principally, as documented in one analyst's book, presenting two ghostwritten books as his own – the first by an established author married to a former law firm colleague using a box of family history documents, and the second compiled from the work of his speechwriters, and apparently partially copied to a yellow legal pad for perceived authorship (I couldn't bring myself to research whether or not he personally authored his picture book); oversaw a near 50% increase in the number of people on food stamps; presided over decades high unemployment (labor force participation rate, as opposed to the purported unemployment rate, which only includes those who've looked for a job in the last six months, and improves as job seekers give up) while claiming job gains in absolute rather than relative numbers because purported gains were significant losses when accounting for population increases and cyclical gains, achieved the 4th lowest economic growth rate (inflation adjusted gross domestic product) of any American President ever, despite interest rates held at emergency levels, and overseeing more spending than all 43 other American Presidents combined, adding about $33,300 in debt to every single citizen (($19.9 trillion - $10.6 trillion)/310 million citizens), including through the purchase of a perceived international legacy by instead of having allies pay for their defense, spending $5.5 trillion over the 8 years to cover 72% of the expenses of the 28 nation NATO alliance, therefore structurally requiring generations of Americans pay higher taxes while receiving lower benefits; appointed people who would hold interest rates at 0% for an unprecedented period of time, which transfers money out of the bond market to artificially inflate it the stock market, and in conjunction with doubling of the debt, structurally arranged an eventual substantial market drop in which many will lose retirement savings; oversaw the lowest household income in over two decades (CPI adjusted real median income, which is inflation adjusted accounting for purchasing power relative to income); delivered a plane load of cash contingent upon the release of hostages, despite threatening parents with jail for terrorism financing if they paid their since beheaded son's ransom; along with his inner circle, personally oversaw a kill list and drone strikes, shown in leaked classified documents to have killed up to 90% civilians, a rate demonstrating intent to violate International Criminal Court War Crimes statute Article 8 section E1 prohibiting "Intentionally directing attacks against the civilian population as such or against individual civilians not taking direct part in hostilities"; supported importing less expensive skilled labor to displace highly paid American workers; elevated tensions around the globe, especially with Russia and China; repeatedly explained of the person who would have been the first female President "she'll say anything and change nothing"; was deemed by the Chair of a United States Congress oversight committee "the most corrupt president of modern times"; was revealed, at exceptional undue personal cost by Edward Snowden, as using surveillance to steal ideas and strategy from countless heads of state, with a particularly shameful example being the electronic theft of the United Nations Secretary General's meeting notes before their meeting; openly arranged similar work against members of Congress who opposed the

Iran nuclear deal; had an appointee and former staffer openly surveil a United States Senate oversight committee; authorized NSA to tap the leading search engine's email servers and use the leading social network to covertly install software on hundreds of thousands of people's computers to download their computers' files; authorized all domestic communications stored by the NSA be available in an unrestricted unauditable manner through software such as XKeyScore and Prism to employees in sixteen different federal agencies, with such communications in the past having been used, for example, to run a traffic stop for purchasing marijuana, and by the IRS to identify tax evaders, in each instance using "parallel construction", where a judge is lied to about the origin of the information, with the federal Privacy and Civil Liberties Board Chair telling the U.S. Senate agencies are "entitled to poke around", the FISA surveillance court having written "There is no question that the government is knowingly acquiring Internet transactions that contain wholly domestic communications", a government lawyer telling FISA surveillance court they can't document warrantless use of American's communications because "If we require our agents to write a full justification every time — think about if you wrote a full justification every time you used Google. Among other things, you would use Google a lot less", and Snowden stated "I, sitting at my desk, could wiretap anyone, from you or your accountant, to a federal judge or even the President", all despite the fact that there has never been a valid legal authorization to access any communication between an American citizen anywhere and any other party without a warrant, because in a constitutional government with equal branches, no Presidential order can be interpreted as changing the constitution's fourth amendment, informed by hard won wisdom, requiring an individual evidence based probably cause warrant, and no court can be legally setup where a person is never allowed the sixth amendment right to confront accusations; classified use of surveillance to prevent oversight by the constitutionally equal judicial branch; prosecuted more leakers and whistleblowers than every other President combined; supported the preparation and distribution of fraudulent character assassinations on opponents, including the one released for Edward Snowden, based on an invasive investigation going back to the time of his birth, concluding he is a "serial fabricator and exaggerator" who was "disgruntled... reprimanded [for] conflicts", despite turning up almost no questionable information, and being known for his even tempered honesty, and whose prosecution for the crime of revealing criminal activity is fundamentally contrary to the concept of law; attacked the current President in conjunction with the Deputy National Security Advisor, National Security Advisor, CIA Director, Chief of Staff, recent Senate Democratic Intelligence Chair, and Senate Democratic Leader, by monitoring the communications of his family for at least the year before office and putting the strategic campaign insights into spreadsheets distributed to dozens of high level allies, with one member of Congress who reviewed the surveillance stating "This is information about their everyday lives... from morning until night...", with those involved viciously insisting on the opposite of the truth even after it became verifiable; the current President explained of him "he tapp[ed] my phones during the very sacred election process… Bad (or sick) guy!" and "With all of the illegal acts that took place in [his] administration, there was never a special counsel appointed."; The British King conducted similar but lesser activities against politically active colonists, leading to the fourth amendment, and laws prohibiting this President's conduct including 18 U.S.C. § 595 which prohibits "a person employed in any administrative position by the United States [from using] his official authority for the purpose of interfering with, or affecting, the nomination or the election of any candidate for the office of President, Vice President, Presidential elector, Member of the Senate, Member of the House of Representatives", 18 U.S. Code § 2511 prohibiting "any person who... intentionally intercepts, endeavors to intercept, or procures any other person to intercept or endeavor to intercept, any wire, oral, or electronic communication...". I reference this President's conduct because after my essays in 2008 and 2009 were released that ultimately passed legislation in the United States for equality, healthcare, and financial reform, and I was identified by him as President in 2016, I was added to his extensive personal surveillance list. I was absolutely astonished when I became aware of this in 2010, when after hearing unique ideas I talked about in private communications repeatedly come out of his mouth shortly after I presented them privately, I made statements intended to provoke responses if he really did access transcripts of my communications, and multiple times received the reaction that could only occur when being monitored, which I identified by utilizing date restricted keyword searches relevant to my statements on Google, leading me to shockingly identify at that time, most of the same people now known to have conducted similar activities against the current President. I explained before the release of my last book, that it would do a historically unprecedented amount of good for the world, and when it was electronically filed with the U.S. Copyright Office on May 13th 2011, at a time when the media explained this President could lose re-election for being out of ideas, a copy of my book's text was pulled for him as soon as I received the filing receipt in my email, and he had the media publish my ideas and my corresponding results as his own. Such theft is additionally in violation of 18 U.S. Code § 1031 prohibiting a person who "knowingly executes, or attempts to execute, any scheme or artifice with the intent... to defraud the United States... to obtain money or property by means of false or fraudulent pretenses". I still sent him a properly printed copy of my book with a flattering note, ignoring his previous actions so that he would support my U.S. Senate campaign, along with two written follow-ups, and instead of receiving any correspondence, he made certain to prevent my public rise during my 2012 U.S. Senate campaign, by having his renowned plagiarist had a stroke half smile too stupid to know better Vice President – who thought corruption would allow him in 2016 to be the dumbest and oldest person ever elected President — put full early unnecessary White House fundraising and endorsements behind a disfavored incumbent, with support from the current senior U.S. Senator from New York, who at the time thought I might run against him because my family also had a New York address. One U.S. Senator recently commenting on this President's use of the NSA stated "The government can use overwhelming force and power to engage in political espionage". There's absolutely nothing more cowardly than attacking in a group, and to do so using the resources of the most powerful offices in human history, is cowardly and disturbed beyond comprehension. I must note that I could never have imagined such conduct was possible, and that I've never made an even slightly similar claim against anyone ever. Unfortunately, for those who request it, the only indisputable proof that could ever be available would be signed confessions from him and his inner circle – a few of whom, tragically for us all, are still in place – since the systems providing communications access are specifically designed for untraceable unauditable use. If anyone ever claims that there could ever have ever been, or could ever be, a legitimate reason to surveil me, that person is an astonishingly mentally ill fraud, and such a reason would have been manufactured as part of the attack as well as for liability coverage during surveillance. However, I was never able to find any indication that any attempt was made to create the perception of legitimate surveillance. There was no limit to elaborate nature of fabrications under this President, who is publicly known to have authorized taking Russian hacking software, modifying it, and using it for hacks to blame on Russians. It is also possible that this President, given his feminine nature, the way he bites his lip, and frolicked with his male personal assistant, may have been running LoveInt on me, an NSA term for employees who read their love interests communications, and his primary interest was in taking my communications up to the residence for some private time. As a penalty for these activities against America and humanity, this President is being provided a lifetime of around a million dollars a year in taxpayer provided income, staff, office space, and protection. If this President doesn't publicly acknowledge what I've stated to be true, or allows any of what I've written to be disputed, he should have his original nose put back on with a Pinocchio extension. These activities could have resulted in me providing anonymous strategic guidance to the current President during his campaign, the value of which could be derived from the content of this book. "It is for us the living," Lincoln said during his Gettysburg address, "to be dedicated here to the unfinished work which those who have died here have thus far so nobly advanced… that a government of the people, by the people, for the people, shall not perish from this Earth."

Of lesser note, I sent many corporate executives my last book, along with my earlier book on business, in early 2012, to add endorsements to my first book, and a second round of endorsements to my last book, and while many have since recited portions of my work in the media, none have acknowledged me, creating the perception that the brilliance, innovations, creativity, courage, inspiration, language, spirit of public service, and sense of humor that brought them praise in the media, was actually their own. A search engine chairman went so far as to immediately have third rate copies of each of my books written, making large investments according to my book, in areas including artificial intelligence, Wi-Fi phone service, and planetary mining, and meeting with several world leaders about the books, actions I suppose driven by the fact that the patents responsible for nearly all of his company's revenue are expiring next year. In another instance, a CEO who got his job after offering the company's dying founder half his liver, and had been too timid to come out as gay his entire career, shortly after receiving my books, decided to utilize them to suddenly present himself as a civil rights and thought leader, I suppose to compensate for his failure in running a company that has unable to produce an innovation and whose products continuously degrade in quality. I'm very happy to have anyone promote my work, as long as I'm acknowledged as the source.

In other instances, I've had people create fake identities, including one purporting to be me, that they've used for negative postings to social media, news article comment sections, and contacting and discouraging supporters, including such activities by an unhinged religious bigot, and in one recurring instance, by a goldmining aspiring John Hinkley Jr. who tries to purchase being a fifth rate clone of me that additionally used an anonymous social media profile with a profile picture of a blown off head to quote me a communication of interest made long ago to his too stupid to revoke his resources now husband. I've wondered, given the level of stalking I've received, if some of my attackers secretly harbor obsessive crushes. If you see or receive any communications, postings, or articles, that encourage you to do anything other than support me, it's exceptionally unlikely they're real.

The past indicates there will be mentally ill frauds salivating over the idea of stealing from me and or attacking me. It's a waste of my gifts at the expense of humanity to have to deal myself with those working against me, and so others must utilize every resource necessary to provide their full support and neutralize attackers by all legal means necessary. Anyone who worked or works against me, or continues to associate with such people, is self-identifying as of absolutely toxic character. I know I have far more untapped value to provide the world, but that requires everyone bring all necessary resources in support of me.

To those who have or would think of doing anything other than supporting me, please recognize that the world will be far better off without your participation and take action accordingly.

INVESTORS

In compliance with public solicitation regulations, the distribution of this book may not at any time be inferred to be a solicitation or offering. If any offering were to occur, formal offering documents are to be relied upon exclusively. Limited investment opportunities may or may not be available, for inventions as disclosed in this book, including: the energy and motor systems; services based on the corresponding disclosures in this book, providing a website platform, a content monetization platform allowing demographic targeting of advertisements to unregistered readers, and an early stage health records collaboration platform; an algorithmic trading hedge fund based on the corresponding disclosure; and later on, books and music. Reading and understanding this text allows for the formation of insights necessary to have an informed opinion. If you want to know what I look or sound like, you can find pictures, interviews, and videos online, including on my website, with further personal information available in my last book that was endorsed by world leaders. Standard legal agreements should be expected to be provided by me, and you can of course personally suggest changes, but I have little interest in correcting terms prepared by others' lawyers who are financially incentivized with billable hours to create the perception of loyalty and protection by trying to subtly slip in investor advantages, which will slow or prevent access to these limited opportunities that I expect will make a few extremely rich. While I am always open to reading thoughts presented by others, I've rarely had my thinking anything but diminished by others' thinking, regardless of intent or capacity, it was my recognition of this that made my inventions possible, and I expect it almost impossible for anyone else to adequately understand the required physics, electrical engineering, mechanical engineering, and global patent law required to have a fully informed opinion, therefore in order to protect all investors, it doesn't make sense to ever have others thinking formalized into a position providing any authority over my work, though I will provide adequate transparency, and if you make an adequately substantial investment and offer enough value and require a title, I don't see any problem with a seat on a board of advisors. If you're interested in investing, and want to find out if I'm interested in raising capital, please contact me using the information provided on the operating company website listed on the cover page.

Prior restrictions on public solicitation and sale of unregistered private securities to existing networks of wealthy investors have reinforced wealth inequality that harms economic fundamentals, discriminates against women and minorities with restricted access to such networks, and reduces entrepreneurs' access to capital allowing investors to ask for more than common non-voting equity which logically makes a company more likely to lose its founder and fail, so supporting and encouraging public solicitation is critical to progress and prosperity. If I do solicit or accept investment, while recent legislation makes public solicitation permissible, it is subject to regulations and compliance adequately burdensome that it took the Securities Exchange Commission four years to finalize such rules after ordered to do so by Congress, with the rules appearing designed to ensure accidental non-compliance, with some entrepreneur and investor unfriendly states illegally trying to override federal law allowing for public solicitation, therefore the terms of the offering, and investor verification and management, may be provided through a third party. Public solicitation laws include: regulation CF, allowing for crowdfunding up to $1 million USD; the more burdensome regulation A+, allowing for crowdfunding up to $50 million USD; and regulation D 506(c) allowing for any amount to be raised from accredited investors, but with some states – in opposition to entrepreneurs and investors as well as in support of previously cited wealth inequality and discrimination – blocking the purchase of securities by investors in their state if the offering wasn't registered prior to public solicitation, among other burdens. All applicable laws will be followed at all times.

To disclose risk factors, in my assessment, there may be four primary risk factors related to the operating and intellectual property holding companies. The first primary risk factor is the validity and enforceability of the patents and thus licensing fees, and to resolve that concern, I read countless patents filed by top lawyers, and read countless patent related court cases, in order to identify every way that a patent has previously been invalidated or diminished, and worked backwards in finalizing my own. I assessed that in the cases I read, invalidation or lack of enforceability have occurred very roughly 40% of the time from prior art anticipating the invention with anticipation not being possible for the world's first self-powered motors and generators, 40% of the time from claims not being supported by the disclosure usually as a result of the patent lawyer filing the patent not understanding the inventions as well as the inventor, 10% of the time from the disclosure or claims not covering all alternative implementations as a result of a lack of creativity and or time commitment by the inventor, 5% of the time from invention not being described in sufficient detail to be constructed, 5% of the time from claims not ensuring infringement by a single entity. In consideration of these factors, I worked myself beyond exhaustion for a long time revising my patents until I considered them stronger than any that I've ever read, and impervious to endless attack with unlimited resources. The second primary risk factor is if I don't maintain absolute control in all significant decisions related to the patents, only for the reason that to secure the patents over their lifetime in each country requires mastery to the extent humanly possible at the intersection of physics, electrical engineering, mechanical engineering, and patent law in over 150 economically significant countries and in ten languages, and even if someone else could be found with this capacity, that person wouldn't have my commitment. The third primary risk factor is that none of the inventions functions or provides benefits as described, and while I don't have prototypes of each to provide validation, in my mind I'm able construct and watch run exceptionally complex systems, while adding, removing, and modifying components, to achieve optimal performance, and I'm as certain as I am that I exist that all of my inventions work, but there is no need to rely on my assessment, as I've provided 100% disclosure, and each investor should make his or her own assessment of whether or not at least one of the inventions works. The fourth primary risk factor, is that funding is not available at required times, in required amounts, on acceptable terms, and to minimize this risk, companies generally solicit multiple competing financing offers in parallel.

As for due diligence, there is no active due diligence participation required by me, because the companies have no revenues, expenses, or debts, and I have no meaningful professional references to check given I've worked about half of the time for myself and the other half of the time for people who never adequately understood either or both me or my work, however meaningful due diligence can be performed by reviewing my 100% disclosure of the inventions and potentially reading my life story as disclosed in my last book.

An intellectual property based startup is very different from a traditional startup, because large fixed fees are due on unchangeable deadlines set by international patent offices, and so if an investor is unable to appreciate that, given the circumstances, I have to simply move on. Any opportunity to invest could be forever gone at any moment, especially given that the businesses will logically become entirely self-funding after orders are taken using commercial prorotypes, and if I were you, I would immediately contact me to ask if I were accepting investment, and if so, on the same day any opportunity were presented, electronically sign the standard investment document provided, and wire money. This is not a pressure tactic, but a simple reality created by market dynamics and the laws put in place around the international patent system. The investors in the operating company who contribute the most value, and or that I like the most, will likely be able to participate in intellectual property holding company, particularly if previous assistance was provided when and as needed, but the intellectual property holding company is not expected to take investment until after commercial prototypes have been produced and offered for sale. If you think any valuation provided is too high, but you think the company could be a top technology company, I remind you that it doesn't matter if you wildly overpaid to get shares in many top technology companies before they went public, since you still could have made around ten to over one thousand times return if you held on to and now sold you shares.

I will also note that there may be someone in your organization who lobbies against me, perhaps maliciously and obsessively so, for one or both of two reasons. First, as a result of my being gay – Fortune 500 CEOs include 32 females, 11 Latinos, 10 Asians, 3 blacks, and only 1 gay who offered half his liver to the dying founder before getting the job, so it's important to remember that despite progress, gay, lesbian, bisexual, and transgender people remain the most professionally discriminated against group in America, where professional discrimination presents in the form of going after someone for something else, and if there is nothing else, creating something, which in a professional context generally gets cited as conflicts, performance problems, or unrelated issues. Second, as a result of that person identifying as having little value, and wanting to feel important, by attacking someone that person thinks is incredibly important. Any person who coordinates thinking against me is toxic to you and your organization, who if you don't fire immediately, aside from deceiving you into damaging yourself, demonstrates toxic character that may ultimately destroy you and or your organization. If you're a limited partner in an investment fund, and don't see that fund reaching out to offer me terms I find acceptable, it is in your interest to eliminate allocations and exposure to those funds.

Any person who wants to be certain to impress me, and obtain my interest, can do so by reading in full one or more of my motor and generator patents, and demonstrating a thorough understanding of each paragraph.

If you would like to know more about how I will build my team, you're welcome to read the chapter on employment opportunities. I've been identifying people to hire since I was a teenager, as well as more recent exceptional candidates I've identified who are relevant to the energy and motor systems, and who will fill in any gaps in my experience. While I've so far only managed within and built organizations of limited size, this is only a result of not having both adequate opportunity and resources to build my own large scale organization, and I have demonstrated unprecedented leadership ability – as detailed in the chapter with the accomplishments section — which is of course infinitely more important than having previously run a large organization. I've been exceptional at recruiting at scale, as demonstrated when I was effectively responsible for my fraternity's recruiting, and it was the only time every single one of our dozens of offers was accepted.

If you're hesitant to quickly put money in on my terms, including any significant question about me, whether or not I can hire and retain the best possible team to build an organization that maximizes revenues and valuation, or you need to see competing offers to close a transaction because you don't trust your judgment, such limited thinking demonstrates a lack of capacity that will prevent you from having a basic understanding of my work. If the roles were reversed, I had money I was looking to invest, and I saw this work, I would drop everything in my life to make an assessment, which would result in me immediately investing as much as I could afford on any terms provided.

If you're interested in positive impact, this may be the only investment opportunity ever where you're very literally saving the world.

Intellectual Property Holding Company

My operating and intellectual property holding companies are logically the best investments available going forward in the areas of energy and motor systems. I have upcoming patent issuance fees due to each of around 150 countries, which I've read can end up costing around $1 million per patent, which if filed for up to 18 inventions totals up to an estimated $18 million, and additionally need funds to speed adoption, support production, and provide for a global legal war chest, for an ideal total of a billion dollars. Therefore, it would be logical to sell to an investor, who has the immediate liquidity and intent to immediately transact, a small piece of the intellectual property holding company, at at least a thousand times discount to the valuations established in the valuation chapter, to cover expenses. Investors are effectively in control of the success of their investment, because there are so many ways they can help increase licensing revenue and sales. I must point out that I worked on Wall Street, and I am aware of the financing scams where a claim is secured on a company, and then financial distress is created to seize the assets at a discount, and I will under absolutely no circumstances, allow a valid legal claim on the assets of the intellectual property holding company. I note that any prospective investors would dramatically improve their position with me, if the relationship begins by sending me a large no conditions attached grant check, to the address listed on the website for the operating company, thereby demonstrating your willingness to write a check, financial capacity, and exceptional genuine support for my work. Your support will benefit every future generation that inhabits this Earth, and I never forget people who support me.

I may be willing to sell the rights to several non energy and motor system pending patents – not included those I am releasing for free as disclosed in the philanthropy chapter — which I could never find the time to properly produce, and would personally like to immediately purchase units of including: the music format which could be worth over a billion USD, if licensed at 5% of each sales price of a song, with the largest digital music store having sold 25 billion songs; bed environment insulation which I expect with the proper retail network would generate at least hundreds of millions in USD of sales; and the air and water distribution system.

Energy & Motor System Production & Licensing Management Company

I may be open to revenue secured debt financing with no recourse to assets, or the sale of shares, to assist with production and sales of the energy systems, in the event that initial orders, grants, or purchase offers are inadequate. The goal of the operating company is to be the leading provider of my self-powered generator systems, has the primary competitive advantage of being run by me, will be to market ahead of my licensees as a result of having a minimum of a two year exclusive license, and also will receive compensation for promoting and managing the licensing of the intellectual property holding company, which of course is an arrangement absolutely required to be cancelable, or it would allow a claim on my intellectual property holding company. I don't intend to produce the motor systems, since self-powered motors are only meaningfully relevant in transportation vehicles, and a core component of transportation system manufacturers' business model is the custom specifications of their engines, and I expect all will be manufacturing their own self-powered motors. In order to be fair to all licensees, the licensing terms provided to the operating company are identical to those that are provided to competitors, with the exception of a limited time exclusive license for the generator systems. Additionally, I may be open to an individual venture capitalist or investment banker, who has an exceptional verifiable relevant personal track record, leading a private capital raise.

PageRock

PageRock is the world's best architected service for creating fully featured websites without technical knowledge. This service, and its predecessors, have been a long standing interest, given I started my first website hosting company in middle school. It implements a software application development system I invented as disclosed in the corresponding section, that significantly removes the need for programmers during web, mobile, and desktop application development, thus having broad implications for the software industry beyond website creation and hosting, and given it is the most robust and easy to use, it could ultimately support a company a valuation of a couple hundred billion USD. PageRock also owns the pending patent for the advertising system I invented which allows publishers to offer advertisers demographic targeting of unregistered audience members, providing advertisers a better alternative than search engines and social media, which is a market already providing tens of billions a year in revenue, and may be worth a couple hundred billion when accounting for the companies that make their money almost exclusively from targeted online advertising are worth around a trillion dollars, and this takes back a large chunk of that revenue for content producers, while providing possibly the only way for newspapers to meaningfully survive, thus allowing for a high royalty rate. There are large well funded competitors in the website development area, which often use too many chefs to make a soup, and none were able to use my capacity to develop a sophisticated and harmonious technological architecture, which provides unprecedented ease of use and flexibility to clients. PageRock currently scales to any number of users, so all that's needed is money to purchase advertising to bring in new clients. I may be interested in revenue secured debt with no asset recourse or very favorable equity terms. PageRock isn't a distraction from my other activities since I've already built the service to be automated and to automatically scale.

Automated Algorithmic Trading Investment Fund

I have an automated algorithmic trading hedge fund that I created from scratch, utilizing the corresponding invention disclosed in the additional inventions section. I however had to remove all funds to pay for this work, so it is temporarily inactive. The legal structure is that same as all investment funds, with two legal entities, an LLC that manages an LP, with the trading account is held in the name of the LP. Trades are selected then executed by connecting my computer system to the broker's computer system. An extremely simplified explanation of the system, is that it evaluates asset performance histories and makes selections. It's not high frequency, manipulative, or disruptive in any way. I understand investors will be curious to know the specific details of my algorithm, but like many or all trading algorithms, the algorithm can't be wholly disclosed without potentially making it less effective, and disclosure might not even be wholly accurate since adjustments may be made over time. My father has been an advisor to exceptionally successful institutional investment managers for about four decades, and my thinking in developing the system was significantly informed by what I learned listening to him.

However, I may never open the fund to investors because independent investment funds are most aggressively policed, often without little or no cause, because such investors have limited defense resources, and make for easy headlines that distract attention from regulators failing to fulfill their obligation to deter the misconduct of financial organizations of market significance. This is not to be misconstrued as my being against large financial institutions, as I have worked at more than one, with much of my family in the industry, and those are valuable when providing capital for socially important companies such as mine. Furthermore, regulators are actively opposed to algorithmic trading, seeking to legally require unrestricted access to all trading code, when even the NSA can't keep its information secure. It's appropriate to note at this time, as I was the only person in the world to predict the Flash Crash of May 2010, in my Copyright Office certified essay to Congress, *Legislation for the Creation and Protection of Real Wealth*, included in my last book, that the only code to regulate is at the exchanges, to ensure their code implements safeguards to block manipulative, disruptive, and dangerous cascade inducing trading.

My most recent live tests have provided what I consider very positive indicators, however I have next to no track record of the software working while live trading, since shortly after I completed fully functional code, and adjusted its performance based on observations when running live, I had to take my money out to continue working on these inventions. I make absolutely no direct or implied guarantee you won't lose money, including the possibility of losing all of it, and so you must have the ability to sustain a full loss of your funds in order for me to accept your investment. I also reserve the right to return your money at any time should the fund become a distraction or not meet my performance expectations. However, my assessment based on expected performance after live runs is that it will perform exceptionally well. I note that because this system doesn't absolutely require manual intervention, neither this nor my other projects take away from each other, and in fact may inform and enhance each other. To find out if I am open to investors, you can contact me using the information on the cover page.

Books and Music

I have written about half of an original music album and about half of each of two fiction books. I have strategically designed the album, and one of the fiction books, to be the top selling book and album ever produced, deriving their framework from an analysis of the highest quality and best selling classical and contemporary works. The album and the fiction book are both subtly educational, and the fiction book may resolve global conflicts permanently.

To get a sense of others reception of my writing, a few quotes from reviews of my last book include, King Norodom Sihamoni who wrote "I am confident that this book will surely attract public attention on the important task of building a peaceful and prosperous world for all." King Mohammed VI who wrote "I should take this opportunity to wish you success in all you set out to accomplish." Second Lady Jill Biden who wrote "It is the kindness of people like you that continue to renew my confidence about what we together can accomplish." and Prime Minister José Maria Pereira Neves who wrote my work is "A pointing of horizons and goals to which we must be aware... the quest for harmony and a blend of attitudes that could reach the heights of the global and total dignity of human beings."

To get a sense of my voice and musical ability, you can listen to *The Fallout of Love*, which was licensed for shows on MTV, VH1, and Discovery Networks, ranked in the top 1% against major label releases through TuneCore by 227 independent reviewers, and is available on my website as a free download. I have included independent reviewers' comments below.

"This song is spectacularly done. Instruments are perfect. The vocals are perfect. The lyrics are perfect. All adding up to a great sound."

"When he started singing I got chills. Such a beautiful song! The lyrics are just breath taking."

"As I was listening to the lyrics was thinking of my father that passed 2 years ago. With the lyrics he seemed to be talking about a love he as missing or someone special."

"I love the tone to his voice. It makes my heart melt."

"This song is pretty much perfect. The lyrics are brilliant and the vocals are outstanding."

"I really hope that this song gets played often on the radio, I think it will make people feel better."

If you put this song in a movie, which is available for free on my website in high definition format, I disclaim all rights to compensation, and you'll receive my appreciation, as I simply don't have the time to arrange such placement otherwise. I'd consider editing a song for a current top selling artist, to make it more articulate and timeless, in exchange for opening your shows for some period or similar compensation. I can only make you more articulate, and would never write another artist a full song.

I have written and published my previous books entirely independently, because I don't consider any book to have the possibility of being historically significant if it has many contributors, as almost all books do when released through large publishers, even when only one author is named, who in some cases, has barely contributed to the book, and because I have little regard for bestseller status as it is overwhelmingly meaningless given the most prestigious bestseller lists rankings only use a week of sales, so be a #1 bestseller, a person or publisher has to do little more than purchase a few thousand copies of his or her book over the course of a slow week. Modern publishing companies have created countless bestselling authors who are neither authors nor bestsellers. However, I did want my books to have a broader reach among the general population, and found that trying to replicate the infrastructure, relationships, and industry knowledge of a major publisher is prohibitively time consuming and expensive. Therefore, to achieve my previously stated goals, I am going to release my future music and books through major publishers. If you have a track record that shows you provide me the infrastructure and platform to make my works all time best sellers, please contact me.

MANUFACTURERS

If interested in being a manufacturer I use to fulfill orders, or that I recommend to others, please mail me at least one functional energy system, ideally one per invention, along with a price quote per unit with individual costs itemized, including shipping and relevant taxes. I realize this incurs a modest upfront investment on the part of the contract manufacturer, but the payoff is obviously potentially massive. The ideal contract manufacturer will have a strong record of high volume technically sophisticated manufacturing, strong financial resources and technical abilities, and the ability to ship units to individual purchasers. Manufacturers will be favored who minimize costs per unit of output, minimize size, and produce units that don't require service. I'd also greatly appreciate but don't require a letter explaining how easy it was to develop units using my book, which will not encourage me to try to reduce your production cost, but is simply for others to see external validation of my designs.

EMPLOYMENT

I have detailed below the roles that I expect will be needed to support the self-powered energy and motor systems. If potentially interested, it is important to read this entire book to understand the inventions, their impact, and the organization being built around the inventions. It would be additionally beneficial to know more about me personally by reading my last book. The attributes the have produced my successes, in descending order of importance, have been intelligence, creativity, work ethic, courage, honesty, and heart, and I generally value the same in others, though not all roles require such attributes, and some may require additional attributes. While I will always have final say on important decisions, every person involved must have a record of continuously contributing value independently with little supervision, as I can't have people involved taking energy or time away from me, waiting on me instead of contributing value, or providing me tasks. I also don't have time to invest in people who aren't loyal to me. If you're a metaphorical rock star, even if it seems I have no specific role for you, please send me your resume anyway. Resumes should be sent to the email address provided on the website specified on the cover page. Please ensure your resume clearly reflects which accomplishments are verifiably and directly attributable to you. You're welcome to include any additional materials that establish your value, including a work portfolio and recommendations, but for security reasons I don't open email attachments beyond PDFs on my mobile phone, so it's best to provide such materials through online links. You will be appropriately compensated, and there is absolutely no greater opportunity in this world to support technology critical to the future of humanity.

1. **Board of Advisors:** the Board of Advisors is expected to consist of former senior government officials such as heads of state, and energy secretaries, as well as top energy executives, each of which will receive compensation solely for revenue contribution. I don't feel a need to add names for prestige, credibility, or advice, only people who will personally and independently substantially expand licensing revenue and enforcement, but who may be too senior in their careers to reasonably be my employee, for example, a former President or King. Current government employees and their families are not legally eligible to receive compensation that could be perceived as related to influencing an official act.

2. **Financing lead**: the financing lead will have strong relevant experience, and work to ensure optimal financial resources, through means including ensure grant making organizations are fully supporting this work, and arranging and closing other financing deals at the lowest possible cost and on the best possible terms.

3. **Patent lead:** the patent lead will be responsible for managing the processing of 18 patents at the patent offices across the 152 countries that are signatories to the WIPO PCT, obtaining translation services and local counsel when necessary, and later maintaining all periodic payments and filings related to maintenance of rights. The ideal candidate will have extensive patent litigation experience and a background in electrical engineering and or physics.

4. **Operating lead:** the operating lead will be responsible for driving global adoption of the inventions among all relevant entities, including energy producers and auto manufacturers, by ensuring all potential licensees are aware of the value of the inventions, and managing the receipt and processing of orders, licensing agreements, and fees. It seems ideal for the operating lead to have an exceptional record of arranging billions of dollars in international patent licensing and sales to large institutions, including governments, perhaps in energy or high priced consumer electronics, as well as having demonstrated exceptional engineering skill, including a Ph.D., masters, or bachelors in engineering, or equivalent experience, and a having a law degree with patent expertise. The operating lead is expected to have a proven ability to build an organization including hiring, retaining, and leading a team of exceptional people achieving exceptional results. If you are currently in a similar executive role at a relevant company, having your organization license, produce, and fully embrace my inventions, would probably be the biggest advantage a candidate could have, as it demonstrates authentic interest and support. There may be additional openings for operating leads in other companies related to inventions other than the energy and motor systems.

5. **Operational strategists:** operational strategists will devise and execute specific detailed plans for ensuring licensing in each country by each relevant organization and the support of populations and their corresponding political leaders.

6. **Office manager:** I expect to hire an office manager to arrange for employees to be undistracted in their focus, including by arranging or providing home laundry, cleaning, grocery delivery, as well as office food delivery and stocking.

7. **Marketers and publicists:** This role is to maximize global awareness of the inventions superiority, by arranging news coverage, by means including identifying relevant individuals, providing information, and following up to establish a relationship and obtain favorable coverage.

8. **Independent sales people including lawyers and lobbyists:** Anyone may receive commission from the sales of units, as explained in the section detailing commission opportunities, as long as such commission is in compliance with applicable laws. I may also be interested in hiring formal sales representatives with extensive sales experience, ideally in selling large contract energy systems to governments and or licensing deals to large manufacturers. I expect a large number of opportunities in this area will be available until the units reach market saturation. Your job is to ensure large payments, from which you will receive your compensation.

9. **Prototype developers:** prototype developers will work to develop, using CAD software and 3D printing, fully optimized versions of the systems for each potential application, with a focus on maximizing durability, output, and aesthetics while minimizing cost, for applications including miniaturized versions embedded in portable consumer electronics. The ideal individuals will have an extensive engineering background, including a Ph.D., masters, or bachelors in engineering, or equivalent experience, with a track record of responsibility for producing similar successful mass market technical products.

10. **Manufacturing lead:** I will be directly taking orders and overseeing manufacturing of some units, and would like a manufacturing lead to manage and optimize the production with the contract manufacturer, and later delivery of units. This ideal individual will have demonstrated exceptional engineering skill, including a Ph.D., masters, or bachelors in engineering, or equivalent experience, as he or she will be responsible for ensuring each component, as well as the overall system, is optimized for price, durability, size, weight, and aesthetics. The ideal individual will have had personal responsibility for running billions of dollars in technology product manufacturing.

11. **Communications director:** I have always done all of my own writing. However, I expect there will be many communications that will need to be managed, and I don't have the time to do so myself. If interested in the position, please provide me as many writing samples as you have, where you were the exclusive author.

12. **Brand ambassadors:** one or more top celebrities working to promote inventions, preferably with an established credible relevant interest, and since there is no cost associated with doing so, first publicly endorsing my work for free to demonstrate authentic interest.

13. **Video producer:** build short videos capturing stories of customer production, sales, and impact.

14. **Security specialists:** if necessary, you will be enabling unimpeded operations to ensure success on behalf of humanity as described in the philanthropy chapter. For physical security, relevant credentials may include former Secret Service, Navy Seals, and SWAT, with the same level of background check required as conducted for Secret Service agents, and a pension that only becomes available much later, if there hasn't been a serious breach of security. For operational security, investigators relevant credentials would include a track record demonstrating the ability to identify and block anyone who would try to operate against America and humanity by working against me or my inventions.

15. **Lawyers:** in order to secure the inventions and corresponding philanthropic contributions, if you're a lawyer who passionately supports my work, and will legally brawl on behalf of my inventions and licensing fees, should there ever be an attempted theft, non-compliance, or challenge, in any country, please send me a note, in the unlikely event your assistance would be beneficial.

BOARD MEMBERSHIPS

If potentially interested in a position with my organization, please see the chapter on employment opportunities.

If you would like me on your board of directors or trustees, you're welcome to send me an offer, using the contact information provided on the cover page. I am most interested in joining boards of organizations with billions in existing revenue from energy and motor systems, and that have licensed my inventions.

EVENT INVITATION

I intend to have an annual event celebrating progress, showcasing invention beneficiaries, and any philanthropic distributions, accompanied by drinks and live music. I expect the event to always be in New York City near the United Nations headquarters, which is within walking distance of where I was born, and within driving distance for the majority of my family and friends. The event date and time is subject to change and must be confirmed on the website listed on the cover page, but it expected to be from 7pm to 10pm on the Saturday on or immediately after my birthday, June 21st. Therefore, the first event is expected to be in New York City on Saturday June 23rd, 2018. I had been the social chair of the top fraternity at my university and more than once was told by attendees they had the night of their life. If you would like to join, please send a short email to the address provided on the website listed on the cover page, indicating your interest, ensuring your identity is easily verifiable, and optionally providing online links containing information about you. Those invited will be added to the entry list and receive an email notification with the exact event location. I don't expect to ever send out invitations to anyone, regardless of how important, because doing so would be prohibitive given the size and international scope of such a guest list.

ADDITIONAL INVENTIONS

Each of the following inventions improves a large number of peoples' lives, and therefore I may be interested in licensing, selling, or receiving investment in these inventions. Some I am using, including the patents for advertisement targeting, automated algorithmic trading, software application creation, and electronic health records collaboration, and so may be only available for investment, as detailed in the chapter on investment opportunities, while others I couldn't possibly ever find the time to produce myself, and would be happy to license or possibly sell. I expect to maintain a single licensing agreement covering all inventions available for license, which will be available for download at a website listed on the cover page.

CONTENT MONETIZATION AND DEVELOPMENT

INVENTOR JONATHAN BANNON MAHER

TECHNICAL FIELD

[0001] Embodiments of the invention relate to the fields of content monetization and development.

ABSTRACT

[0002] Systems, methods, apparatuses, and computer programs encoded on a computer storage medium, allow for matching advertisements with content, with advertisement and content formats including text, image, audio, video, and animation, by means which may include expected demographic of content audience, content words, and visitor locations, and may provide the option to the advertiser to review a listing of the individual content that was matched for presentation of the advertisement, where the advertiser may disable presentation for selected content, as well as add presentation for additional content, and set bids on individual content, allowing content publishers to provide advertisers an unprecedented level of targeting, as well as more relevant content if producing content supporting desired advertiser targeting, all while providing the content audience more relevant advertisements.

REFERENCE TO RELATED DOCUMENTS

[0003] This application is provided the benefit and priority date of United States Patent and Trademark Office provisional patent application number 62/522,659, filed June 20th 2017 by inventor Jonathan Bannon Maher, which is incorporated herein in its entirety.

BACKGROUND

[0004] This section is intended to introduce the reader to various aspects of the art that may be related to various aspects of the present techniques, which are described and or claimed. This discussion is believed to be helpful in providing the reader with background information to facilitate a better understanding of the various aspects of the

present disclosure. Accordingly, it is understood that these statements are to be read in this light, and not a citation of prior any art.

[0005] There several current methods for targeting advertisements, including the overwhelmingly untargeted method of print and online news media, where a category is selected, the overwhelmingly untargeted method of television and radio networks, where a category or show is selected, those of search engines, which display advertisements relevant to the keywords of the search, and those of social networks, which allow advertisers to target advertisements to user profile demographics. Countless content producers, such as news organizations, have in recent years been driven to financial distress or bankruptcy, as a result of being unable to offer advertisers either the active targeting of search engines, or the passive demographic targeting of social networks.

BRIEF DESCRIPTION OF THE ILLUSTRATIONS

[0006] Illustrations are presented by way of example, and not by way of limitation, in the figures of the accompanying drawings, and embodiments may not contain all components, may contain additional components, and may contain functionally similar components.

[0007] FIG. 1 is an embodiment of an example of a screen allowing a user to sign in to the system.

[0008] FIG. 2 is an embodiment of an example of a screen providing account information with a link to add new content, a list of created content, a link to add a new advertisement, and a list of created advertisements.

[0009] FIG. 3 is an embodiment of an example of a screen allowing a user to submit content to the system and specify targeting factors including expected audience demographics.

[0010] FIG. 4 is an embodiment of an example of a screen displaying the form to submit an advertisement with targeting of content based on factors including expected audience demographics.

[0011] FIG. 5 is an embodiment of an example of a screen displaying a user created advertisement with aggregate performance statistics, specified targeting, and corresponding individual content along with a bid and performance statistics.

[0012] FIG. 6 is an embodiment of an example of a screen displaying content with an advertisement, including the ability for a logged in user to enable or disable display of their advertisements on the content.

[0013] FIG. 7 is an embodiment of an example of a screen displaying information including demographics the advertisers on the platform are currently targeting, and the amounts they are willing to pay to for content matching that targeting.

DETAILED DESCRIPTION

[0014] It is understood that, as in any engineering or design project, the development of any actual implementation will include numerous implementation specific decisions made to achieve the developers' specific goals, such as compliance with business related and system related constraints, which may vary from one implementation to another. It is understood that such a development effort might be complex and time consuming, but is nevertheless a routine undertaking of design, fabrication, and manufacture for those skilled in the art having the benefit of this disclosure. The disclosed steps may be read as prefaced by "In some embodiments, including one complete embodiment, ", may be executed or performed in other orders or sequences, and are not limited to the order and sequence shown and described, which are provided to enable ease in constructing an embodiment, and along with each components of each step, may be removed, modified, combined, or rearranged, and other steps and or step components may be added, without departing from the scope of this disclosure and or invention. Although embodiments of the invention have been described and illustrated in the disclosed implementations, it is understood that the present disclosed subject matter, including apparatuses, methods, specification, and illustrations, has been made only by way of example, not by way of limitation, and the methods and apparatuses may be used in other systems, and that numerous changes and optimizations in the details of implementation of the invention and or embodiment are made without such modifications departing from the spirit and scope of this disclosure and or embodiments of the invention. Although the disclosure has been shown and described with respect to one or more embodiments, features of the disclosed embodiments can be combined and rearranged in various ways, and changes including equivalent alterations, substitutions, modifications, and additional efficiencies will of course occur to someone of ordinary skill in the art without departing from the spirit and scope of this disclosure and or invention. In particular regard to the various functions performed by the described components, the terms used to describe such components are intended to correspond, unless otherwise indicated, to any component which performs the specified function of the described component, or is functionally equivalent to the described component, even though not structurally equivalent to the disclosed structure which performs the function in the implementations described in this disclosure. In addition, while a particular feature of the disclosure may have been provided with respect to only one of several embodiments, such feature may be combined with one or more other features of other embodiments as may be desired and advantageous for any given or particular application. In some instances, well-known circuits, structures and techniques have not been shown in detail in order not to obscure the understanding of this disclosure. Articles in this disclosure such as "a" "an" and "the" may allow for both singular and plural forms. Verbs in this disclosure such as "is" may be read as "may be". Conjunctions in this disclosure such as "or" as used herein may be interpreted as inclusive or meaning any one or any combination, where "A, B or C" means "any of the following: A; B; C; A and B; A and C; B and C; A, B and C". Relational terms in this disclosure, for example first and second, top and bottom, left and right, are to distinguish one entity or action from another, and may not necessarily require or imply a relationship, or order between, such entities or actions. The disclosure includes the best mode contemplated by the inventor, a completely described specific embodiment, along with optional components and alternative embodiments to best suit the implementer, measurements in imperial and metric units to support universal understanding, and dramatically exceeds claims support requirements and enablement requirements by allowing for selection and or construction of the required components to be carried out easily, quickly, and routinely by persons of ordinary skill in the art, who are provided the additional benefit of utilizing readily available commodity components whenever possible. The present disclosure includes material protected by copyrights, and the owner of the copyrights hereby reserves all rights, but with authorization for publication as required by government patent offices. Various embodiments of the present invention may provide all, some or none of the disclosed technical advantages.

[0015] The computer code descriptions disclosed, in order to provide comprehensive enabling disclosure, rather than utilizing flow charts, which according to Patent Cooperation Treaty 11.11a are prohibited from containing "text matter, except a single word or words, when absolutely indispensable, such as... a few short catchwords indispensable for understanding", are provided in a text only format where the number of arrows preceding a line indicate logical block level, semicolons indicate a new segment of a logical block, and periods indicate the closure of one or more logical blocks. It is understood that any computer code representations in this disclosure are merely illustrative, rather than restrictive. While code may be written in nearly any computer language, including Java and C++, the illustrative computer code descriptions were derived from code written the Python language, which may be run through the Python interpreter, with appropriate supportive libraries, which at the time of disclosure, may run on nearly any computer, for example one with an Intel or AMD processor, running a current version of Linux, Windows, or Mac OS. All code components may read as if prefaced by "In some embodiments, including one complete embodiment, ". In some embodiments, functionality may be modified, rearranged, excluded, and added. To provide more fundamental computer system details, in some embodiments, the functionality associated with the disclosed computer code descriptions may be referred to as a script, module, software, software application, or code, and can be written in any form of language, including compiled, interpreted, declarative, or procedural, able to be deployed in any form suitable for use in a computing environment, including as an independent or integrated program, module, component, or subroutine, for execution by the computer system, implemented on one or more independent or integrated computers, utilizing a central processing unit in the form of one or more general or special purpose microprocessors, in conjunction with digital electronic circuitry, which may include special purpose logic circuitry such as a field programmable gate array or application specific integrated circuit, with the computer controlled by and operatively coupled to tangibly embodied software and or firmware, which may include code that creates an environment for code execution, including individual or combined use of processor firmware, a protocol stack, a database management system, and an operating system, where such software and or firmware may exist in one or more parts in memory on one or more computers, and is encoded on one or more tangible non transitory software carriers, such as individual or combined use of a random or serial access device or substrate, a semiconductor memory device, transient or persistent random access memory, a magnetic, magnetic optical, or optical disk, or encoded on an artificially generated transmitted signal, for example, optical, electrical, or electromagnetic, transmitted using a sending and a receiving apparatus, where the interaction between the user and the software may be implemented by operatively coupling, to the local implementing computer, or a local computer connected to one or more remote computers through a local or wide area network, a display device which may implement fluid crystals or light emitting diodes, a keyboard, and a pointing device.

[0016] The inventor retains absolutely no liability for any implementation of this invention, and the invention is implemented exclusively at the risk and liability of the implementer.

[0017] In some embodiments, including one complete embodiment, unprecedented demographic advertisement targeting of an unregistered audience is provided, where content and advertisements may be in formats including text, image, audio, video, and animation, and where content is tagged with demographics of the expected audience, allowing advertisements to target a combination of demographics, content keywords, and visitor location, where when an advertisement is created, all available content is scanned for a match with the desired demographic criteria, matching content is selected for the advertisement, where the advertiser may for maximum control optionally disable presentation on individual and or groups of content, as well as add presentation to items or groups of content, and set bids on individual content and or content groups, therefore providing unprecedented targeting by offering both the active targeting of search engines, which provides advertisements to individuals actively seeking information, and the passive targeting of social networks, by targeting audience demographic, therefore offering unprecedented targeting exceeding that offered by search engines, social networks, news outlets, radio stations, and television networks, while additionally providing maximum transparency and verification of advertisement placement, because unlike social networks which for privacy reasons can't reasonably allow an advertiser to review everyone's profile who's been served an advertisement, there is no privacy issue with presenting the individual content where the advertisement has been presented and the number of times a presentation resulted in a click, and therefore publishers are provided increased advertising rates through unprecedented targeting, users more relevant and interesting advertisements, advertisers more relevant content if publishers produce content supporting desired advertiser targeting, and potentially providing a full reversal of adopting media outlets declining financial condition, and a restoration of the reporting resources necessary to the security of a free people.

[0018] The system is designed to allow for both a platform that allows content publishers and advertisers to sign up to have content matched with advertisements, as well as for content publishers to implement it independently for their own exclusive use directly with their advertisers.

[0019] **In some embodiments, including one complete embodiment, create data store structures to hold information.** In some embodiments, including one complete embodiment, data structures are created in a suitable data store, for example where the data store is a relational database management system and data store structures are tables, by performing actions comprising:
> create a data store structure for accounts, with fields including an auto incrementing integer primary key, a record creation timestamp, a username, and a password.

> create a data store structure for sessions, with fields including a record creation timestamp, a session ID, and an account ID.
> create a data store structure for publishers, with fields including an auto incrementing integer primary key, a record creation timestamp, an associated account ID, a name, a domain name, and a content index URL.
> create a data store structure for advertisements, with fields including an auto incrementing integer primary key, a record creation timestamp, an associated account ID, an advertisement title, advertisement text, an uploaded file, the uploaded file name, a link to advertisers content to send the a user to, content, category, sub category, demographic fields including minimum and maximum age, sexual orientations, ethnicities, interests, countries, whether or not the advertisement has been indexed against content, whether or not its active, the default bid price per click, and the default bid price per impression, the total budget, and the start and end dates of presentation.
> create a data store structure for content, with fields including an auto incrementing integer primary key, a record creation timestamp, an associated account ID, an associated publisher ID, the title, author, content, keyword tags, category, sub category, demographic fields including minimum and maximum age, gender, sexual orientations, ethnicities, interests, countries, a file, and a file name.
> create a data store structure for content words, with fields including an associated content ID, and the word.
> create a data store structure for content matched to advertisements, with fields including an associated advertisement ID, an associated content ID, whether or not advertisement presentation is enabled for the content, whether or not the advertiser individually selected the content, the bid per click, and the bid per presentation.
> create a data store structure for locations associated with IP addresses, with fields including the IP address, the associated latitude, longitude, country, state or province, and city.
> create a data store table for content presentations, with fields of the content ID, a record creation timestamp, and the IP address or other identifier for the individual audience member.
> create a data store structure for advertisement clicks with fields of a record creation timestamp, advertisement ID, content ID, and the IP address or other unique identifier of the visitor.

[0020] In some embodiments, when the system is integrated with an existing data store, adjustments may be made to existing data structures to support the disclosed data structures and functionality.

[0021] **In some embodiments, including one complete embodiment, create the software code to support the functionality of the system.** In some embodiments, including one complete embodiment, the advertisement is inserted into the content being presented to users, by means which may include serving the advertisement file to the publisher who inserts it into the stream being presented to users, where the server is designed to allow advertisements to be presented within content including text, image, audio, video, and animation, as well as in native desktop and mobile applications, where the method of communicating with the server, and integrating the advertisements, is customized to the computer language and or requirements of the application. In some embodiments, including one complete embodiment, functionality is provided to the user, including the ability to insert, select, update, and delete data for each of the data tables, through either the application server and user interface, or through the user interface of the data management software. In some embodiments, including one complete embodiment, during the presentation of content, its presentation is recorded, and an advertisement is selected for presentation from those matched to the content, as well as optionally retrieving the geographic location of the audience member and restricting the advertisement served to one of without location targeting or one where the audience member is within the bounds of the specified target location, recording the presentation of the advertisement, and if the advertisement is clickable and clicked, recording the click of the advertisement and redirecting the user to the advertiser's designated location.

[0022] In some embodiments, including one complete embodiment, in reference to FIG. 1, an example user interface is presented that allows the user to sign in or create an account by entering and email address 1000 and password 1001 then pressing the start button 1002.

[0023] In some embodiments, including one complete embodiment, in reference to FIG. 2, an example screen is presented having a title 2000, providing a link to add new content 2001, providing a list by title of all recorded content 2002, providing a link to a screen to add a new advertisement 2003, a list of advertisements linked by title 2004, and a link to sign out 2005.

[0024] In some embodiments, including one complete embodiment, in reference to FIG. 3, an example user interface is provided to add and edit content while specifying expected audience demographic information, where the screen provides the title of the organization offering this platform 3000, fields to input the content including title, contents, category, and a file upload field to allows the upload of a content file if applicable, which may be in formats including text, image, audio, video, and animation 3001, fields for content targeting by advertisers including tags and words 3002, fields to specify demographic targeting, by gender, minimum age, maximum age, sexual orientation, ethnicities, interests, book categories, music categories, and movie categories 3003, fields to specify visitor location targeting by country, state or province, and city 3004, and submit button 3005 which when pressed saves the content to the server to then be stored and matched with advertisements.

[0025] In some embodiments, including one complete embodiment, in reference to FIG. 4, an example user interface is provided to allow advertisers to create an advertisement and target it to specific content, providing fields to specify targeting of content by categories, tags, and words 4000, fields to specify demographic targeting by attributes including gender, minimum age, maximum age, sexual orientation, ethnicities, interests, book categories, music categories, and movie categories 4001, fields to specify visitor location targeting by country, state or province, city, or other means of location targeting including postal code 4002, fields to specify campaign start date, end date, and default bid per click, and bid per presentation, and optionally a budget 4003, a checkbox indicating whether or not to target individual content 4004, a file upload field allowing the upload of an advertisement file, which may be in formats including text, image, audio, video, and animation 4005, and a submit button 4006 which saves the advertisement to the server to be stored and matched with content so that the advertiser can enable and disable display on selected content, and so that the system can run the advertisement

[0026] In some embodiments, including one complete embodiment, in reference to FIG. 5, an example user interface presents a list of content selected for the advertisement, and to allow for advertisers to optionally target individual content, where when restricting the list of content presented to those targeted for the advertisement, may provide the advertisement title, targeting, actual and expected aggregate performance statistics including the visitors to the selected content in the last 30 days, the percent of the time the advertisement is expected to be presented based on competing bids, the average click through rate if applicable of the targeted content, and the total expected clicks 5000, a listing of individual content with performance statistics that has been targeted for the advertisement, including presentations and clicks, display of the bid amount with the ability to increase and decrease the individual bid, as well as a button or link for enabling and disabling presentation within the content 5001, and a link to all content in the system for advertisers to search through content and enable presentation of the advertisement within additional content 5002.

[0027] In some embodiments, including one complete embodiment, a screen presenting content in the data store is made available, where the display may include a list of all content titles with a link to the individual content.

[0028] In some embodiments, including one complete embodiment, in reference to FIG. 6, an example of the display of individual content is provided, where the display includes an advertisement 6000, the title of the content 6001, the body of the content 6002, a list of any advertisements by the signed in user with the option to enable or disable the presentation of each advertisement with that content 6003.

[0029] In some embodiments, including one complete embodiment, in reference to FIG. 7, an example is presented of a screen provided to an administrator, to see what demographics the advertisers on the platform are currently targeting, and the amounts they are willing to pay to for content matching that targeting, to provide real time feedback to content producers about what content is most profitable to produce, including a list of advertisers keywords in descending order of bid 7000, a list of advertisers keywords in descending order of count 7001, and a list of advertisers targeted demographics by percentage 7002.

[0030] In some embodiments, including one complete embodiment, the code to provide a user interface to the system, and respond to requests, provides functionality comprising:
> import a data management library and initialize a connection to a data store containing the previously disclosed data structures;
> create a function to handle all server get requests;
>> initialize a variable to hold the server response;
>> if the requested action is to start a new session, in reference to example FIG. 1, return a screen allowing for a user to create a session by means which may include submitting an email address, username, or phone number and password;
>> if the requested action is to display the users account, in reference to example FIG. 2, return a screen providing a list of the user's content, a link to a form to add content, a list of the user's advertisements, and a link to a form to add a new advertisement, and the option to sign out;
>> if the requested action is to add or edit content, in reference to example FIG. 3, generate the content submission form, and display each form field including those that allow for content entry and targeting, as provided for in the previously disclosed corresponding data structures, including age minimum, age maximum, gender, ethnicity, sexual orientation, interests, etc.;

>> if the requested action is to add or edit an advertisement, in reference to example FIG. 4, generate the advertisement submission form and display each form field including those that allow for content targeting, as provided for in the previously disclosed corresponding data structures, including age minimum, age maximum, gender, ethnicity, sexual orientation, interests, etc.;
>> if the request is to return to an advertisement to a logged in user that user created, in reference to example FIG. 5, return the specified targeting for the advertisement, retrieve through data store queries and calculations the specified aggregate performance statistics, including percent of time advertisement is expected to be presented based on competing advertisers bids, average click through rate, expected clicks, display a list of content where presentation has been either enabled or disabled along with the option to enable and disable display, and the bids, display statistics, and clicks for individual content with the option to increase and decrease the bids, and a link to a list of content which may be presented as a result of a search to select further content for presentation of the advertisement;
>> if the requested action is to display a list of content, display to the user a list of all user retrievable content by retrieving from the data store specified content fields including title;
>> if specific content was requested, in reference to example FIG. 6, parse from the request the content identifier, retrieve the content from the data store, display that content to the user, record the presentation of the content to content presentations table in the data store, and if the user is logged in, list any advertisements associated with the account, with a link to enable or disable the advertisement with the content, and record that the content was presented;
>> if the requested action is to display a list of advertiser targeting, in reference to example FIG. 7, return aggregate statistics including demographics the advertisers on the platform are currently targeting, and the amounts they are willing to pay to for content matching that targeting, including a list of advertisers' keywords in descending order of bid, a list of advertisers' keywords in descending order of count, and a list of advertisers targeted demographics by percentage.
>> if the requested action is to present an advertisement, based on a advertisement ID previously generated by a post to the server, retrieve the advertisement from the data store, record that the advertisement was presented to the data store table holding advertisement presentations, optionally including for billing purposes whether or not it was targeted to specified content and or the bid, and return the advertisement, including if applicable a header specifying the content type of the advertisement, which may include text, image, audio, video, or animation;

>> if the requested action is to redirect a user who has selected an advertisement, parse from the requesting URL the content ID and advertisement ID, then retrieve from the data store the advertisement's redirect location, record the visitor's IP address or unique identifier, to not bill more than one click per visitor, and return or redirect the visitor to the advertisers specified location;
> create a function to respond to all server post requests;
>> create variables holding the posted data including the posted action;
>> if the action is to start a session, check if the provided email address, username, or phone number is in the data store, and if not create an account, and if it is, verify the password is correct for the account with the provided email address, if a new account was created, or an existing account with a matching password was found, create and associate with the account a session ID and store it to the user interface or user interface software to include with requests;
>> if the post includes a session ID, retrieve the account ID associated with the session from the data store sessions table;
>> if the action is to submit content, record the users account ID, and the content including all submitted fields and files to the corresponding data store structure;
>> if the action is to submit an advertisement, record the users account ID, and the advertisement including all submitted fields and files to the corresponding data store structure, and insert the advertisement ID into the Advertisement Locations table;
>> if the request is to retrieve an appropriate advertisement, take the posted data, and check if the publisher ID and content ID exist in the data store, and if not parse the posted content, including targeting information, and save it to the data store so it can be associated with advertisements, then check if the specified content ID has specific advertisements targeted to it, and if so select one to provide, by means which may include randomly selecting among the highest bidding advertisements, or selecting a random advertisement, and return the advertisement or an identifier that can be used to retrieve the identified advertisement;
>> if the request is to enable or disable presentation of advertisement with content, record the association to the advertisement content data store structure;
>> if the request is to increase or decrease the bid of an advertisement associated with content, record to the bid change to the advertisement content data store structure;
>> if the request is to sign out, disassociate the session ID or expire the session from the user interface client;

[0031] In some embodiments, there may be a policy stating that advertisements found to be abusive in content or targeting will be disabled, for example, targeting content about a specific individual or organization with negative advertisements about that person or organization.

[0032] **In some embodiments, including one complete embodiment, add content along with demographics to the data store.** In some embodiments, including one complete embodiment, content is added and the content that is tagged, may be in any format, including text, image, audio, video, and animation. If the content is not text, a description of the content may be saved for targeting, such as a transcript of the audio. In some embodiments, including one complete embodiment, content with associated demographics is added to the data store in the corresponding table through the data store's included management interface, with the content demographics including but not limited to those in the previously disclosed data structure: expected minimum and maximum age, gender, location, sexual orientation, and content keywords, interest keywords, category, sub category. For example, it's reasonable to assume content discussing prostate cancer will be reviewed by older males interested in prostate cancer information, thus allowing it to be tagged with attributed of male and 40-80, with targeting of the words "prostate" and "cancer", thus allowing for prostate cancer drug manufacturers upon submitting targeting criteria to have their advertisement set to present with that content, who will pay immensely more than for simple category based targeting, and who may be able to use the platform to target multiple advertising mediums all at once, including online and broadcast. The demographic tagging of content, and advertiser selection and deselection of content, may in some cases seem somewhat tedious or time consuming, but it's not expensive, as it can be performed by any modestly intelligent person including an intern or contractor. Content producers and advertisers who have a problem with that level of individualized targeting have the option of continuing to utilize less effective and less profitable targeting.

[0033] In some embodiments, the demographic data is embedded in the content, and the content is posted to the server to retrieve a matched advertisement, through software providing functionality comprising:
> initialize a variable to hold the content being presented including related information such as demographic targeting;
> include the script that posts the content to the server then inserts a targeted advertisement;
> specify the demographic targeting criteria;
> create a place holder for the advertisement;
> identify the beginning of the content and present the content to the user;
> identify the end of the content;

> post the content and targeting information to the server, retrieve a matched advertisement, and insert it into the advertisement placeholder.

[0034] In some embodiments, a server side script continuously or periodically runs that retrieves the content from the publisher along with its demographics, where the script provides functionality comprising:
> import a data store access library and initialize a variable to hold the data store connection;
> run the script on a continuous loop;
>> establish a connection to the data store if none has been established or the connection was lost;
>> retrieve publishers content index addresses from the data store;
>> iterate through each publisher, retrieve the content index, and check each content ID to see if it exists in the data store, or if its update timestamp is newer than the content in the data store for that content ID, and if either condition is true, retrieve the content, parse it, and save it to the data store content table, including overwriting the record for that ID if it already exists;
>> pause program execution for a moment to not drain publisher server resources.

[0035] In some embodiments, demographic data embedded in content is sent to the server through client side code.

[0036] **In some embodiments, including one complete embodiment, write a script that continuously or periodically indexes any unindexed content words, and matches advertisements to content utilizing content demographics and advertisement specified targeting.** In some embodiments, including one complete embodiment, a script is executed to continuously run in the background, adding individual words appearing in the content or the content description to a data structure containing content IDs and words, so that content can be targeted where certain words appear, and additionally any advertisement has any specified demographics searched for in content tagged with those demographics, and if the content demographics and words match those specified in the advertisement, an entry is created in Content Advertisements data store structure linking the Advertisement ID with the Content ID, so that advertisement is presented with the matching content, with the software providing functionality comprising:
> import the data store library and initialize a connection to the data store containing the previously disclosed data structures;

> continuously run in the background to index all content and match advertisements to content;
> create variable to hold whether or not unindexed content records were found, and the index of the current record;
> loop while records are found;
>> select the record ID and the Content text from Content records that haven't been indexed;
>> if there aren't records, set records found to false to end the loop, but if records were found, proceed;
>>> create an array of each unique word in the content text, and insert each into a data store structure storing individual content words for targeting, with the Content ID and the word, then update the record as indexed;
>>> check if the content contains tags indicating the demographic content hasn't been parsed from the content, if such tags exist, parse the demographic data and update the content record;
>>> create variable to hold whether or not unindexed advertisement records were found, and the index of the current record;
>> match content to advertisements by retrieving unindexed advertisements, then retrieving the content that has demographics that match the advertisement, record each matching content ID, then if the advertisement includes word targeting, select the content IDs that contain that word, from the content word table, and recording any content IDs that match both content and word targeting to the advertisement content table.

[0037] When new advertisements are added to the system, an administrator can review the targeting saved with the advertisement in the data store, and search for and manually associate any content relevant to the requested targeting that hasn't been tagged with that targeting, in order to ensure maximum relevant reach.

[0038] **In some embodiments, including one complete embodiment, write a script to be included in the content to retrieve and insert a matching advertisement and optionally send the content with demographics to the server.** In some embodiments, including one complete embodiment, client side code such as JavaScript is embedded in the webpage, and allows various types of advertisements to be dynamically inserted, such as text, images, audio, video, or animation, and allows the content of the page to be posted to the server with embedded demographics in case the content hasn't already been indexed, and may prevent

advertisements from being presented and counted when crawled by a search engine for indexing purposes, providing functionality comprising:
> post the content to the server, and read the generated advertisement ID, where posting of the content allows applications to integrate an advertisement when content can't be otherwise accessed by the server.
> retrieve and insert into the advertisement placeholder specified in the display, the advertisement to display and its associated link, by utilizing the previously retrieved advertisement ID.

[0039] **In some embodiments, including one complete embodiment, offer unprecedented targeting to advertisers, and advertisement relevance to users.** In some embodiments, including one complete embodiment, once the system has been built, and the content providers have integrated the advertisements, advertisers can be offered unprecedented targeting and additional targeted content, publishers can increase advertising rates, audiences can see more relevant and interesting advertisements, and in the case of news organizations, recapture lost revenues to restore the reporting resources necessary to the security of a free people.

CLAIMS

What is claimed is:

1. A system for targeted advertising, with the invention comprising:
content and advertisements in formats including but not limited to text, image, audio, video, and animation;
associating actual or expected audience demographic information with said content;
associating targeted audience demographic information with said advertisement;
matching said advertisements to said content based on said demographic targeting of said advertisement and actual or expected demographics of said content audience.
2. Claim 1 further comprising:
providing the ability to enable or disable presentation of said advertisement on individual and or groups of said content.
3. Claim 1 further comprising:
providing the ability to set the bid of said advertisement presented with individual and or groups of said content.

4. A non-transitory computer-readable recording medium holding stored instructions, which when executed by one or more processing devices, cause the one or more processing devices to implement a method comprising:
storing contents and advertisements in formats including but not limited to text, image, audio, video, and animation;
associating actual or expected demographic data with said contents;
associating targeted demographic data with said advertisements;
matching said advertisements to said contents based on the demographic targeting of said advertisements and the actual or expected audience demographics of said contents.

5. Claim 4 further comprising:
providing the ability to enable or disable presentation of said advertisement with said content.

6. Claim 4 further comprising:
providing the ability to set the bid to present said advertisement with said individual and or groups of content.

7. A system for content development, with the invention comprising:
displaying current advertisers requested content targeting, individually and or in aggregate, to allow for development of profitable content.

8. A non-transitory computer-readable recording medium holding stored instructions, which when executed by one or more processing devices, cause the one or more processing devices to implement a method comprising:
displaying current advertisers requested content targeting, individually and or in aggregate, to allow for development of profitable content.

FIG. 1

1000 – **Account**

1001 – Email Address: [ ]
Password: [ ]

1002 – [Start]

FIG. 2

2000 – **Account**

2001 – Content [New]
2002 – Election Outcome Uncertain
September Polling Results

2003 – Advertisements [New]
2004 – Maher for America

2005 – [Sign out]

FIG. 3

3000 – **Content**

3001 – Content
Category: [ ]
Title: [ ]
Description: [ ]
Text: [ ]
[ ]
[ ]
File [Upload]
3002 – Content Targeting
Tags: [ ]
Words: [ ]
3003 – Demographic Targeting
Gender o Male o Female
Age Minimum: [ ]
Age Maximum: [ ]
Ethnicities: [ ] White [ ] Black [ ] Latino [ ] Asian
Sexual Orientation: o All o Straight o Gay
Interests: [ ]
Book Categories: [ ]
Movie Categories: [ ]
Music Categories: [ ]
3004 – Visitor Location Targeting
Countries (two letter codes): [ ]
States/Provinces: [ ]
Cities: [ ]

3005 – [Submit Content]

FIG. 4

Advertise

4000 – Content Targeting

Categories: [ ]

Tags: [ ]

Words: [ ]

4001 – Demographic Targeting

Gender o Male o Female

Age Minimum: [ ]

Age Maximum: [ ]

Ethnicities: [ ] White [ ] Black [ ] Latino [ ] Asian

Sexual Orientation: o All o Straight o Gay

Interests: [ ]

Book Categories: [ ]

Movie Categories: [ ]

Music Categories: [ ]

4002 – Visitor Location Targeting

Countries (two letter codes): [ ]

States/Provinces: [ ]

Cities: [ ]

4003 – Campaign

Default Bid per Click: [ ]

Default Bid per Display: [ ]

Budget: [ ]

Start (MM/DD/YY): [ ]

End (MM/DD/YY): [ ]

4004 – Target Individual Content

o Yes o No

4005 – Advertisement File Upload [Select File]

4006 – [Submit Advertisement]

FIG. 5

Advertisement

5000 – Maher for America

Targeting	Gender: Male, Female Age: 18-35 Interests: Politics
Recent Content Visitors	1,417,220
Expected Advertisement Presentation Percentage:	60%
Expected Click Through Rate:	1.5%
Expected Clicks:	12,750

5001 – Selected Content & Statistics

Title	Presentations	Clicks	Bid per Click	Bid CPM	Status
Election Uncertain	1,220	43	$0.80 [+/-]	$0.40 [+/-]	Enabled [Disable]
September Polling Released TV News Episode	308,070		$0.50 [+/-]	$0.25 [+/-]	Enabled [Disable]

5002 – [all content]

FIG. 6

Content

6000 – [Seeking Campaign Volunteers]

6001 – Election Outlook Uncertain

6002 – With the general election under way …

6003 – [Advertisement 1] [Enable / Disable]
[Advertisement 2] [Enable / Disable]

FIG. 7

Advertisers' Targeting

[7000] *Advertisers' keywords by bid amount*

2.80 Cancer
0.87 California Election
0.17 Viagra

[7001] *Advertisers keywords by count*

417 Election
280 John Smith
66 Oil
17 Cancer

[7002] *Advertisers demographic targeting*

Male 60%
Female 40%
Average Minimum Age: 23
Average Maximum Age: 58

SIMPLIFIED DEVELOPMENT OF COMPUTER APPLICATIONS

INVENTOR JONATHAN BANNON MAHER

TECHNICAL FIELD

[0001] Embodiments of the invention relate to the field of computer software application and website development.

ABSTRACT

[0002] Systems, methods, apparatuses, and computer programs encoded on a computer storage medium, allow for the development of fully featured computer applications including software and websites using any non-technical spoken language, consisting of core and optional components including a template, content, an integration function, a processor function, a settings file, a menu file, and a translation function, providing for a unique advancement in the separation and integration of components, and the robust processing of submitted data without custom programming.

REFERENCE TO RELATED DOCUMENTS

[0003] This application is provided the benefit and priority date of United States Patent and Trademark Office provisional patent application number 62/522,667, filed June 20th 2017 by inventor Jonathan Bannon Maher, which is incorporated herein in its entirety.

BACKGROUND

[0004] This section is intended to introduce the reader to various aspects of the art that may be related to various aspects of the present techniques, which are described and or claimed. This discussion is believed to be helpful in providing the reader with background information to facilitate a better understanding of the various aspects of the present disclosure. Accordingly, it is understood that these statements are to be read in this light, and not a citation of prior any art.

[0005] Developing custom software applications and websites currently generally requires technical expertise. Furthermore, current methods of website and software development either integrate the programming of the front end displayed to the user, and the backend supporting display and processing of interactions with the front end, or allow for separation of the front end from the back end, but in all cases, require custom programming to create customized robust software applications and websites.

BRIEF DESCRIPTION OF THE ILLUSTRATIONS

[0006] An embodiment of the invention is illustrated by way of example, and not by way of limitation, in the figures of the accompanying drawings, and embodiments may not contain all components, may contain additional components, and may contain functionally similar components.

[0007] FIG. 1 is an example of an embodiment of a screen produced by the integration function utilizing the template, settings, and content.

[0008] FIG. 2 is an example of an embodiment of a screen produced by the integration function displaying a form utilizing the template, settings, and content.

[0009] FIG. 3 is an example of an embodiment of a screen produced by the integration function displaying a table utilizing the template, settings, content, and a data source.

DETAILED DESCRIPTION

[0010] It is understood that, as in any engineering or design project, the development of any actual implementation will include numerous implementation specific decisions made to achieve the developers' specific goals, such as compliance with business related and system related constraints, which may vary from one implementation to another. It is understood that such a development effort might be complex and time consuming, but is nevertheless a routine undertaking of design, fabrication, and manufacture for those skilled in the art having the benefit of this disclosure. The disclosed steps may be read as prefaced by "In some embodiments, including one complete embodiment, ", may be executed or performed in other orders or sequences, and are not limited to the order and sequence shown and described, which are provided to enable ease in constructing an embodiment, and along with each components of each step, may be removed, modified, combined, or rearranged, and other steps and or step components may be added, without departing from the scope of this disclosure and or invention. Although embodiments of the invention have been described and illustrated in the disclosed implementations, it is understood that the present disclosed subject matter, including apparatuses, methods, specification, and illustrations, has been made only by way of example, not by way of limitation, and the methods and apparatuses may be used in other systems, and that numerous changes and optimizations in the details of implementation of the invention and or embodiment are made without such modifications departing from the spirit and scope of this disclosure and or embodiments of the invention. Although the disclosure has been shown and described with respect to one or more embodiments, features of the disclosed embodiments can be combined and rearranged in various ways, and changes including equivalent alterations, substitutions, modifications, and additional efficiencies will of course occur to someone of ordinary skill in the art without departing from the spirit and scope of this disclosure and or invention. In particular regard to the various functions performed by the described components, the terms used to describe such components are intended to correspond, unless otherwise indicated, to any component which performs the specified function of the described component, or is functionally equivalent to the described component, even though not structurally equivalent to the disclosed structure which performs the function in the implementations described in this disclosure. In addition, while a particular feature of the disclosure may have been provided with respect to only one of several embodiments, such feature may be combined with one or more other features of other embodiments as may be desired and advantageous for any given or particular application. In some instances, well-known circuits, structures and techniques have not been shown in detail in order not to obscure the understanding of this disclosure. Articles in this disclosure such as "a" "an" and "the" may allow for both singular and plural forms. Verbs in this disclosure such as "is" may be read as "may be". Conjunctions in this disclosure such as "or" as used herein may be interpreted as inclusive or meaning any one or any combination, where "A, B or C" means "any of the following: A; B; C; A and B; A and C; B and C; A, B and C". Relational terms in this disclosure, for example first and second, top and bottom, left and right, are to distinguish one entity or action from another, and may not necessarily require or imply a relationship, or order between, such entities or actions. The disclosure includes the best mode contemplated by the inventor, a completely described specific embodiment, along with optional components and alternative embodiments to best suit the implementer, measurements in imperial and metric units to support universal understanding, and dramatically exceeds claims support requirements and enablement requirements by allowing for selection and or construction of the required components to be carried out easily, quickly, and routinely by persons of ordinary skill in the art, who are provided the additional benefit of utilizing readily available commodity components whenever possible. The present disclosure includes material protected by copyrights, and the owner of the copyrights hereby reserves all rights, but with authorization for publication as required by government patent offices. Various embodiments of the present invention may provide all, some or none of the disclosed technical advantages.

[0011] The computer code descriptions disclosed, in order to provide comprehensive enabling disclosure, rather than utilizing flow charts, which according to Patent Cooperation Treaty 11.11a are prohibited from containing "text matter, except a single word or words, when absolutely indispensable, such as... a few short catchwords indispensable for understanding", are provided in a text only format where the number of arrows preceding a line indicate logical block level, semicolons indicate a new segment of a logical block, and periods indicate the closure of one or more logical blocks. It is understood that any computer code representations in this disclosure are merely illustrative, rather than restrictive. While code may be written in nearly any computer language, including Java and C++, the illustrative computer code descriptions were derived from code written the Python language, which may be run through the Python interpreter, with appropriate supportive libraries, which at the time of disclosure, may run on nearly any computer, for example one with an Intel or AMD processor, running a current version of Linux, Windows, or Mac OS. All code components may read as if prefaced by "In some embodiments, including one complete embodiment, ". In some embodiments, functionality may be modified, rearranged, excluded, and added. To provide more fundamental computer system details, in some embodiments, the functionality associated with the disclosed computer code descriptions may be referred to as a script, module, software, software application, or code, and can be written in any form of language, including compiled, interpreted, declarative, or procedural, able to be deployed in any form suitable for use in a computing environment, including as an independent or integrated program, module, component, or subroutine, for execution by the computer system, implemented on one or more independent or integrated computers, utilizing a central processing unit in the form of one or more general or special purpose microprocessors, in conjunction with digital electronic circuitry, which may include special purpose logic circuitry such as a field programmable gate array or application specific integrated circuit, with the computer controlled by and operatively coupled to tangibly embodied software and or firmware, which may include code that creates an environment for code execution, including individual or combined use of processor firmware, a protocol stack, a database management system, and an operating system, where such software and or firmware may exist in one or more parts in memory on one or more computers, and is encoded on one or more tangible non transitory software carriers, such as individual or combined use of a random or serial access device or substrate, a semiconductor memory device, transient or persistent random access memory, a magnetic, magnetic optical, or optical disk, or encoded on an artificially generated transmitted signal, for example, optical, electrical, or electromagnetic, transmitted using a sending and a receiving apparatus, where the interaction between the user and the software may be implemented by operatively coupling, to the local implementing computer, or a local computer connected to one or more remote computers through a local or wide area network, a display device which may implement fluid crystals or light emitting diodes, a keyboard, and a pointing device.

[0012] The inventor retains absolutely no liability for any implementation of this invention, and the invention is implemented exclusively at the risk and liability of the implementer.

[0013] Embodiments of the invention are the first to allow for fully separated non-repeating re-useable custom template components that can be assembled by an integrator, using assembly instructions specified in the content, with user interaction handled by a processor, using processor instructions specified in the content, thereby requiring no custom software development to develop full software applications. More specifically, this is accomplished by providing an integrator that takes a template, a content file that optionally specifies template components, which optionally specify one or more sources of data, optional menu and settings data, with the integration function generating a unified output presented to the user, and a processor function that handles data submissions by adding or modifying data sources and records, and that can be extended through optional custom libraries. The content text, menu text, and settings text, are all written in standard natural language, means that no programming or technical expertise is required to produce robust software applications. Additionally, rather than having to define one template per application screen, the system provides the advantage of unprecedented separation and simplicity of the template structure, by creating individual non-repeating template components that can be re-used across the entire application, with components specified in the content. This is the first invention to allow an entirely non-technical user to create fully featured applications and websites in a generic text editor by writing in basic non-technical text format.

[0014] Embodiments are equally applicable to desktop software, mobile device software, and web based software, and may run either as an integrated whole, or server side, where the integration is completed before the screen is presented to the user, and or client side, where the data is integrated with the template through a client side implementation of the integration and or processing functions.

[0015] **In some embodiments, including one complete embodiment, create an optional settings file.** In some embodiments, including the one complete embodiment, settings text is maintained in plain text format, that optionally specifies settings to be integrated by the processor function, and or the integrator function into the template in specified corresponding placeholders before the output is displayed to the user, with example settings comprising:
> application title;
> application logo image file name;
> application background color;
> database IP address;
> database name;
> the database username;
> database password.

[0016] **In some embodiments, including one complete embodiment, create optional menu text.** In some embodiments, including one complete embodiment, menu text is contained in a text format containing linked application screens, and link titles, to be displayed in the application menu, where for example when displaying in HTML, the format for menu items is one item per line, with the link, then the title, where example menu text contains titles and corresponding links comprising:
> / home
> /about/ about
> /order/ order
> /contact/ contact

[0017] **In some embodiments, including one complete embodiment, create content text.** In some embodiments, including one complete embodiment, content text is provided, which contains data that will be integrated by the integrator into template components for display to the user, where the content text may specify plain text, one or more template items such as a form or a table, which specifies one or more data sets, where data sets may be in any format, which may include CSV, JSON, and SQL statements that return RDBMS objects containing data sets, and provide data rows that can be integrated into a template item specified surrounding the data set.

[0018] In some embodiments, including one complete embodiment, an example of a content file is where the first text block is identified as plain text content that will be displayed as presented to the user because it is not specified inside a template component, a second text block is data to be displayed through a template component as identified by

enclosing tags that indicate which template component the data is to be integrated into, where for example, to display values inside the template's table template, if the format is HTML, the tag around the data could be "<template:table>" and "</template:table>", with data in formats including CSV, JSON, or a Structured Query Language data request to the data source specified in the previously disclosed settings file returning data, where the data is integrated for display into the specified template component. In some embodiments, including one complete embodiment, the example text has content comprising:
> a first line containing the text "Welcome to our Company".
> a tag indicating the start of a table, for example "<template:table>".
> a first data record containing the separated data to be displayed in columns within the table, for example "joe.jpg","CEO","Joe Smith".
> a second data record containing the separated data to be displayed in columns within the table, for example "bob.jpg","CFO","Bob Jones".
> a tag indicating the end of a table, for example "</template:table>".

[0019] **In some embodiments, including one complete embodiment, create content text specifying display in a form template.** In some embodiments, including one complete embodiment, a content file specifies data fields and settings for a form, to be processed by the integrator, into the form template, for display to the user, with the text comprising:
> a title.
> a tag indicating the start of a table template data is to be integrated into.
> the name of a first data type, followed by the name of the corresponding data store field, followed by the display title of the data field, for example "text credit_card_number Credit Card Number".
> the name of a first data type, followed by the name of the corresponding data store field, followed by the display title of the data field, for example "text credit_card_number Credit Card Expiration".
> the name of a first data type, followed by the name of the corresponding data store field, followed by the display title of the data field, for example "text amount Amount".
> the text "processor_extension" followed by the name of an external processor to pass the data to, for example "credit_card_processor.py", to indicate that the submitted form data should be passed to a processor extension to process a credit card transaction.
> a tag indicating the end of data to be integrated into the table.

[0020] **In some embodiments, including one complete embodiment, create a template file.** In some embodiments, including one complete embodiment, a template

is provided which may contain sub templates, which may contain any level of nested sub templates, for any graphical user interface element, where each sub template within the main template may be integrated with any data set by the integration function, and the template may be in any computer code format, including HTML as used in the example template, and contains enclosed sections for each graphical user interface component, such as tables, forms, and images, where the enclosed sections are identified by starting and ending tags, such as "<template:item>" and "</template:item>", and "item" is a name for the template item. In some embodiments, including one complete embodiment, within the template, the location of the values to be inserted is identified with the name of the value to be inserted identified by a method such as a double underscore on either side of the value to be replaced __value__. In some embodiments, including one complete embodiment, the template contains functionality comprising:
> specify the content type of the template, for example "<html>".
>> provide a screen title place holder, for example "<title>__title__</title>".
>> initialize the main screen body, for example "<body>".
>> define the menu template, for example "<template:menu><table>";
>>> define the menu item template including placeholders for label and value, for example "<template:menu_item><tr><td><a href="__link__">__value__</a></td></tr></template:menu_item>";
>> close the menu template, for example "</template:menu>".
>> define the table template, for example "<template:table><table>";
>>> define the table header tow template, for example "<template:table_header><tr>";
>>>> define the table header row item template including a value placeholder, for example "<template:table_header_item><td><b>__value__</b></td></template:table_header_item>";
>>>> close the header row item template, for example "</tr></template:table_header>";
>>> close the header row template, for example "</tr></table_header>".
>>> define the table row template, for example "<template:table_row><tr>"
>>>> define the table row item template, for example "<template:table_row_item><td>__value__</td></template:table_row_item>";
>>>> define the table row administration options item template including update and delete, for example "<table:table_row_administration><td><a href="/?action=update&id=__value__">update</a> <a href="/?action=update&id=__value__">delete</a></td></table:table_row_administration>";
>>> close the table row template, for example "</table:table_row>".

>> close the table template, for example "</template:table>".
>> define the form template, for example "<template:form><table>"
>>> define the form row template, for example "<template:form_row><tr>"
>>>> define the form row item template including a place holder for field name and a user input box, for example "<template:form_input><td><input type=text name=__name__>__value__</td></template:form_input>";
>>>> close the form row item template, for example "</template:form_row_input>".
>>> define a hidden field for email to send form submission notification to and define a placeholder value, for example "<input type=hidden name=email_notification value=__value>";
>> close the form template, for example "</template:form>".
>> close the template body definition, for example "</body>".
> close the template definition, for example "</html>".

[0021] **In some embodiments, including one complete embodiment, optionally create a translation function.** In some embodiments, including one complete embodiment, a content translator is provided, where the complete separation of the content from the rest of the application allows for the content to be easily run through an external translation service to display content in any language. The computer code function may take three parameters, the text, the source language and the destination language. In some embodiment, the translation results are stored locally so as to reduce calls to an external translation service. In some embodiments, including one complete embodiment, software providing translations provides functionality comprising:
> define a function to translate content that accepts parameters including source language, destination, and text to translate;
>> encode the text to translate for transit;
>> prepare a call to retrieve a translation of the text that includes the address of the translation service, the source and destination languages, and the text to translate;
>> read the translated text provided to the translation address;
>> return the translation result to the requesting function.

[0022] **In some embodiments, including one complete embodiment, create the software code providing the integrator function and the processor function.** In some embodiments, including one complete embodiment, an integrator function is created, which is computer code, written in any language, that reads in the template content and replaces values in the template, with text from the content file, and any specified text data and data from data sources and data stores, for example tables in a relational database management system, and returns the resulting output for display to

the user, where the integrator can either run on the client side, for example through a scripting language like JavaScript, or native mobile application code, or the server side, in nearly any computer language. In some embodiments, including one complete embodiment, computer code is created that processes all user data submissions, accessed by passing form values by the server, where the processor takes each field and corresponding value submitted by the form, and as specified in the submission, may save them to a data store table, write them to a file, and email the submitted data, with the email address and file name specified in hidden fields submitted as part of the form. In some embodiments, the integrator and or processor is accessed as an internally integrated function within a mobile or desktop application. In some embodiments, including one complete embodiment, an action may be specified in a hidden field in the form, along with a corresponding code file name, to further utilize an external code file for processing, where such code files may be provided as a library and or custom developed, and if a corresponding code file is found, fields and data are passed to an instance of the class contained in the code file to perform additional functions, such as processing a charge on a credit card that was submitted with the form. It is therefore possible using basic text to create a custom order form on a website, have it save to an orders table, and have a credit card charge processed. In some embodiments, including one complete embodiment, the computer code is executed on an Internet connected server, started from the command line, that binds to the IP address of the server to provide an interface through a web browser over standard HTTP or HTTPS ports. In some embodiments, user interface rendering and interaction may be performed by means including native application code, or rendering user interfaces inside the application from a format such as HTML. In some embodiments, including one complete embodiment, this functionality is provided through software code providing functionality comprising:
> optionally import the previously created translation library;
> define a function to display the user interface;
>> define an array to hold the content segments to be integrated;
>> define an array to hold the template components to be integrated;
>> define a variable to hold the compiled output to display to the user;
>> define a variable and set it to any additional action specified with the request;
>> read the template definition to a variable;
>> initialize a variable for the application title;
>> initialize a variable for the application background color;
>> read the optional application settings text and set variables for the application including title and background color to those settings;

>> replace the template placeholder values for application title and background color equal to those provided in the settings file;
>> initialize an array of template item components, and split the template into an array of components, utilizing the established tag that distinguishes the start of a template component;
>> iterate through each item in the template array;
>>> set the name of the template item to the name specified in the template identification tag;
>>> set the template array item, at index of the template item name, equal to the content of the current item, between the start and end tags;
>> if there is menu text specified, integrate the menu items' titles and links into the menu item template, and then integrate the integrated menu items into the menu template, and replace the menu in the template with the integrated menu;
>> read in the content file for the specified screen;
>> create a content array, by splitting the text in the content file by template tag;
>> for each item in the content array, identify any template component for the content, remove the template tags from the content, and insert the content into the appropriate template component placeholders;
>>> if the content type is a form, retrieve the form template, split the form data into rows of field type, field name, and field label;
>>>> create a variable to hold the integrated form rows;
>>>> iterate through each row in the form;
>>>>> integrate form fields into the form template row place holders, integrate the form rows into the form, and add the integrated form row to the complied form rows variable;
>>>> integrate the integrated form rows into the form template, and add the integrated form to the variable holding the output to be returned to the user;
>>> if the content type is a table, retrieve the table template, split the table data into rows;
>>> create a variable to hold the compiled table rows;
>>>> iterate through each row of the table
>>>>> add a table row to the variable holding compiled table rows by replacing template components with the corresponding data;
>>>> integrate the table rows into the table, and add the integrated table to the output;
>>> if there is no content type associated with the content item, simply add the content item text to the output;
>> delete all remaining unintegrated template items from the output;
>> display the complied output screen to the user.

> define a function to process user interactions;
>> import the appropriate data store library;
>> create an account ID variable, and if a session ID is provided, retrieve and store the associated account ID to the account ID variable;
>> read in the values posted from the user interface, which may include the requested action, the ID and file name of the record to acted upon;
>> if the action is to start, verify that the accounts table exists in the data store provided in the application settings, and if not send a command to the data store that creates the accounts table with fields of email address and password, as well as the sessions table with fields of account ID and session ID, then check if the provided email address is listed in accounts, and if not, create an account, and if it is, verify the password is correct for the account with the provided email address, then if a new account was created, or an existing account with a matching password was found, create and associate with the account and record a session ID, record the account ID and session ID to the sessions table, and provide it to the user interface to include with requests;
>> if a record ID number is provided, and the action isn't to delete the record, default the action to update;
>> establish a variable to hold the content;
>> check if the file name contains an extension, for example ".db", that indicates it is a table name in the data store;
>> if a data store table was specified, set the data store table name to act upon equal the provided file name less the extension indicating it is a database table;
>> ask the data store if the structure exists, and if the data store structure doesn't exist, execute a command to create it with the fields provided by the form, by iterating through each field to build a create data structure statement, such as a create table SQL statement, then execute the statement on the data store;
>> if the requested action is to insert the submitted data as a data store record, build an insert statement, such as an SQL insert statement, by iterating through the field value pairs of the form, and if applicable include the field for account ID and its associated value, and execute the statement on the database;
>> if the requested action is to update the record in the database, build and execute an update statement, such as a SQL update statement, by iterating through the form fields and using the ID provided by the form as the ID of the record in the database;
>> if the action is to delete, build and execute a delete statement, such as a SQL delete statement, to delete from the specified data structure the record with the matching ID;
>> proceed if the file name does not indicate a database table;
>> if the specified file doesn't exist, create it and write a header row to it of the list of form fields;

>> create a record as a text string by iterating through the form values;
>> if the action is insert, read in the specified file if it exists and determine the ID number for the new record, otherwise start with a record ID of 1, and create the file with a header row consisting of a comma separated list of field names;
>> if the action is to update the record, read in all records in the file, and replace the one with the ID with the new record with the same ID;
>> if the action is to delete the record, read in all the records, then overwrite the file with all records except the one with the specified ID;
>> if a processor extension was specified in the form, create an instance of the named class and call the associated name function with the submitted data fields and values;
>> if a notification email address was specified, create a text string for the message body containing the form field value pairs, and call the email library to send the message;
>> redirect the user back to the submitting page or display a message such as one indicating the requested actions have been performed.

[0023] **In some embodiments, including one complete embodiment, build an optional extension to the processor.** In some embodiments, including one complete embodiment, an optional extension to the processor function is implemented, which may contain and execute external code. If the content file specifies an action in a hidden field, and that action corresponds to a code file, that named code file and associated class and main function will be passed the submitted data. In some embodiments, including one complete embodiment, an example of an extension to the processor provides functionality comprising:
> define a class to extend the processor by executing credit card transactions;
>> define a function to extend the processor by executing credit card transactions that accept as parameters the form fields and values;
>>> import the credit card processing company's application programming interface (API);
>>> execute the credit card transaction through the imported API.

[0024] **In some embodiments, including one complete embodiment, run the code to start the application.** In some embodiments, including one complete embodiment, once the code is complete, it may be started in the manner corresponding to the language used and or type of application created. In some embodiments, where the code is written in Python, and saved as "application.py", the application may be started by typing at the command line "python application.py" and pressing enter.

[0025] **In some embodiments, including one complete embodiment, verify displayed output of main screen.** In some embodiment, including one complete embodiment, in reference to FIG. 1, an example of output provided by the integrator to the user is displayed combining the menu, settings, content or translation of content, and template in the manner previously described, with title 1000, menu 1001, plain text 1002, an integrated table 1003, and additional links 1004.

[0026] **In some embodiments, including one complete embodiment, view displayed output of form.** In some embodiments, including one complete embodiment, in reference to FIG. 2, an example of a generated form is displayed as a result of the integration function bringing together the content text, menu text, and settings text, where the generated order form may now be filled out and submitted to the processor function, including title 2000, integrated form 2001, and menu 2002.

[0027] **In some embodiments, including one complete embodiment, review form processing to file.** In some embodiments, including one complete embodiment, an example of the data text that results from the form data being posted to the processor would be displayed, where the first row in the file is the names of the fields, and is only added when data is added, if the file doesn't exist, where the first text line contains sample titles for the data columns of "Name,Address,Product", the second text line contains sample values for the data columns of "Joe Smith,1 Birch Street,Annual Membership", and the third test line contains additional sample values for the data columns of "Bob Jones,34 Main Street,Annual Membership".

[0028] **In some embodiments, including one complete embodiment, review form processing to data table.** In some embodiments, including one complete embodiment, an example of an alternate embodiment and representation of form data storage in a database management system is contained in Data Table Orders, and if the table doesn't exist, it is created with the field names that are specified in the content, which for example contains sample titles for the data columns of "Name,Address,Product", sample values for the data columns of "Joe Smith,1 Birch Street,Annual Membership", and sample values for the data columns of "Bob Jones,34 Main Street,Annual Membership".

[0029] **In some embodiments, including one complete embodiment, verify display of processed form data.** In some embodiments, including one complete embodiment, an example of content text to display data for orders that were submitted by the generated order form, with data comprising:

> a tag specifying use of the table template component in presenting the data, for example "<template:table>".
> the name of the file containing formatted data to insert into the template, for example "Orders.csv".
> a closing tag for the template indicating there is no more data to include, for example "</template:table>".

[0030] In some embodiments, including one complete embodiment, in reference to FIG. 3, the output described may reassemble the output an example of output to the user as a result of the integration of previously described components in the example code in Displayed Output Orders Received, where the first displayed component is the title of the application taken from the settings file 3000, the second displayed component is the menu generated for the application from the menu file 3001, the third displayed component is the plain text title set by the user to display above the table from the content file 3002, the fourth displayed component is the data from the orders file displayed after integration with the table template 3003.

CLAIMS

What is claimed is:

1. A system providing for computer application user interface creation, in contrast to current methods utilizing one template per screen rather than per application, with the invention comprising:
a user interface template containing reuseable individual user interface component templates;
application settings, which may include including menu items and data store location and credentials;
application content, able to specify template components the content should be integrated into, where content may be composed of one or more components including text, a referenced data file, a data store query;
integrating said application settings and said application content into said application template and displaying the result to the user.

2. A non-transitory computer-readable recording medium holding stored instructions, which when executed by one or more processing devices, cause the one or more processing devices to implement a method comprising:
a memory which is able to store a series of characters including template information, content information, and data in a structured format;

a memory controller which combines said template and said content along with any associated data stored in memory;

a display which is operatively connected to said memory for displaying the integration of said template, said content and any associated data.

3. A system providing computerized processing of the submitted computer data, in contrast to current methods requiring pre-defined structures, with the improvement comprising:

a data submission processor that accepts the submission of data, and optionally emails said data, optionally passes said data to another data processor, and optionally records said data to a data store including automatically creating if none exists a data store structure comprised of submitted fields, or adjusting said data structure to add new submitted fields, and adding the new data record.

4. A machine for processing input from a user interface comprising:

a memory that is able to store data structures and associated records consisting of characters, and add fields to said data structures in memory if they are submitted by the user and don't already exist, and is able to add a data record to memory.

FIG. 1

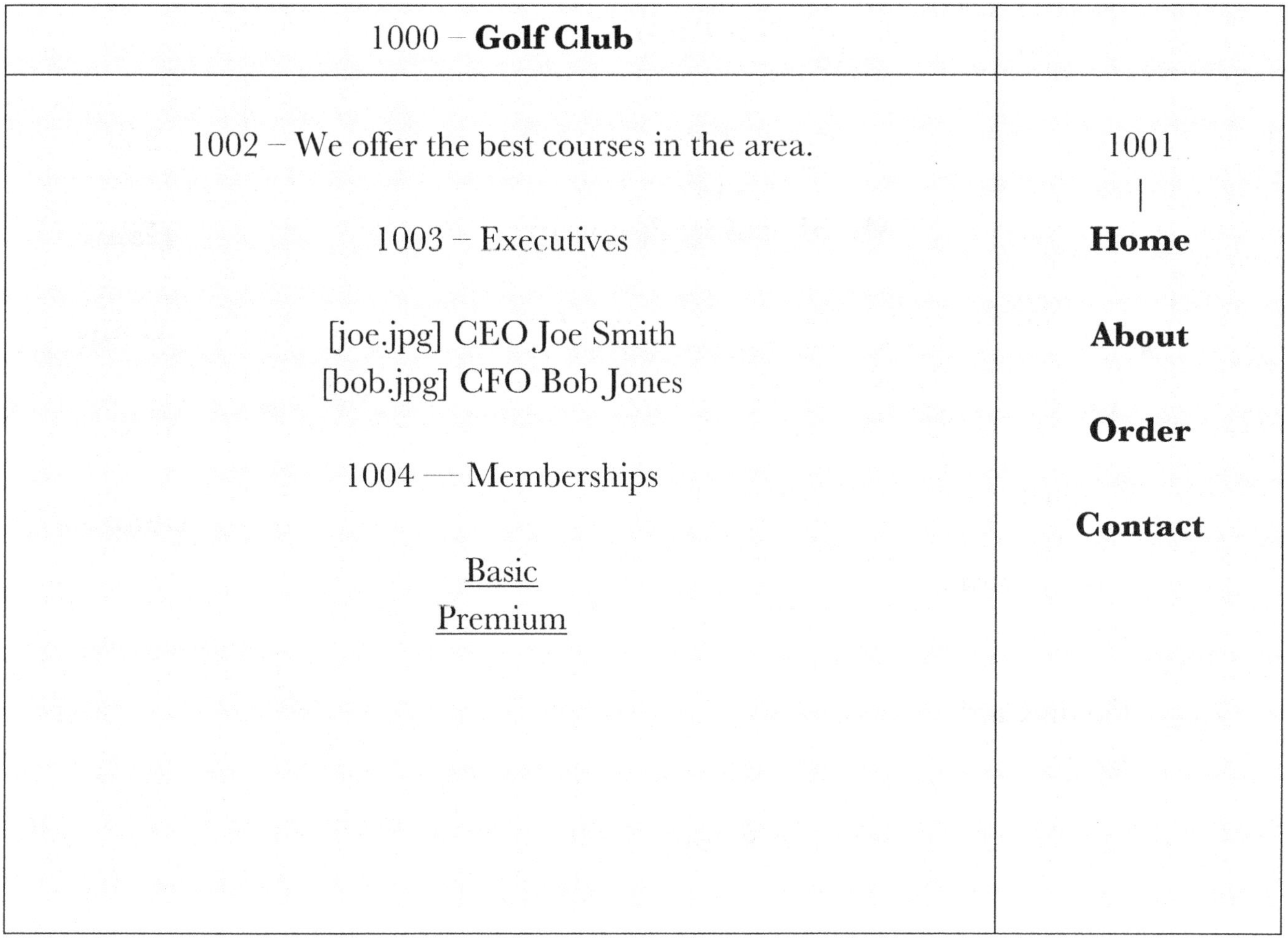
1000 – Golf Club
1001
Home
About
Order
Contact
1002 – We offer the best courses in the area.
1003 – Executives
[joe.jpg] CEO Joe Smith
[bob.jpg] CFO Bob Jones
1004 — Memberships
Basic
Premium

FIG. 2

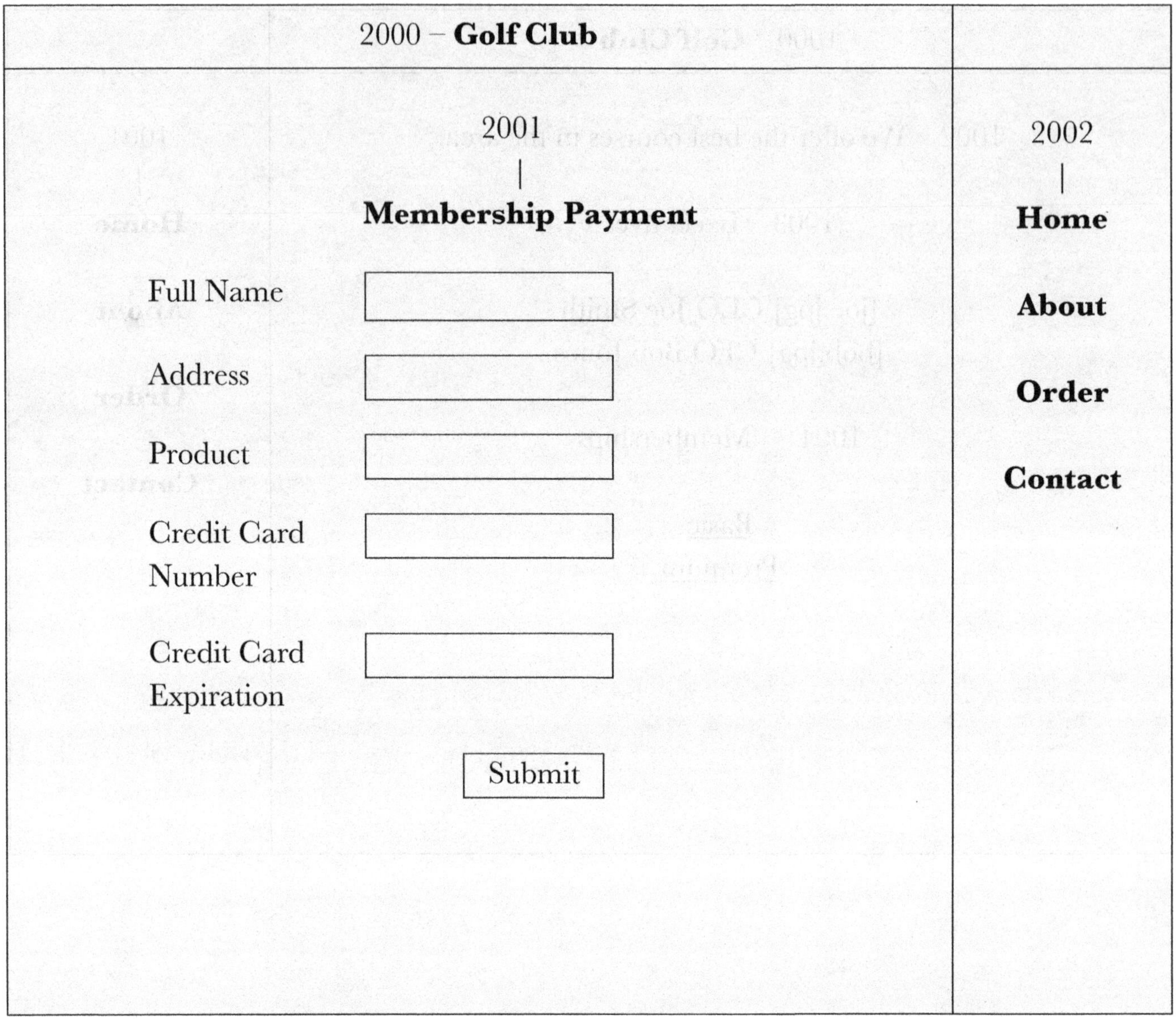
2000 – Golf Club
2001
Membership Payment
2002
Home
About
Order
Contact
Full Name
Address
Product
Credit Card Number
Credit Card Expiration
Submit

FIG. 3

3000 — **Golf Club**

3002 — **Membership Payment**

3003

Full Name	Address	Product
Joe Smith	1 Birch Street	Annual Membership
Bob Jones	34 Main Street	Annual Membership

3001

Home

About

Order

Contact

AUTOMATED ALGORITHMIC ASSET SELECTION AND TRADING

This system is not being disclosed because doing so is not socially beneficial.

AUTOMATED REMOVAL OF TARGETED TISSUE

INVENTOR JONATHAN BANNON MAHER

TECHNICAL FIELD

[0001] Embodiments of the invention relate to the field of autonomous medical devices and cancer.

ABSTRACT

[0002] Systems, methods, apparatuses, and computer programs encoded on a computer storage medium, provide for minimally invasive precision suction cutting of tissue, by providing a thin profile flexible extendable rotatable arm that contains an imaging device accompanied by any necessary illuminating device, a cutting device, and a suction tube, collectively allowing for a single incision to be made while allowing unrestricted movement inside the incision, where the apparatus may be autonomously controlled by computer software continuously comparing the image feed to specifications of tissue to be removed, and moving the apparatus to areas of matched tissue within a specified distance, while performing matched extractions until all specified tissue has been removed, allowing for example, without using chemotherapy or radiation, and while potentially leaving all healthy tissue intact, the automated removal of brain cancer.

REFERENCE TO RELATED DOCUMENTS

[0003] This application is provided the benefit and priority date of United States Patent and Trademark Office provisional patent application number 62/522,681, filed June 20th 2017 by inventor Jonathan Bannon Maher, which is incorporated herein in its entirety.

BACKGROUND

[0004] This section is intended to introduce the reader to various aspects of the art that may be related to various aspects of the present techniques, which are described and or claimed. This discussion is believed to be helpful in providing the reader with background information to facilitate a better understanding of the various aspects of the

present disclosure. Accordingly, it is understood that these statements are to be read in this light, and not a citation of prior any art.

[0005] There are currently several methods for eliminating undesirable tissue from a human body. One method involves targeted radiation to kill cells. Another method may require multiple incisions for the insertion of multiple manually controlled instruments, including cutting devices and laparoscopic cameras, which that can't be effectively flexed inside the patient to reach difficult areas, therefore increasing the risk of undesirable tissue being left behind. Another method is laparoscopic morcelation, where tissue is shredded inside the patient during extraction, which is known to spread cancerous cells, sometimes including more dangerous cancer cells that had not yet emerged. Deep brain tumors are currently only partially accessible through large openings in the skull and brain, as well as by toxic chemotherapy drugs, and radiation, which destroy substantial amounts of healthy tissue, rendering permanent damage.

BRIEF DESCRIPTION OF THE ILLUSTRATIONS

[0006] An embodiment of the invention is illustrated by way of example, and not by way of limitation, in the figures of the accompanying drawings, and embodiments may not contain all components, may contain additional components, and may contain functionally similar components.

[0007] FIG. 1 is an embodiment of the apparatus that allows for removal of undesired tissue.

[0008] FIG. 2 is an example of an embodiment of the settings file that specifies the identification and removal of undesired tissue to be read in by the operating software.

[0009] FIG. 3 is an embodiment of a computer code embedded flow diagram, which provides the software code for image processing.

[0010] FIG. 4 is an embodiment of a computer code embedded flow diagram, which provides the software code that controls the identification and removal of undesired tissue.

DETAILED DESCRIPTION

[0011] The disclosure is related to fields including automated brain surgery. It is understood that any reference to a person skilled in the art, recognizes that at the time of filing, there is no one else skilled in the art of this particular field, or a closely related field. Given the extraordinary nature of the disclosure, regardless of how full, clear, concise and exact the disclosure in enabling the production and use of embodiments of the disclosure, what could be construed to be undue experimentation during production and use, is simply the ordinary effort required in the assembly and use of an embodiment of such a disclosure.

[0012] It is understood that, as in any engineering or design project, the development of any actual implementation will include numerous implementation specific decisions made to achieve the developers' specific goals, such as compliance with business related and system related constraints, which may vary from one implementation to another. It is understood that such a development effort might be complex and time consuming, but is nevertheless a routine undertaking of design, fabrication, and manufacture for those skilled in the art having the benefit of this disclosure. The disclosed steps may be read as prefaced by "In some embodiments, including one complete embodiment, ", may be executed or performed in other orders or sequences, and are not limited to the order and sequence shown and described, which are provided to enable ease in constructing an embodiment, and along with each components of each step, may be removed, modified, combined, or rearranged, and other steps and or step components may be added, without departing from the scope of this disclosure and or invention. Although embodiments of the invention have been described and illustrated in the disclosed implementations, it is understood that the present disclosed subject matter, including apparatuses, methods, specification, and illustrations, has been made only by way of example, not by way of limitation, and the methods and apparatuses may be used in other systems, and that numerous changes and optimizations in the details of implementation of the invention and or embodiment are made without such modifications departing from the spirit and scope of this disclosure and or embodiments of the invention. Although the disclosure has been shown and described with respect to one or more embodiments, features of the disclosed embodiments can be combined and rearranged in various ways, and changes including equivalent alterations, substitutions, modifications, and additional efficiencies will of course occur to someone of ordinary skill in the art without departing from the spirit and scope of this disclosure and or invention. In particular regard to the various functions performed by the described components, the terms used to describe such components are intended to correspond, unless otherwise indicated, to any component which performs the specified function of the described component, or is functionally equivalent to the described component, even though not structurally equivalent to the disclosed structure which performs the function in the implementations described in this disclosure. In addition, while a particular feature of the disclosure may have been provided with respect to only one of several embodiments, such feature may be combined with one or more other features of other embodiments as may be desired and advantageous for any given or particular application. In some instances, well-known circuits, structures and techniques have not been shown in detail in order not to obscure the understanding of this disclosure. Articles in this disclosure such as "a" "an" and "the" may allow for both singular and plural forms. Verbs in this disclosure such as "is" may be read as "may be". Conjunctions in this disclosure such as "or" as used herein may be interpreted as inclusive or meaning any one or any combination, where "A, B or C" means "any of the following: A; B; C; A and B; A and C; B and C; A, B and C". Relational terms in this disclosure, for example first and second, top and bottom, left and right, are to distinguish one entity or action from another, and may not necessarily require or imply a relationship, or order between, such entities or actions. The disclosure includes the best mode contemplated by the inventor, a completely described specific embodiment, along with optional components and alternative embodiments to best suit the implementer, measurements in imperial and metric units to support universal understanding, and dramatically exceeds claims support requirements and enablement requirements by allowing for selection and or construction of the required components to be carried out easily, quickly, and routinely by persons of ordinary skill in the art, who are provided the additional benefit of utilizing readily available commodity components whenever possible. The present disclosure includes material protected by copyrights, and the owner of the copyrights hereby reserves all rights, but with authorization for publication as required by government patent offices. Various embodiments of the present invention may provide all, some or none of the disclosed technical advantages.

[0013] The computer code descriptions disclosed, in order to provide comprehensive enabling disclosure, rather than utilizing flow charts, which according to Patent Cooperation Treaty 11.11a are prohibited from containing "text matter, except a single word or words, when absolutely indispensable, such as... a few short catchwords indispensable for understanding", are provided in a text only format where the number of arrows preceding a line indicate logical block level, semicolons indicate a new segment of a logical block, and periods indicate the closure of one or more logical blocks. It is understood that any computer code representations in this disclosure are merely illustrative, rather than restrictive. While code may be written in nearly any computer language, including Java and C++, the illustrative computer code descriptions were derived from code written the Python language, which may be run through the Python interpreter, with appropriate supportive libraries, which at the time of disclosure, may run on nearly any computer, for example one with an Intel or AMD processor, running a current version of Linux, Windows, or Mac OS. All code components may read as if prefaced by "In some embodiments, including one complete embodiment, ". In some embodiments, functionality may be modified, rearranged, excluded, and added. To provide more fundamental computer system details, in some embodiments, the functionality associated with the disclosed computer code descriptions may be referred to as a script, module, software, software application, or code, and can be written in any form of language, including compiled, interpreted, declarative, or procedural, able to be deployed in any form suitable for use in a computing environment, including as an independent or integrated program, module, component, or subroutine, for execution by the computer system, implemented on one or more independent or integrated computers, utilizing a central processing unit in the form of one or more general or special purpose microprocessors, in conjunction with digital electronic circuitry, which may include special purpose logic circuitry such as a field programmable gate array or application specific integrated circuit, with the computer controlled by and operatively coupled to tangibly embodied software and or firmware, which may include code that creates an environment for code execution, including individual or combined use of processor firmware, a protocol stack, a database management system, and an operating system, where such software and or firmware may exist in one or more parts in memory on one or more computers, and is encoded on one or more tangible non transitory software carriers, such as individual or combined use of a random or serial access device or substrate, a semiconductor memory device, transient or persistent random access memory, a magnetic, magnetic optical, or optical disk, or encoded on an artificially generated transmitted signal, for example, optical, electrical, or electromagnetic, transmitted using a sending and a receiving apparatus, where the interaction between the user and the software may be implemented by operatively coupling, to the local implementing computer, or a local computer connected to one or more remote computers through a local or wide area network, a display device which may implement liquid crystals or light emitting diodes, a keyboard, and a pointing device.

[0014] The inventor retains absolutely no liability for any implementation of this invention, and the invention is implemented exclusively at the risk and liability of the implementer.

[0015] In some embodiments, including one complete embodiment, hardware and software working together provide for the automated precision removal of specific tissue, including cancer, providing for location independent sophisticated surgical procedures, while reducing patient risks by eliminating or reducing the need for chemotherapy and radiation, and by shortening surgery time, as the machine operates at a fixed continuous pace, and eliminating the possibility of physician fatigue, incompetence, distraction, or carelessness. In some embodiments, including one complete embodiment, it's possible that all healthy tissue may be left entirely undamaged, as a single incision may made, and a single instrument may be inserted, which in conjunction with software, can then automatically maneuver itself around inside of the patient to visualize and remove undesired tissue. In some embodiments, including one complete embodiment, the extraction device may be manually controlled by a physician through a graphical interface from anywhere in the universe. In some embodiments, the hardware may be detached with the wires attached to a hand held controller. In some embodiments, including one complete embodiment, when the device is manually operated, the images of removed tissue are recorded to a log that can be used by the machine to later make the same tissue removal decisions without the physician, in what is commonly referred to as machine learning and artificial intelligence, therefore potentially allowing every person in the world to access the world's top doctor in any given procedure, in exactly the same manner every single time, and at a fraction of cost.

[0016] **In some embodiments, including one complete embodiment, design and produce a thin flexible mechanical arm along with a connected head unit and base unit.** In some embodiments, including one complete embodiment, a thin profile flexible rotatable disposable mechanical arm is designed and constructed to be controlled by wires attached to motors, to be inserted into a patient, with the insertion end having a head unit, securing a fiber optic illuminated camera line or alternative imaging apparatus, a cutting blade, and a suction tube connected to a surgical suction machine, with the blade mounted to cut tissue in front of the suction tube, so that the tissue is slightly elevated for cutting by the suction, and once cut is immediately removed through suction, thus effectively removing the possibility that undesirable cells are ever spread, where designing the unit to utilize wires for motion control, rather than direct control with motors, allows for disposal of the wires and mechanical arm so every patient receives an unused sterile arm. In some embodiments, including one complete embodiment, the arm and head units do not contain bolts or nuts to hold components together or in place, and instead rely on structuring the components so that they are held in place by slightly bending connecting metal components of sufficient strength that the bend can not come accidentally undone.

[0017] In some embodiments, including one complete embodiment, in reference to FIG. 1, a flexible mechanical arm is constructed with base 1000 having wire holding loops

1001 1002 1003 1004 1005, and base arm 1006 welded to base 1000, and middle arm 1007 connected to base arm 1006, with base arm hook 1008 passed through middle arm hole 1009 and bent inward to hold the two together, and middle arm 1007 connecting to head unit 1005 with hook 1012 passing through arm hole 1010 and head unit port 1003 then bent back to hold the head unit in place, with base arm having extended above the hook and hole arm extension wire hole 1024, where middle arm 1007 has extended holes 1010 1011 for use by wires connected to the linear actuators, with middle arm hook 1013 connected to head hole 1013 with middle arm hook 1013 bent back to hold in place head hole 1016, with middle arm having extended above the hook and hole arm extension wire hole 1025, where head 1017 has extended holes 1015 1016 for use by wire connected to the linear actuators, where head 1017 is a hollow metal cube head component with a square opening with blade 1018 held in place as it moves by left slot 1019 and right slot 1020, with blade 1018 having on its backside a painted rubber coating to wipe the camera, with camera line holder 1021, fiber optic light cable holder 1022, and suction tube holder, 1023 with the tube then connecting to a suction machine. In some embodiments, including one complete embodiment, base arm 1006 is 1 foot (.3 meters) long, middle arm 1007 is 1.5 inches (.038 meters) long, allowing for tissue of approximately 3 inches (.076 meters) in diameter to be cleared, and the arm components inserted have a thin core diameter of 0.25 inches (0.006 meters), and the head blade cuts an area of 0.1 square millimeters at a time. In some embodiments, multiple arm lengths are created and assembled to allow for the internally flexing arm range approximately matches the diameter of the disuse to be removed, as determined by pre surgery imaging. In some embodiments, other means of constructing a robotic arm may be used. In some embodiments, the head unit may be any shape including round, square, triangular, or not have definite boundaries. In some embodiments, an additional blade may be integrated into the head, which may be controlled by one or more additional linear actuator connected wires, to allow the head cut through, rather than extract, tissue, in order to move to a location of interest.

[0018] In some embodiments, where cutting and suturing surgery is performed, such as heart surgery, in order to prevent anything more than a small incision, thereby replacing open heart surgery, a head may be used in place of head 1019 that has a vertically retractable blade in place of a horizontal cutting blade 1021, and the addition of a medical stapler 1023 using dissolvable staples, where cutting blade 1021 is vertically retractable rather than horizontally retractable, and a medical stapler 1023 is provided, which carries a the number of staples required for the procedure, and will eject a staple each time an electrical charge is applied, where the unit may be operated manually, or autonomous code may be written for various procedures.

[0019] In some embodiments, for use in dental applications, the head may contain a drill bit in place of a blade – which is rotated back and forth, by means which may include wires connected to linear actuators or equivalents, or by a miniature motor mounted directly in the head – a needle for injecting anesthesia, a pressurized water tube for area cleaning, a pressurized air tube for area drying, a needle for extruding filler material, and a light for hardening the material.

[0020] In some embodiments, including one complete embodiment, the previously specified components utilizing the previously specified designs are modeled in Computer Aided Design (CAD) software. In some embodiments, including one complete embodiment, in order to minimize incision size, and make the parts as precise and small, and so 3D metal printed. In some embodiments, after having accounted for the details and specifications of the designs, graphite molds may be made to cast the metal components. The molds may be manually etched, or may be setup in a CAD (Computer Aided Design) file, which is then will then processed by a CNC (Computer Numerical Control) machine to etch each mold. While there are many versions of CNC machines, for many the design process may allow for a part to be designed in a CAD (Computer Aided Design) program, which may export the design to the vendor neutral Initial Graphics Exchange Specification (IGES) format, developed by the United States Air Force for parts manufacturing, and may have the CNC machine's software convert the designs to G-Code instructions, which may control the movement of the CNC machine tools. The selected metal may then be melted and poured into the graphite molds.

[0021] In some embodiments, including one complete embodiment, in reference to FIG. 1, after the arm components have been created, they are assembled as previously specified, with connecting nuts securing bolts, or their equivalents, are welded in place so they can't be lost inside the patient.

[0022] **In some embodiments, including one complete embodiment, select and attach motorized controllers.** The motorized controllers will need to allow the device to rotate forward and backward, move into and out of the patient, flex the mechanical arm inside the patient, and rotate the unit head inside the patient. Stepper motors are commonly used in robotic applications and allow for movements of exact degrees or distance referred to as steps, by having multiple coils, with each coil activation providing a measured amount of rotation, and when the coils are activated in sequence, the stepper motor can move forward or backward precisely. A stepper motor linear actuator is a linear actuator that uses a stepper motor to precisely extend and retract a rod.

[0023] In some embodiments, including one complete embodiment, in reference to FIG. 1, rear flange mounted stepper motor 1100 rotates the unit at 1.6 degrees per step and whose axle is welded to stepper motor 1101, stepper motor rear flange mounted threaded linear actuator 1101 allows the device to move in and out of the patient with a stroke length of 6 inches, and whose threaded metal piston is welded to the embodiment base 1000, stepper motor threaded linear actuator 1102 is welded to base 1000 to control the mechanical arm, stepper motor rear flange mounted threaded linear actuator 1105 is welded to base 1001 to control the head, and stepper motor threaded linear actuator is welded to base 1001 to control the blade.

[0024] In some embodiments, other devices that provide for control of the arm may be used, which may include standard motors, or hydraulics, and may be mounted along the length of the arm.

[0025] **In some embodiments, including one complete embodiment, attach tubing, near infrared light, and camera components.** In some embodiments, including one complete embodiment, in reference to FIG. 1, fiber optic camera line 1700 is connected to computer 1500 at one end, and to the head unit holder 1021 secured compressing the holder, fiber optic light cable 1701 is connected at one end to a light source provider 1702 and at the other end to head unit holder 1022 secured compressing the holder, and suction tubing 1601 is connected on one end to a surgical suction machine 1600, with the other end attached to head by placing it through suction tube holder 1023 secured by compressing the holder.

[0026] In some embodiments, which may include one complete embodiment, when utilizing image recognition on individual cells, a magnifying lens may be added to one or both ends of the fiber optic camera line.

[0027] In some embodiments, the imaging device may be of any type which can produce an image that can be processed including sonar or radar. In some embodiments, in place of or in addition to a camera line, radar and or sonar may be used to capture densities, with the software checking densities ranges and or shapes against those for specified for undesired tissue.

[0028] **In some embodiments, including one complete embodiment, sharpen head unit blade, optionally add cauterizer, and attach to head unit.** In some embodiments, including one complete embodiment, in reference to FIG. 1, the head unit cutting blade 1017 will need to be sharpened, and then attached to head unit, with wires attached to the blade. In some embodiments, including one complete embodiment, in reference to FIG. 1, the cutting blade may be heated to cauterize cut tissue to prevent bleeding. The head unit cutting blade 1017, may be heated in a manner similar to a soldering iron, with wire wound around flat electrically non-conductive thermally conductive block which may be aluminum nitride ceramic, shaped to fit inside the compartment of the cutting blade compartment, and again encased in a block of the same material, so that the heating component fits inside the blade compartment. Wire may be wound around the thermal block, after the sheath of that wire has been removed for only the portion of the wire that will be wound around the block. The wire may then be connected to relay board, or may be connected to a power source of adequate strength so as to cause adequate heat for cauterizing.

[0029] **In some embodiments, including one complete embodiment, attach wires between motorized controllers and corresponding components.** In some embodiments, including one complete embodiment, wire is used to connect the mechanical arm components to the linear actuators, where utilizing a long wire looped through the arm and bolted on either end between the bolts on the linear actuator, prevents potential loss of the wire and any associated components inside a patient. In some embodiments, which may include one complete embodiment, sterilized galvanized soft annealed mild steel wire defined in ASTM A641 with a Class 1 Coating at 12 gauge (.106 inches in diameter) that will withstand force up to 209 pounds, may be used. In some embodiments, including one complete embodiment, wire is looped through each mechanical arm hole, with the ends secured together by compressing them between two commodity bolt nuts secured to form a tight loop on the threaded end of the corresponding linear actuator.

[0030] In some embodiments, including one compete embodiment, in reference to FIG. 1, arm wire 1200 is looped through arm extension hole 1024, and through base loop 1005, then compressed between two nuts on linear actuator 1102; arm wire 1201 is looped through arm extension hole 1024, then compressed between two nuts on linear actuator 1102; head wire 1202 is looped through arm extension holes 1024 and 1025, and through base loop 1005, then compressed between two nuts on linear actuator 1102; head wire 1203 is looped through arm extension holes 1024 and 1025, then compressed between two nuts on linear actuator 1103; cutting wire 1201 is looped through arm extension holes 1024 and 1025, then compressed between two nuts on linear actuator 1104.

[0031] **In some embodiments, including one complete embodiment, connect multi channel relay board to motorized controllers and computer.** The multi channel relay board may be rated to support the amps and volts of the purchased linear actuators and purchased suction machine. In some embodiments, because Ethernet may provide faster throughput than USB, serial, and firewire, the relay board is an Ethernet relay board, which is available at a static IP address, attached to the computer through an Ethernet crossover cable, and accessed to through a computer protocol which may include telnet or secure shell. In some embodiments, including one complete embodiment, in reference to FIG. 1, relay board 1400 may have 16 channels, to support the motors that operate the flexible mechanical arm and tools, as well as the suction machine 1600.

[0032] In some embodiments, including one complete embodiment, in reference to FIG. 1, the electrical wires from the linear actuators are stripped as necessary and connected to correspond with the software code in FIG. 2, that operates the relays that control the motors, where vertical control actuator 1101 first positive wire may be connected to relay 1401 and relay 1404 and first negative wire is connected to relay 1402 and 1403 and second positive wire is connected to relay 1405 and 1408 and second negative wire is connected to relay 1406 and 1407; rotational control actuator 1100 first positive wire is connected to relay 1409 and 1412 and first negative wire is connected to relay 1410 and 1411 and second positive wire is connected to relay 1413 and 1416 and second negative wire is connected to relay 1414 and 1415; arm forward actuator 1102 first positive wire is connected to relay 1417 and 1420 and first negative wire is connected to relay 1418 and 1419 second positive wire is connected to relay 1421 and 1424 and second negative wire is connected to relay 1422 and 1423; head linear actuator 1103 first positive wire is connected to relay 1425 and 1428, first negative wire is connected to relay 1426 and 1427, second positive wire is connected to relay 1429 and 1432, and second negative wire is connected to relay 1430 and 1431; cutting linear actuator 1104 first positive wire is connected to relay 1433 and 1436, first negative wire is connected to relay 1434 and 1435, second positive wire is connected to relay 1433 and 1436, second negative wire is connected to relay 1434 and 1435; suction machine 1600 cut power wire positive wire is attached to relay 1437 and negative wire is connected to relay 1438.

[0033] **In some embodiments, including one complete embodiment, create images of tissue or cellular components.** In some embodiments, including one complete embodiment, in reference to FIG. 1, images are taken of each tissue component, to be used by image recognition software. Photos may be produced from extracted cancer cells from biopsies or from cadavers immediately after death. Images may be captured with the snapshot function, using command line snapshot command, and may then in an image editor such as Photoshop, isolating the relevant cells, while maintain the size of the cut out but otherwise save with the same settings such as resolution as the original image.

[0034] Matching may also be done by shape. An outline of an object within a shape, such as a cancer or virus, can be trace, the export image outline from an illustration program after tracing the outline that supports Scalable Vector Graphics format, open the file in a text editor, and take the coordinates of the outline. The scale of the original image should be maintained. May also write out the x,y coordinates in sequence of the points that make up the border. Coordinates should be on the same line as a sequence of x,y values, for example "x,y x,y x,y x,y"

[0035] **In some embodiments, including one complete embodiment, develop a settings file containing the images and colors of components to recognize.** In some embodiments, including one complete embodiment, in reference to FIG. 1, a setting file similar to FIG. 2 may be created, to list image file names and or colors of tissue components to recognize, and then whether or not to sort the component. In some embodiments, define shapes to look for in image, by providing the geometric coordinates of the border and contents of a specific component. If utilizing a library of images not created by the system operator, the operator may adjust camera color balance and light until a photo from the camera matches the cell sizes and color and contrast of library of images.

[0036] In some embodiments, including one complete embodiment, the settings file may be that in FIG. 2, where text line 2000 lists an image carcinoma_1.jpg remove .85; text line 2001 lists an image carcinoma_2.jpg remove .85; text line 2002 lists an image carcinoma_3.jpg remove .85; text line 2003 lists an image carcinoma_4.jpg remove .85; text line 2004 lists an image lymphoma_1.jpg remove .85; text line 2005 lists an image lymphoma.jpg remove .85; text line 2006 lists an image lymphoma_3.jpg remove .85; text line 2007 lists an image sarcoma_1.jpg remove .80; text line 2008 lists an image sarcoma_2.jpg remove .80; text line 2009 lists an image sarcoma_3.jpg remove .80; text line 2010 lists an image sarcoma_4.jpg remove .80; text line 2011 contains the a sequence of x,y coordinates that form the outline of a component to remove; text lines

2012 and above colors are displayed to sort out of the blood, which are shades of green displayed by cells when exposed to an imaging agent such as Indocyanine Green.

[0037] **In some embodiments, including one complete embodiment, write the software code to process images and control the embodiment.** In some embodiments, the software functionality described may reassemble the functionality explained in the example code in FIG. 3 and FIG. 4, where FIG. 3 is an image processing library, and FIG. 4 controls the embodiment. The software and device doesn't have to remove every single cancerous cell to provide benefit, as a third of people have cancerous cells by the time they die of old age, and various other treatments can be used to remove residual cells. However, because of the level of non-invasive precision required when used for brain surgery, and it is ideal to remove as many undesired cells as possible, and given the software code is as similarly complex as that to create an automated driving car, the code will of course evolve over time, as all software does, to allow the device to operate with greater precision and coverage. Adjust percent match based on correct and incorrect identifications. Library used to detect faces in images, can be implemented in a similar manner to the image recognition library, and provided with images of check if in image. In some embodiments, code could be simply extended to include a [x,y,z] matrix of the positions of known healthy cells, and the arm is not moved through those positions again to not press on known healthy areas. on fiber optic before camera image recognition on individual cells, if one cell in group cancerous remove.

[0038] In some embodiments, including one complete embodiment, the library for image processing, provides functionality comprising:

> import image analysis libraries that provide identification of image pixel colors, extracting and comparing shapes in images, and comparing two images overall similarity.

> define a class for image processing;

>> create variables holding the minimum and maximum percent of image expansion for matching, the image expansion step percent, and the image rotation degree step;

>>> create a function that determines whether or not a larger image contains a smaller image.

>>> initialize an array to hold every possible cut out of the large image of the size of the smaller image.

>>> initialize a variable to hold the current small image degree rotation.

>>> iterate though each image degree rotation while rotation is less than 360 degrees, with all images always being circles since a microscope provides circles both for live images and stored images, which also happens to allow for easy image rotation.

>>>> iterate through each image percent size between the previously specified image size minimum and maximum, utilizing the specified step size.

>>>> generate the image using the determined rotation and sizing.

>>>> append the image to the modified images array.

>>>> increment the image step size.

>>>> increment the image degree rotation.

>>> iterate through each image in the modified images array.

>>>> create variables holding the dimensions of the large image and the small image.

>>>> initialize top left starting point x and y variables to iterate through large image.

>>> loop through each top left x and y point within the large image that and take a cut out the size of the small image.

>>>> if the cut out of the large image matches the small image at or above the previously specified match threshold, return that a match was found.

>>>> increment the y position.

>>> increment the x position.

>>> if no match was found, return false.

>> create a function to check if an image contains a color;

>>> load the individual pixel colors from the image into an array using the corresponding previously imported library;

>>> initialize variables holding the number of pixels processed and found;

>>> iterate through each image pixel, increment the pixel counter, and if the pixel color matches the specified color, increment the pixel found counter;

>>> if the percent of pixels matching the specified color are above the previously set threshold return true, otherwise return false;

>> create a function to determine if the image contains a shape;

>>> create a variable specifying if the shape was found in the image;

>>> initialize a coordinate pairs array;

>>> split the coordinate pairs from the line in the settings file into an array;

>>> iterate through each coordinate pair;

>>> split the coordinate pair into x and y;

>>>> append the x,y coordinates to the coordinates array;

>>>> generate a polygon from the coordinates using the corresponding imported library;

>>>> generate an array of all polygons in the image using the corresponding imported library.

>>> iterate through each polygon found in the image;

>>>> check if the polygon in the image matches the polygon specified in the settings file within the threshold previously specified and if so set the found variable equal to true;

>>> return the variable indicating whether or not the shape was found in the image.

[0039] In some embodiments, including one complete embodiment, the software to operate the system provides functionality comprising:
> import the created image processing library.
> import a library for three dimensional arrays.
> import a library for system resource access.
> import a library to connecting to the relay.
> import a library to pause program execution.
> import a library for accessing http resources.
> create variables holding the relay on off values.
> create variables holding the relay board connection, IP address, username, and password.
> create variables holding the state of each relay on the relay board.
> create variables holding the step motor step millimeters, degrees, and seconds.
> initialize variables to hold the settings and log file.
> create variables holding the pixel minimum percent match for extraction, and the minimum image and shape percent match.
> initialize variables to hold the current color, shape, or image matching against, the current degree rotation, current degree flexion, and whether or not a match has been found in the current position.
> initialize a three dimensional array to hold the status of the tissue at each point.
> create variables for the values indicating tissue is good or bad.
> initialize variables to hold the current x, y, and z of the head unit.
> read the lines of the settings file into an array.
> open the log file.
> create variables holding the camera IP address, username, and password.
> create a function to update the relay board, first establishing a connection to the relay board, if it has not been initialized, or has been dropped, then creating a string sequence of relay states, and sending the states to the relay board.
> create a function that when called moves the embodiment in by providing power to the corresponding relay attached step motor in the sequence that will move it as desired by one full step sequence.
> create a function that when called moves the embodiment out by providing power to the corresponding relay attached step motor in the sequence that will move it as desired by one full step sequence.

> create a function that when called rotates the embodiment forward by providing power to the corresponding relay attached step motor in the sequence that will move it as desired by one full step sequence.
> create a function that when called rotates the embodiment backward by providing power to the corresponding relay attached step motor in the sequence that will move it as desired by one full step sequence.
> create a function that when called moves the embodiment arm forward by the degrees in the function parameter by providing power to the corresponding relay attached step motor in the sequence that will move it as desired by full step sequences until the desired degrees have been achieved.
> create a function that when called moves the embodiment arm backward by the degrees in the function parameter by providing power to the corresponding relay attached step motor in the sequence that will move it as desired by one full step sequence.
> create a function that when called moves the embodiment head forward by the degrees specified in the parameter by providing power to the corresponding relay attached step motor in the sequence that will move it as desired by full step sequences until the desired angle is achieved.
> create a function that when called moves the embodiment head backward by the degrees specified in the parameter by providing power to the corresponding relay attached step motor in the sequence that will move it as desired by one full step sequence.
> create a function that when called turns the suction machine on then operates the motor connected to the blade wire to move the blade back and forth to cut the tissue, then turns the suction machine off.
> create a function to take a snapshot from the camera;
>> access the current camera image;
>> write the current camera image to a file so that its accessible to the both the system and the user interface.
> define a function to process the tissue at the current location;
>> retrieve the current camera image;
>> initialize a variable specifying if a match was found;
>> iterate through each setting;
>>> create a variable for whether or not to remove the tissue if there's a match and set it to the value from the setting;
>>> create a variable holding the setting match string;
>>> create a variable for the match method and with a default of image recognition;
>>> if the match string contains coordinates set the recognition method to shape;
>>> if the match string doesn’t contain an image file or coordinates set the recognition method to color;

>>> split the setting by spaces to an array;
>>> if the recognition mode is color, send the image to the function that checks if the color in the setting exists in the image;
>>> if the recognition mode is shape, send the image to the function that checks if the shape in the setting exists in the image;
>>> if the recognition mode is image, check if the image in the setting exists in the image.
>>> if a match was found, save the image to an incrementally numbered file and record the time, file name, and setting, to the log file;
>>> record the status of the tissue at the current x, y, z as bad;
>>> if the tissue match is to be removed according to the setting, call the function to suction cut the tissue.
>> if the no match was found, record the status of the tissue at the current x, y, z as good.
> create a function that allows the embodiment to extract tissue in an automated manner through image recognition, where if moving horizontally on x, and a point before that horizontally is clear x with the same y and z, skip=true to next point, if moving vertically and point before vertically, skip to next point, also for rotating, point array is of .2 mm increments as determined by stepper motors;
>> create variable that holds whether or not a match was found by retrieving the result of the function to process tissue at the current location;
>> while a match was found move in, while incrementing the y position of the embodiment, and then setting whether or not a current match was found by calling the function to process tissue at the current location;
>>> move the embodiment head forward 90 degrees;
>>> continue while the current arm rotation degree is less than one revolution, then return the embodiment to zero degrees of rotation;
>>> proceed while the y position is not equal to the point of origin, and call the function to process the tissue at that location;
>>>> proceed while the current arm flexion degree is less than one revolution, then return the embodiment to zero degrees of rotation, and call the function to process the tissue at that location.
> create a function to provide manual operation of the embodiment.
>> if the action requested is to move in call the function to do so.
>> if the action requested is to move out call the function to do so.
>> if the action requested is to flex forward call the function to do so.
>> if the action requested is to flex backward call the function to do so.
>> if the action requested is to rotate forward call the function to do so.

>> if the action requested is to rotate backward call the function to do so.
>> if the action requested is to move the head forward call the function to do so.
>> if the action requested is to move the head backward call the function to move in.
>> if the action requested is to suction cut call the function to do so.
> create a function to be called upon program execution;
>> retrieve any command line arguments, including the settings file name, the log file name, and any requested embodiment action;
>> if the setting file name was named snapshot, then call the function that saves a picture of the current embodiment camera image;
>> proceed if a snapshot was not specified;
>> open the log file to record unit operation;
>> if the request to the embodiment was to perform a manual action, send that action to the function that performs manual actions;
>> if the embodiment is to be operated autonomously, retrieve the settings file, and call the function to begin autonomous operation.

[0040] **In some embodiments, including one complete embodiment, access and control the unit operation software through a web browser.** In some embodiments, including one complete embodiment, the device and its operation are controlled and monitored through a web browser, accessible at the IP address of the computer running the software, where the software to display the user interface and accept commands is in FIG. 5, and the screen generated by the software in the web browser is represented in FIG. 6. By allowing for all monitoring and control through a web browser, the device can easily be controlled from anywhere in the world, allowing the best surgeon possible to perform an operation, even if the person is on the other side of the world or another planet. Allowing image based remote operation, can be done by simply providing a surgeon the IP address. The computer screen also allows autonomous mode to be selected and monitored as images are consistently provided to the screen.

[0041] In some embodiments, including one complete embodiment, the software to provide the user interface provides functionality comprising:
> import the libraries for http post and get.
> import the library for system resource access.
> define a function to provide control of the system though http and a web browser that is executed when the server is called through http;
>> create a variable holding any command provided with the http request;
>> if there was a command, build the command line call to the operating script;
>> if the command is autonomous, append autonomous to the command line call;

>> if the command isn't autonomous, append manual to the command line call then append the user command;
>> execute the command line call;
>> create an array of available commands;
>> create a variable to hold the display output;
>> for each command, generate a link that will call this application through http with the command and add the link to the output;
>> add the last image retrieved by the system to the user display;
>> display the output the user.

[0042] **In some embodiments, including one complete embodiment, optionally validate the unit.** In some embodiments, it may be wise for any manufacturer as well as practitioner, to use a produced unit for testing and practice, by a means with may include injecting a piece of meat with green Jello, and removing only the Jello, and or by using the body of an animal that died naturally from cancer.

[0043] **In some embodiments, including one complete embodiment, save lives.** In some embodiments, including one complete embodiment, a contrast agent may be administered to the patient to identify cancerous cells, such as Indocyanine green, which makes cancer cells fluorescent green under an appropriate light and has a half-life of 4 minutes.

[0044] A skull clamp may be mounted on the device and connected to the patient to hold the patient in place. The patient may then be anesthetized, with a skull access point created, with a slice created to the point of the tissue to be removed, the slice held open with an expansion clamp, to the width of the mechanical arm, thereby potentially leaving all healthy tissue intact when the expansion clamp is removed. The head unit of the device may be positioned at top center of the tissue to be removed.

[0045] Before removing the device, in order to avoid the possibility of leaving any cancer cells that may be stuck to the device are or head, slip a sterilized plastic tube, sliced down the middle, around the full length of the device, including the head, so that when the device is retracted, only the sterile tube touches the healthy incision tissue.

[0046] Of additional note, there is no present technical limitation prohibiting a software controlled machine from being created, using this design as a template, to use magnetic resonance imaging to determine tumor location, apply anesthesia, create the initial incision, and place this device at the start point, to ultimately make operations run from start to finish with the push of a button.

CLAIMS

What is claimed is:

1. A method for removing tissue with the invention comprising:
a mechanical arm, with one or more flexible segments, controlled through connected wires, or directly or indirectly by any other means that provide motion, with a head unit that contains components which may include one or more blades, drills, or lasers, and cauterizing or burning tools, and suction tubes, which may be provided suction to remove tissue and or tissue residue.
2. An apparatus for removing tissue with the invention comprising:
a mechanical arm, with one or more flexible segments, controlled through connected wires, or directly or indirectly by any other means that provide motion, with a head unit that contains components which may include one or more blades, drills, or lasers, and cauterizing or burning tools, and suction tubes, which may be provided suction to remove tissue and or tissue residue.
3. A system for the automated removal of tissue with the invention comprising:
computer software the performs image processing, by means which may include color recognition, visual, sonar, or radar image match recognition, which may use a range of combinations of rotated and resized input or output images, to identify tissues to be removed, then controlling the motors that control the mechanical arm components, to move to the position of the tissue and remove it.
4. The method of claim 5, a mechanical arm, with one or more flexible segments, controlled through connected wires, or directly or indirectly by any other means that provide motion, with a head unit that contains components which may include one or more blades, drills, or lasers, and cauterizing or burning tools, and suction tubes, which may be provided suction to remove tissue and or tissue residue.
5. A non-transitory computer-readable recording medium holding stored instructions, which when executed by one or more processing devices, cause the one or more processing devices to implement a method comprising:
computer software the performs image processing, by means which may include color recognition, visual, sonar, or radar image match recognition,

which may use angled and resized input or output images, to identify tissues to be removed, then controlling the motors that control the mechanical arm components, to move to the position of the tissue and remove it.

a mechanical arm, with one or more flexible segments, controlled through connected wires, or directly or indirectly by any other means that provide motion, with a head unit that contains components which may include one or more blades, drills, or lasers, and cauterizing or burning tools, and suction tubes, which may be provided suction to remove tissue and or tissue residue.

6. A system for recording the positions of tissue identified to contain or not contain identified tissue, with the invention comprising:

maintaining a record of the x, y, and z coordinates of matches and or non-matches.

7. A non-transitory computer-readable recording medium holding stored instructions, which when executed by one or more processing devices, cause the one or more processing devices to implement a method comprising:

maintaining a record of the x, y, and z coordinates of matches and or non-matches.

FIG. 1

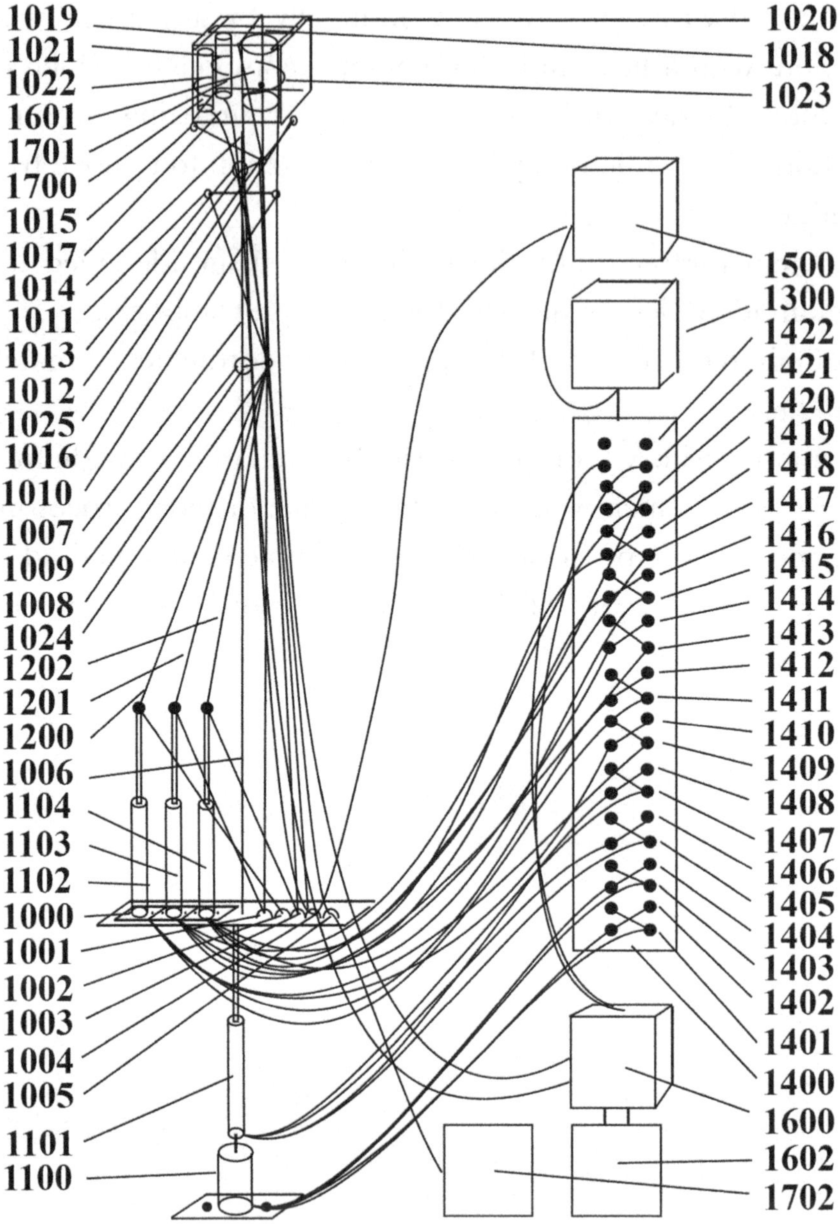

Jonathan Bannon Maher

FIG. 2

2000 – carcinoma_1.jpg remove .85
carcinoma_2.jpg remove .85
carcinoma_3.jpg remove .85
carcinoma_4.jpg remove .85
2001 – lymphoma_1.jpg remove .85
lymphoma.jpg remove .85
lymphoma_3.jpg remove .85
2002 – sarcoma_1.jpg remove .80
sarcoma_2.jpg remove .80
sarcoma_3.jpg remove .80
sarcoma_4.jpg remove .80
2003 – inflammation.jpg .80
inflammation_2.jpg .80
inflammation_3.jpg .80
2004 – infection.jpg .80
infection_2.jpg .80
infection_3.jpg .80
2005 – damage.jpg .80
damage_2.jpg .80
damage_3jpg .80
2006 – 0,0 10,10 10,20 20,20 20,10
2007 – 5A6351 remove
636F57 remove
526F35 remove
3B5323 remove
78AB46 remove
4A7023 remove
458B00 remove
66CD00 remove
76EE00 remove
7CFC00 remove
7FFF00 remove
7F9A65 remove

FIG. 3

3000 – import image analysis libraries that provide identifying image pixel colors, extracting and comparing shapes in images, and comparing two images overall similarity

```
import cv2
from shapely import geometry
from PIL import Image
import ImageChops
```

3001 – define a class for image processing

```
class image_processing():
```

3002 – create variables holding the minimum and maximum percent of image expansion for matching, the image expansion step percent, and the image rotation degree step

```
    image_size_minimum = 0.8
    image_size_maximum = 1.2
    image_size_step = .02
    image_degree_step = 2
```

3003 – create a function that determines whether or not a larger image contains a smaller image

```
    def image_contains_image(image_large, image_small):
```

3004 – initialize an array to hold every possible cut out of the large image of the size of the smaller image

```
        image_large_array = []
```

3005 – initialize a variable to hold the current small image degree rotation

```
degree = 0
```

3006 – iterate though each image degree rotation while rotation is less than 360 degrees, with all images always being circles since a microscope provides circles both for live images and stored images, which also happens to allow for easy image rotation

```
while degree < 360:
```

3007 – iterate through each image percent size between the previously specified image size minimum and maximum, utilizing the specified step size

```
    size = image_size_minimum
    while size <= image_size_maximum:
        width = image_large.width*size
        height = image_large.height*size
```

3008 – generate the image using the determined rotation and sizing

```
        image_large_adjusted = img.rotate(
            degrees, expand=True)
        image_large_adjusted = img.resize(
            (width,height), PIL.Image.ANTIALIAS)
```

3009 – append the image to the modified images array

```
        image_large_array.append(
            image_large_adjusted)
```

3010 – increment the image step size

```
size += image_size_step
```

3011 – increment the image degree rotation

```
degree += image_degree_step
```

3012 – iterate through each image in the modified images array

```
for large_adjusted in large_image_array:
```

3013 – create variables holding the dimensions of the large image and the small image

```
    large_width = large_adjusted.width
    large_height = large_adjusted.height
    small_width = image_small.width
    small_height = image_small.height
```

3014 – initialize top left starting point x and y variables to iterate through large image

```
    x = 0
    y = 0
```

3015 – loop through each top left x and y point within the large image that and take a cut out the size of the small image

```
while (x + small_width < = large_width)
    while (y + small_height < = large_height):
        large_cutout = large_adjusted.crop(
```

```
(x, y, small_width, small_height))
```

3016 – if the cut out of the large image matches the small image at or above the previously specified match threshold, return that a match was found

```
if ImageChops.difference(
    large_cutout, image_small) >=
    match_threshold:
    return True
```

3017 – increment the y position

```
y = y+1
```

3018 – increment the x position

```
x = x+1
```

3019 – if no match was found, return false

```
return False
```

3020 – create a function to check if an image contains a color

```
def image_contains_color(
    image_source, color, image_match_threshold)
```

3021 – load the individual pixel colors from the image into an array using the corresponding previously imported library

```
pixels = image_source.load()
```

3022 – initialize variables holding the number of pixels processed and found

```
pixel_count = 0
pixel_found_count = 0
```

3023 – iterate through each image pixel, increment the pixel counter, and if the pixel color matches the specified color, increment the pixel found counter

```
for x in range(image_source.width):
    for y in range(image_source.height):
        pixel_count += 1
        pixel_color = pixels[x,y]
        if pixel_color == color:
            pixel_found_count += 1
```

3024 – if the percent of pixels matching the specified color are above the previously set threshold return true, otherwise return false

```
if pixel_count > 0:
    if (pixel_found_count/pixel_count) >
        image_match_threshold:
        return True
return False
```

3025 – create a function to determine if the image contains a shape

```
def image_contains_shape(
    coordinates, image_large):
```

3026 – create a variable specifying if the shape was found in the image

```
found = False
```

3027 – initialize a coordinate pairs array

```
cordinates_array = []
```

3028 – split the coordinate pairs from the line in the settings file into an array

```
coordinates = coordinates.split(";")
```

3029 – iterate through each coordinate pair

```
for coordinate in coordinates:
```

3030 – split the coordinate pair into x and y

```
    coordinate = coordinate.split(",")
    x = coordinate[0]
    y = coordinate[1]
```

3031 – append the x,y coordinates to the coordinates array

```
    coordinates_array.append([x,y]]
```

3032 – generate a polygon from the coordinates using an the corresponding imported library

```
    coordinates_polygon = geometry.Polygon(
        [[p.x, p.y] for p in points])
```

3033 – generate an array of all polygons in the image using the corresponding imported library image_polygons = cv2.findContours(image_large)

3034 – iterate through each polygon found in the image

for image_polygon in image_polygons:

3035 – check if the polygon in the image matches the polygon specified in the settings file within the threshold previously specified and if so set the found variable equal to true

```
if image_polygon.almost_equals(
    coordinated_polygon, decimal=
    match_threshold]):
    found = True
```

3036 – return the variable indicating whether or not the shape was found in the image

return found

FIG. 4

4000 – import the created image processing library

```
from image_processing import image_processing
```

4001 – import a library for three dimensional arrays

```
import numpy as np
```

4002 – import a library for system resource access

```
import sys
```

4003 – import a library to connecting to the relay

```
import telnetlib
```

4004 – import a library to pause program execution

```
import time
```

4005 – import a library for accessing http resources

```
import urllib
```

4006 – create variables holding the relay on off values

```
off = 0
on = 1
```

4007 – create variables holding the relay board connection, IP address, username, and password

relay = None
relay_ip = "127.0.0.1"
relay_username = "admin"
relay_password = "admin"

4008 – create variables holding the state of each relay on the relay board

relay_1_motor_vertical_1_1 = off
relay_2_motor_vertical_1_2 = off
relay_3_motor_vertical_2_1 = off
relay_4_motor_vertical_2_2 = off
relay_5_motor_rotational_1_1 = off
relay_6_motor_rotational_1_2 = off
relay_7_motor_rotational_2_1 = off
relay_8_motor_rotational_2_2 = off
relay_9_motor_arm_1_1 = off
relay_10_motor_arm_1_2 = off
relay_11_motor_arm_2_1 = off
relay_12_motor_arm_2_2 = off
relay_13_motor_head_1_1 = off
relay_14_motor_head_1_ 2 = off
relay_15_motor_head_2_1 = off
relay_16_motor_head_2_2 = off
relay_17_motor_blade_forward = off
relay_18_motor_blade_backward = off
relay_19_motor_suction_state = off

4009 – create variables holding the step motor step millimeters, degrees, and seconds

motor_step_millimeters = 0.1
motor_step_degrees = 1.6
motor_step_seconds = 0.2

4010 – initialize variables to hold the settings and log file

settings = None
log_file = None

4011 – create variables holding the pixel minimum percent match for extraction, and the minimum image and shape percent match

pixel_minimum_percent_match = .05
match_threshold = .85

4012 – initialize variables to hold the current color, shape, or image matching against, the current degree rotation, current degree flexion, and whether or not a match has been found in the current position

match_found = True
image_current = None
shape_current = 0
color_current = 0
degree_rotation_current = 0
degree_flexion_current = 0

4013 – initialize a three dimensional array to hold the status of the tissue at each point

tissue_coordinates_array = np.zeros((1000,1000,1000))

4014 – create variables for the values indicating tissue is good or bad

tissue_good = 1
tissue_bad = -1

4015 – initialize variables to hold the current x, y, and z of the head unit

```
tissue_current_x = 0
tissue_current_y = 0
tissue_current_z = 0
```

4016	–	read the lines of the settings file into an array

```
settings_file_name = sys.argv[1]
settings_file = open(settings_file_name, "r")
settings_file_lines = settings_file.readlines()
settings_file.close()
```

4017	–	open the log file

```
log_file = open("log.txt","w+")
```

4018	–	create variables holding the camera IP address, username, and password

```
camera_ip = "127.0.0.2"
camera_username = "admin"
camera_password = "admin"
```

4019	–	create a function to update the relay board, first establishing a connection to the relay board, if it has not been initialized, or has been dropped, then creating a string sequence of relay states, and sending the states to the relay board

```
def update_relay():

    if not relay:
        relay = telnetlib.Telnet(relay_ip, 23)
        relay.read_until(b"User Name: ")
        relay.write(
            relay_username.encode('ascii') + b"\n")
        relay.read_until(b"Password: ")
        relay.write(relay_password.encode('ascii') + b"\n")
```

```
command =
   str(relay_1_motor_vertical_1_1)
command+=
   str(relay_2_motor_vertical_1_2)
command+=
   str(relay_3_motor_vertical_2_1)
command+=
   str(relay_4_motor_vertical_2_2)
command+=
   str(relay_5_motor_rotational_1_1)
command+=
   str(relay_6_motor_rotational_1_2)
command+=
   str(relay_7_motor_rotational_2_1)
command+=
   str(relay_8_motor_rotational_2_2)
command+=
   str(relay_9_motor_arm_1_1)
command+=
   str(relay_10_motor_arm_1_2)
command+=
   str(relay_11_motor_arm_2_1)
command+=
   str(relay_12_motor_arm_2_2)
command+=
   str(relay_13_motor_head_1_1)
command+=
   str(relay_14_motor_head_1_2)
command+=
   str(relay_15_motor_head_2_1)
command+=
   str(relay_16_motor_head_2_2)
command+=
   str(relay_17_motor_blade_forward)
command+=
   str(relay_18_motor_blade_backward)
command+=
```

```
    str(relay_19_motor_suction_state)
  relay.write(
    command.encode('ascii') + b"\n")
```

4020 – create a function that when called moves the embodiment in by providing power to the corresponding relay attached step motor in the sequence that will move it as desired by one full step sequence

```
def move_in():
  relay_1_motor_vertical_1_1 = on
  update_relay()
  time.sleep(motor_step_seconds)
  relay_1_motor_vertical_1_1 = off
  relay_2_motor_vertical_2_1 = on
  update_relay()
  time.sleep(motor_step_seconds)
  relay_2_motor_vertical_2_1 = off
  relay_4_motor_vertical_2_2 = on
  update_relay()
  time.sleep(motor_step_seconds)
  relay_4_motor_vertical_2_2 = off
  relay_3_motor_vertical_2_1 = on
  update_relay()
  time.sleep(motor_step_seconds)
  relay_3_motor_vertical_2_1 = off
  update_relay()
```

4021 – create a function that when called moves the embodiment out by providing power to the corresponding relay attached step motor in the sequence that will move it as desired by one full step sequence

```
def move_out():
  relay_4_motor_vertical_2_2 = on
  update_relay()
  time.sleep(motor_step_seconds)
```

```
    relay_4_motor_vertical_2_2 = off
    relay_2_motor_vertical_1_2 = on
    update_relay()
    time.sleep(motor_step_seconds)
    relay_2_motor_vertical_1_2 = off
    relay_2_motor_vertical_2_1 = on
    update_relay()
    time.sleep(motor_step_seconds)
    relay_2_motor_vertical_2_1 = off
    relay_1_motor_vertical_1_1 = on
    update_relay()
    time.sleep(motor_step_seconds)
    relay_1_motor_vertical_1_1 = off
    update_relay()
```

4022 — create a function that when called rotates the embodiment forward by providing power to the corresponding relay attached step motor in the sequence that will move it as desired by one full step sequence

```
def rotate_forward(requested_degrees):
    current_degrees = 0
    while current_degrees < requested_degrees:
        relay_5_motor_rotational_1_1 = on
        update_relay()
        time.sleep(motor_step_seconds)
        relay_5_motor_rotational_1_1 = off
        relay_6_motor_rotational_2_1 = on
        update_relay()
        time.sleep(motor_step_seconds)
        relay_6_motor_rotational_2_1 = off
        relay_8_motor_rotational_2_2 = on
        update_relay()
        time.sleep(motor_step_seconds)
        relay_8_motor_rotational_2_2 = off
        relay_7_motor_rotational_2_1 = on
        update_relay()
```

```
    time.sleep(motor_step_seconds)
    relay_7_motor_rotational_2_1 = off
    update_relay()
    current_degrees += motor_step_degrees*4
```

4023 — create a function that when called rotates the embodiment backward by providing power to the corresponding relay attached step motor in the sequence that will move it as desired by one full step sequence

```
def rotate_backward(requested_degrees):
  current_degrees = 0
  while current_degrees < requested_degrees:
    relay_8_motor_rotational_2_2 = on
    update_relay()
    time.sleep(motor_step_seconds)
    relay_8_motor_rotational_2_2 = off
    relay_6_motor_rotational_1_2 = on
    update_relay()
    time.sleep(motor_step_seconds)
    relay_6_motor_rotational_1_2 = off
    relay_7_motor_rotational_2_1 = on
    update_relay()
    time.sleep(motor_step_seconds)
    relay_7_motor_rotational_2_1 = off
    relay_5_motor_rotational_1_1 = on
    update_relay()
    time.sleep(motor_step_seconds)
    relay_5_motor_rotational_1_1 = off
    update_relay()
    current_degrees += motor_step_degrees*4
```

4024 — create a function that when called moves the embodiment arm forward by the degrees in the function parameter by providing power to the corresponding relay attached step motor in the sequence that will move it as desired by full step sequences until the desired degrees have been achieved

```
def flex_forward(requested_degrees):
    current_degrees = 0
    while current_degrees < requested_degrees:
        relay_9_motor_arm_1_1 = on
        update_relay()
        time.sleep(motor_step_seconds)
        relay_9_motor_arm_1_1 = off
        relay_11_motor_arm_2_1 = on
        update_relay()
        time.sleep(motor_step_seconds)
        relay_11_motor_arm_2_1 = off
        relay_12_motor_arm_2_2 = on
        update_relay()
        time.sleep(motor_step_seconds)
        relay_12_motor_arm_2_2 = off
        relay_10_motor_arm_2_1 = on
        update_relay()
        time.sleep(motor_step_seconds)
        relay_10_motor_arm_2_1 = off
        update_relay()
        current_degrees += motor_step_degrees*4
```

4025 — create a function that when called moves the embodiment arm backward by the degrees in the function parameter by providing power to the corresponding relay attached step motor in the sequence that will move it as desired by one full step sequence

```
def flex_backward(requested_degrees):
    current_degrees = 0
    while current_degrees < requested_degrees:
```

```
        relay_12_motor_arm_2_2 = on
        update_relay()
        time.sleep(motor_step_seconds)
        relay_12_motor_arm_2_2 = off
        relay_10_motor_arm_1_2 = on
        update_relay()
        time.sleep(motor_step_seconds)
        relay_10_motor_arm_1_2 = off
        relay_11_motor_arm_2_1 = on
        update_relay()
        time.sleep(motor_step_seconds)
        relay_11_motor_arm_2_1 = off
        relay_9_motor_arm_1_1 = on
        update_relay()
        time.sleep(motor_step_seconds)
        relay_9_motor_arm_1_1 = off
        update_relay()
        current_degrees += motor_step_degrees*4
```

4026 — create a function that when called moves the embodiment head forward by the degrees specified in the parameter by providing power to the corresponding relay attached step motor in the sequence that will move it as desired by full step sequences until the desired angle is achieved

```
def head_forward(requested_degrees):
    current_degrees = 0
    while current_degrees < requested_degrees:
        relay_13_motor_head_1_1 = on
        update_relay()
        time.sleep(motor_step_seconds)
        relay_13_motor_head_1_1 = off
        relay_15_motor_head_2_1 = on
        update_relay()
        time.sleep(motor_step_seconds)
        relay_15_motor_head_2_1 = off
        relay_16_motor_head_2_2 = on
```

```
        update_relay()
        time.sleep(motor_step_seconds)
        relay_16_motor_head_2_2 = off
        relay_14_motor_head_2_1 = on
        update_relay()
        time.sleep(motor_step_seconds)
        relay_14_motor_vertical_2_1 = off
        update_relay()
        current_degrees += motor_step_degrees*4
```

4027 – create a function that when called moves the embodiment head backward by the degrees specified in the parameter by providing power to the corresponding relay attached step motor in the sequence that will move it as desired by one full step sequence

```
def head_backward(requested_degrees):
    current_degrees = 0
    while current_degrees < requested_degrees:
        relay_16_motor_head_2_2 = on
        update_relay()
        time.sleep(motor_step_seconds)
        relay_16_motor_head_2_2 = off
        relay_14_motor_head_1_2 = on
        update_relay()
        time.sleep(motor_step_seconds)
        relay_14_motor_head_1_2 = off
        relay_15_motor_head_2_1 = on
        update_relay()
        time.sleep(motor_step_seconds)
        relay_15_motor_head_2_1 = off
        relay_13_motor_head_1_1 = on
        update_relay()
        time.sleep(motor_step_seconds)
        relay_13_motor_head_1_1 = off
        update_relay()
        current_degrees += motor_step_degrees*4
```

4028 — create a function that when called turns the suction machine on then operates the motor connected to the blade wire to move the blade back and forth to cut the tissue, then turns the suction machine off

```
def suction_cut():
    relay_19_suction_state = on
    update_relay()
    time.sleep(.25)
    relay_17_motor_cut_forward = on
    update_relay()
    time.sleep(.5)
    relay_17_motor_cut_forward = off
    relay_18_motor_cut_backward = on
    update_relay()
    time.sleep(.5)
    relay_18_motor_cut_backward = off
    update_relay()
    time.sleep(.25)
    relay_19_suction_state = off
    update_relay()
```

4029 — create a function to take a snapshot from the camera

```
def snapshot():
```

4030 — access the current camera image

```
    camera_image =
        urllib2.urlopen("http://" + camera_ip +
        "/?username=" + camera_username +
        "&password=" + camera_password + ").read()
```

4031 – write the current camera image to a file so that its accessible to the both the system and the user interface

```
  image_file = open("snapshot.jpg","wb")
  image_file.write(image)
  image_file.close()
```

4032 – define a function to process the tissue at the current location

```
def process():
```

4033 – retrieve the current camera image

```
  camera_image = urllib2.urlopen(
    "http://" + camera_ip + "/?username=" +
    camera_username + "&password=" +
    camera_password + ").read()
```

4034 – initialize a variable specifying if a match was found

```
  match_found = False
```

4034 – iterate through each setting

```
  for setting in settings:
```

4035 – create a variable for whether or not to remove the tissue if there's a match and set it to the value from the setting

```
    remove = False
    if setting.find("remove"):
      remove = True
```

4036 – create a variable holding the setting match string

```
match_string = setting[0]
```

4037 – create a variable for the match method and with a default of image recognition

```
recognition_method = "image"
```

4038 – if the match string contains coordinates set the recognition method to shape

```
if setting.find(",") > -1:
    recognition_method = "shape"
```

4039 – if the match string doesn't contain an image file or coordinates set the recognition method to color

```
if setting.find(",") == -1 and setting.find(".") ==-1:
    recognition_method = "color"
```

4040 – split the setting by spaces to an array

```
setting = setting.split(" ")
```

4041 – if the recognition mode is color, send the image to the function that checks if the color in the setting exists in the image

```
if recognition == "color":
    match_found = image_processing.
        image_contains_color(
        camera_image, color, match_threshold)
```

4042 — if the recognition mode is shape, send the image to the function that checks if the shape in the setting exists in the image

```
if recognition == "shape":
   match_found = image_processing.
      image_contains_shape(
      camera_image, setting[0])
```

4043 — if the recognition mode is image, check if the image in the setting exists in the image

```
if recognition == "image":
   match_found  = image_processing.
      image_contains_image(
         current_image, image_file):
```

4044 — if a match was found, save the image to an incrementally numbered file and record the time, file name, and setting, to the log file

```
if match_found:
      image_file_name =
         "image-" + image_counter + ".jpg"
      image_file = open(image_file_name, "wb")
      image_file.write(camera_image)
      image_file.close()
      log_file_line = str(sys.time()) + " " +
         image_file_name + " " + setting
```

4045 — record the status of the tissue at the current x, y, z as bad

```
      coordinates_array[x,y,z] = tissue_bad
```

4046 — if the tissue match is to be removed according to the setting, call the function to suction cut the tissue

```
if remove:
    suction_cut()
```

4047 — if the no match was found, record the status of the tissue at the current x, y, z as good

```
if not match_found:
    coordinates_array [x,y,z] = tissue_good
```

4048 — create a function that allows the embodiment to extract tissue in an automated manner through image recognition, where if moving horizontally on x, and a point before that horizontally is clear x with the same y and z, skip=true to next point, if moving vertically and point before vertically, skip to next point, also for rotating, point array is of .2 mm increments as determined by stepper motors

```
def autonomous():
```

4049 — create variable that holds whether or not a match was found by retrieving the result of the function to process tissue at the current location

```
    match_found = process()
```

4050 — while a match was found move in, while incrementing the y position of the embodiment, and then setting whether or not a current match was found by calling the function to process tissue at the current location

```
    while match_found:
        move_in()
```

```
tissue_current_y += 1
match_found = process()
```

4051 — move the embodiment head forward 90 degrees

```
head_forward(90)
tissue_x_current += 1
rotation_current = 0
```

4052 — continue while the current arm rotation degree is less than one revolution, then return the embodiment to zero degrees of rotation

```
while rotation_current < 360:
    rotate_forward(45)
    rotation_current +=45
```

4053 — proceed while the y position is not equal to the point of origin

```
    while tissue_current_y >= 0:
        move_out()
        tissue_current_y -= 1
        match_found = process()
```

4054 — proceed while the current arm flexion degree is less than one revolution, then return the embodiment to zero degrees of rotation

```
        while flexion_current < 360:
            flex_forward(45)
            flexion_current += 45
            found = process()
            flex_backward(360)
```

4055 – create a function to provide manual operation of the embodiment

```
def manual(action):
```

4056 – if the action requested is to move in call the function to do so

```
  if action == "move_in":
          move_in()
```

4057 – if the action requested is to move out call the function to do so

```
  if action == "move_out":
    move_out()
```

4058 – if the action requested is to flex forward call the function to do so

```
  if action == "flex_forward":
    flex_forward(motor_step_degrees)
```

4059 – if the action requested is to flex backward call the function to do so

```
  if action == "flex_backward":
    flex_backward(motor_step_degrees)
```

4060 – if the action requested is to rotate forward call the function to do so

```
  if action == "rotate_forward"
    rotate_forward(motor_step_degrees)
```

4061 – if the action requested is to rotate backward call the function to do so

```
  if action == "rotate_backward":
    rotate_backward(motor_step_degrees)
```

4062 – if the action requested is to move the head forward call the function to do so

```
  if action == "head_forward":
    head_forward(motor_step_degrees)
```

4063 – if the action requested is to move the head backward call the function to move in

```
  if action == "head_backward":
    head_backward(motor_step_degrees)
```

4064 – if the action requested is to suction cut call the function to do so

```
  if action == "suction_cut":
    suction_cut()
```

4065 – create a function to be called upon program execution

```
def main():
```

4066 – retrieve any command line arguments, including the settings file name, the log file name, and any requested embodiment action

```
  parameters = sys.argv
  settings_file_name = ""
  log_file_name = ""
  try:
    settings_file_name = parameter[1]
    log_file_name = parameter[2]
```

```
        action = parameter[3]
    except:
        pass
```

4067 – if the setting file name was named snapshot, then call the function that saves a picture of the current embodiment camera image

```
    if settings_file_name == "snapshot":
        snapshot()
```

4068 – proceed if the a snapshot was not specified

```
    else:
```

4069 – open the log file to record unit operation

```
        log_file = open(log_file_name, "w+")
```

4070 – if the request to the embodiment was to perform a manually action, send that action to the function that performs manual actions

```
        if settings_file_name == "manual":
            if action:
                manual(action)
```

4071 – if the embodiment is be operated autonomously, retrieve the settings file, and call the function to begin autonomous operation

```
        else:
            settings = open(settings_file_name,"r").split("\n")
            autonomous()
```

FIG. 5

5000 — import the libraries for http post and get

```
from flask import Flask
from flask import request
from flask import Response
```

5001 — import the library for system resource access

```
import os
```

5002 — define a function to provide control of the system though http and a web browser that is executed when the server is called through http

```
@application.route('/*', methods=['GET'])
def remote():
```

5003 — create a variable holding any command provided with the http request

```
  command = get_args("command")
```

5004 — if there was a command, build the command line call to the operating script

```
  if command:
    call = "python run.py "
```

5005 — if the command is autonomous, append autonomous to the command line call

```
    if command == "autonomous":
```

```
call+= "autonomous"
```

5006 — if the command isn't autonomous, append manual to the command line call then append the user command

```
else:
    call+= "manual " + command
```

5007 — execute the command line call

```
os.execute(call)
```

5008 — create an array of available commands

```
commands = ["autonomous", "in", "out", "forward",
    "backward", "rotate right", "rotate left"]
```

5009 — create a variable to hold the display output

```
output = "<html><meta refresh=5>"
```

5010 — for each command, generate a link that will call this application through http with the command and add the link to the output

```
for command in commands:
    output+= "<a href=\"/?command="
    + command + "\">" + command + "</a>"
```

5011 — add the last image retrieved by the system to the user display

```
output += "<img src=/current_image.jpg>"
```

5012 — display the output the user

return output

FIG. 6

6000 - **Surgical Robot Control**

6001 - Autonomous | In | Out | Rotate Forward | Rotate Backward | Forward | Backward | Head Forward | Head Backward

6002 - [CURRENT CAMERA IMAGE]

COMPUTERIZED ANALYSIS AND SEPARATION OF COMPONENTS IN FLUID

INVENTOR JONATHAN BANNON MAHER

TECHNICAL FIELD

[0001] Embodiments of the invention relate to the fields of medical and industrial devices.

ABSTRACT

[0002] Systems, methods, apparatuses, and computer programs encoded on a computer storage medium, provide for circulation of fluid under a microscope coupled with a computer connected imaging device, to obtain images of the fluid, to identify fluid components using custom image processing software, and optionally sort out specified components using a custom designed sorting device, with the sorted fluid going to a separate fluid container, and the rest of the fluid flowing back to the source or a second fluid container, where fluid flow may be split among any number of embodiments to allow for improved operational speed. Applications include eliminating custom fluid tests and associated costs, as well as when applied to blood, because the sorting is manual, removing without side effects, components including resistant bacteria and viruses, as well as cancer.

REFERENCE TO RELATED DOCUMENTS

[0003] This application is provided the benefit and priority date of United States Patent and Trademark Office provisional patent application number 62/522,685, filed June 20th 2017 by inventor Jonathan Bannon Maher, which is incorporated herein in its entirety.

BACKGROUND

[0004] This section is intended to introduce the reader to various aspects of the art that may be related to various aspects of the present techniques, which are described and or claimed. This discussion is believed to be helpful in providing the reader with background information to facilitate a better understanding of the various aspects of the

present disclosure. Accordingly, it is understood that these statements are to be read in this light, and not a citation of prior any art.

[0005] There are currently two fluid filtration and or separation methods, both of which are rather crude, as neither allow for the individual recognition of fluid components, and are thus only able to mechanically filter out a limited set of matter. The first method is membrane filtration based dialysis, which uses a porous filter to prevent some fluid components from passing through. The second method is rotating tubes of fluid in a centrifuge to separate heaver components from lighter components, where an Apheresis machine passes fluid through a centrifuge, to separate it into components, but when applied to blood, can only remove all white blood cells, not just leukemia cells.

[0006] Fluid analysis methods currently often require separate tests for each component in a fluid sample, requiring many tubes of blood, and include exposing the fluid to various costly reactive agents to estimate the level of a particular component.

BRIEF DESCRIPTION OF THE ILLUSTRATIONS

[0007] An embodiment of the invention is illustrated by way of example, and not by way of limitation, in the figures of the accompanying drawings, and embodiments may not contain all components, may contain additional components, and may contain functionally similar components.

[0008] FIG. 1 is an embodiment of the invention, which allows fluid to circulate under a microscope coupled with a computer connected camera, through a specialized sorting device, and sorted according to image recognition software.

[0009] FIG. 2 is an example of an embodiment of the settings file that specifies the identification and removal of undesired fluid components to be read in by the operating software.

[0010] FIG. 3 is an embodiment of a computer code embedded flow diagram, providing the software that processes the images.

[0011] FIG. 4 is an embodiment of a computer code embedded flow diagram, providing the software that controls the fluid sorting device.

DETAILED DESCRIPTION

[0012] The disclosure is related to fields including automated cancer and virus removal. It is understood that any reference to a person skilled in the art, recognizes that at the time of filing, there is no one else skilled in the art of this particular field, or a closely related field. Given the extraordinary nature of the disclosure, regardless of how full, clear, concise and exact the disclosure in enabling the production and use of embodiments of the disclosure, what could be construed to be undue experimentation during production and use, is simply the ordinary effort required in the assembly and use of an embodiment of such a disclosure.

[0013] It is understood that, as in any engineering or design project, the development of any actual implementation will include numerous implementation specific decisions made to achieve the developers' specific goals, such as compliance with business related and system related constraints, which may vary from one implementation to another. It is understood that such a development effort might be complex and time consuming, but is nevertheless a routine undertaking of design, fabrication, and manufacture for those skilled in the art having the benefit of this disclosure. The disclosed steps may be read as prefaced by "In some embodiments, including one complete embodiment, ", may be executed or performed in other orders or sequences, and are not limited to the order and sequence shown and described, which are provided to enable ease in constructing an embodiment, and along with each components of each step, may be removed, modified, combined, or rearranged, and other steps and or step components may be added, without departing from the scope of this disclosure and or invention. Although embodiments of the invention have been described and illustrated in the disclosed implementations, it is understood that the present disclosed subject matter, including apparatuses, methods, specification, and illustrations, has been made only by way of example, not by way of limitation, and the methods and apparatuses may be used in other systems, and that numerous changes and optimizations in the details of implementation of the invention and or embodiment are made without such modifications departing from the spirit and scope of this disclosure and or embodiments of the invention. Although the disclosure has been shown and described with respect to one or more embodiments, features of the disclosed embodiments can be combined and rearranged in various ways, and changes including equivalent alterations, substitutions, modifications, and additional efficiencies will of course occur to someone of ordinary skill in the art without departing from the spirit and scope of this disclosure and or invention. In particular regard to the various functions performed by the described components, the terms used to describe such components are intended to correspond, unless otherwise indicated, to any component which performs the specified function of the described component, or is functionally equivalent to the described component, even though not structurally equivalent to the disclosed structure which performs the function in the implementations described in this disclosure. In addition, while a particular feature of the disclosure may have been provided with respect to only one of several embodiments, such feature may be combined with one or more other features of other embodiments as may be desired and advantageous for any given or particular application. In some instances, well-known circuits, structures and techniques have not been shown in detail in order not to obscure the understanding of this disclosure. Articles in this disclosure such as "a" "an" and "the" may allow for both singular and plural forms. Verbs in this disclosure such as "is" may be read as "may be". Conjunctions in this disclosure such as "or" as used herein may be interpreted as inclusive or meaning any one or any combination, where "A, B or C" means "any of the following: A; B; C; A and B; A and C; B and C; A, B and C". Relational terms in this disclosure, for example first and second, top and bottom, left and right, are to distinguish one entity or action from another, and may not necessarily require or imply a relationship, or order between, such entities or actions. The disclosure includes the best mode contemplated by the inventor, a completely described specific embodiment, along with optional components and alternative embodiments to best suit the implementer, measurements in imperial and metric units to support universal understanding, and dramatically exceeds claims support requirements and enablement requirements by allowing for selection and or construction of the required components to be carried out easily, quickly, and routinely by persons of ordinary skill in the art, who are provided the additional benefit of utilizing readily available commodity components whenever possible. The present disclosure includes material protected by copyrights, and the owner of the copyrights hereby reserves all rights, but with authorization for publication as required by government patent offices. Various embodiments of the present invention may provide all, some or none of the disclosed technical advantages.

[0014] The computer code descriptions disclosed, in order to provide comprehensive enabling disclosure, rather than utilizing flow charts, which according to Patent Cooperation Treaty 11.11a are prohibited from containing "text matter, except a single word or words, when absolutely indispensable, such as... a few short catchwords indispensable for understanding", are provided in a text only format where the number of arrows preceding a line indicate logical block level, semicolons indicate a new segment of a logical block, and periods indicate the closure of one or more logical blocks. It is understood that any computer code representations in this disclosure are merely illustrative, rather than restrictive. While code may be written in nearly any computer language, including Java and C++, the illustrative computer code descriptions were derived from code written the Python language, which may be run through the Python interpreter, with appropriate supportive libraries, which at the time of disclosure, may run on nearly any computer, for example one with an Intel or AMD processor, running a current version of Linux, Windows, or Mac OS. All code components may read as if prefaced by "In some embodiments, including one complete embodiment, ". In some embodiments, functionality may be modified, rearranged, excluded, and added. To provide more fundamental computer system details, in some embodiments, the functionality associated with the disclosed computer code descriptions may be referred to as a script, module, software, software application, or code, and can be written in any form of language, including compiled, interpreted, declarative, or procedural, able to be deployed in any form suitable for use in a computing environment, including as an independent or integrated program, module, component, or subroutine, for execution by the computer system, implemented on one or more independent or integrated computers, utilizing a central processing unit in the form of one or more general or special purpose microprocessors, in conjunction with digital electronic circuitry, which may include special purpose logic circuitry such as a field programmable gate array or application specific integrated circuit, with the computer controlled by and operatively coupled to tangibly embodied software and or firmware, which may include code that creates an environment for code execution, including individual or combined use of processor firmware, a protocol stack, a database management system, and an operating system, where such software and or firmware may exist in one or more parts in memory on one or more computers, and is encoded on one or more tangible non transitory software carriers, such as individual or combined use of a random or serial access device or substrate, a semiconductor memory device, transient or persistent random access memory, a magnetic, magnetic optical, or optical disk, or encoded on an artificially generated transmitted signal, for example, optical, electrical, or electromagnetic, transmitted using a sending and a receiving apparatus, where the interaction between the user and the software may be implemented by operatively coupling, to the local implementing computer, or a local computer connected to one or more remote computers through a local or wide area network, a display device which may implement fluid crystals or light emitting diodes, a keyboard, and a pointing device.

[0015] The inventor retains absolutely no liability for any implementation of this invention, and the invention is implemented exclusively at the risk and liability of the implementer.

[0016] The disclosure resolves the previously cited deficiencies, by in some embodiments, including one complete embodiment, using commodity components including a microscope coupled with a computer connected imaging device, to feed images of fluid to custom image recognition software, to determine if a camera image matches an image specified, logging matches, and controlling the sorting device to either send the fluid unit to one fluid container, or back to the source or a second fluid container.

[0017] In embodiments when used on blood, because the sorting is entirely mechanical, free from drugs, radiation, or engineered cells, this system eliminates the potential for side effects. In some embodiments, including one complete embodiment, when applied to blood, it may provide the only defense that can be instantly used against any virus, blood cancer, or bacteria, for which there is no known or effective treatment or cure, including Zika, HIV, Bird Flu, or whatever virus next jumps from other animals to humans, and including viruses that mutate and develop resistance to drugs, with embodiments working with unchanged efficiency regardless of viral mutations that can make drugs ineffective. In embodiments when used on blood, given blood cancers are particularly difficult to treat in a targeted non-damaging way because of the difficulty in targeting only the relevant cells, this device can instead be used to recognize and sort out individually blood borne cancer cells such as leukemia, as well as potentially metastasizing cancer cells, making it possible that person with leukemia or HIV may never need to undergo treatment if they periodically have their blood filtered by this system, where when applied to viruses and combined with a drug that flushes out viral reservoirs, it may be possible enough is removed that any residual fails to replicate and dies off. In embodiments when used on blood, blood borne infections, a common cause of death, some strains of which are immune to antibiotics, can now be separated out, while chemotherapy drugs and immunotherapy cells that are no longer needed, may also be separated out. In embodiments when used on blood, it can also be used to isolate only healthy blood components from a donor, for infusion to another person, including for example only white blood cells or platelets.

[0018] In some embodiments, fluid component analysis may be performed, with a drop or test tube of fluid run through the system in place of a fluid stream, and a log made of a count of every type of particle identified in the fluid. There is cost to perform every fluid known analysis is limited and fixed, since the unit is simply counting fluid component matches. When applied to blood, this can allow a person to go into a local drug store and take a finger prick of blood, place it in a machine, and potentially have every aspect of his or her blood health diagnosed.

[0019] In some embodiments, given the only expensive component is a powerful personal computer, when the cost is allocated among many beneficiaries, use is effectively free. Therefore, when applied to blood, because this system provides on demand nearly free blood analysis and comprehensive filtration against a range of conditions, it has the ability to eliminate a significant portion of global healthcare costs over time while improving and saving countless lives.

[0020] **In some embodiments, including one complete embodiment, design and produce the mechanical fluid sorting device.** In some embodiments, including one complete embodiment, in reference to FIG. 1, the fluid sorter 1000 is made of 3D printed metal, with a clear glass top 1001, made of commodity microscope slide glass, to allow clear images to be taken through microscope 1100, using the computer connected camera 1200, and sized and assembled in such a way, that allows it to sit below the microscope in the place of a slide, where the sorter is hollow in the center to allow the fluid to flow through it, with fluid inlet 1002 being 3 millimeters in diameter to connect to a standard intravenous line which may be connected from the patient, fluid outlet 1003 of 3 millimeters in diameter to connect to a standard intravenous line, and fluid outlet 1004 of 3 millimeters in diameter to connect to a standard intravenous line which may be connected back to the patient, which then connects to a standard fluid collection bag, with the dimensions of the sorting device's fluid chamber matching the circular field of view of a microscope at the utilized magnification level of 1000 times which provides a circular field of view with a diameter of around 0.2 millimeters, and therefore the fluid chamber has a diameter of 0.2 millimeters, with a height that allows for individual cells to be visible with little potential for overlap, which may be around 0.2 millimeters, with the sorter having sliding doors on the inlet and outlet, with the sliding doors having gear ridges that connect to the gear on the corresponding motor axle, with ridged sliding door 1005 blocking inlets 1002, and ridged sliding door 1006 blocking inlets 1003 1004, and with motor support rod 1007 welded to the support structure, and welded to each motor support rods 1008 1009, with motor support rod 1008 welded to motor 1200, and motor support rod 1009 welded to motor 1201, that hold the motors in place that connect to the ridged sliding doors, with the motors each having a custom axle gear 1202 1203 designed to snugly interlock with the compartment door ridges. The sorter is then 3d printed and must be printed at such a quality level that allows it to be airtight.

[0021] In some embodiments, the sorting device instead utilizes magnets, where entry of fluid is controlled by motor 1200 operating port blocker to block both out ports, with each magnet being coated or encased in the same material as the slide, where computer controlled motor 1200 with attached gear 1202 moves the port blocker 1002 back and forth, based on image recognition, to allow fluid to enter. The exit is controlled by a motor 1201, controlled by port blocker 1003 to block both out ports. When the computer controlled motor 1201 with attached gear 1204 moves the port blocker 1003 back and forth, based on image recognition, to open the appropriate port for the fluid to exit.

[0022] In some embodiments, instead of a sorting device, a one or more lasers may be aligned over the fluid compartment 1000, and wired the relay control board 1400, in the same manner as specified for the for the motors, to be turned on and off, when the software in FIG. 4 identified that the fluid chamber contains undesirable components, to damage and disable selected fluid components, rather than sort them out.

[0023] In some embodiments, including one complete embodiment, to design the sorters, the previous specifications are modeled using CAD (Computer Aided Design) software. The sorters are then printed by a 3D printer in a high strength metal such as steel or titanium. In some embodiments, the molds may be manually etched, or may be setup in a CAD (Computer Aided Design) file, which is then will then processed by a CNC

(Computer Numerical Control) machine to etch each mold. In some embodiments, after having accounted for the details and specifications of the designs, graphite molds may be made to cast the metal components. While there are many versions of CNC machines, for many the design process may allow for a part to be designed in a CAD program, which may export the design to the vendor neutral Initial Graphics Exchange Specification (IGES) format, developed by the United States Air Force for parts manufacturing, and may have the CNC machine's software convert the designs to G-Code instructions, which may control the movement of the CNC machine tools. The selected metal may then be melted and poured into the graphite molds.

[0024] **In some embodiments, including one complete embodiment, select, acquire, and attach imaging device and microscope.** In some embodiments, including one complete embodiment, in reference to FIG. 1, microscope 1600 has a commodity computer connected camera 1601 secured on its eye piece, with clamp 1006, where the microscope capable of and set to provide magnification of 1000 times, and the camera is a commodity IP camera, therefore allowing multiple units to be easily attached to the same computer, and because latency on currently available commodity IP cameras is 0.15 seconds, while it is 0.75 seconds on currently available commodity USB cameras. In some embodiments, any level of magnification may be used that allows the cells to be visible for computerized recognition. In some embodiments, a fiber optic camera line is run from above the sorting device to the viewing area of a microscope where it is secured to provide images. In some embodiments, a digital microscope is used.

[0025] **In some embodiments, including one complete embodiment, attach motors to control the sorting compartment.** Stepper motors allow for precisely measured movements and are commonly used in robotics applications. In some embodiments, including one complete embodiment, in reference to FIG. 1, motors that consume 12 volts, where motor 1200 is welded to base welded support beam 1005, and has mounted to it gear 1202, to control ridged sliding door 1005, while motor 1201 is welded to base welded support beam 1006 and has mounted gear 1203 to control ridged sliding door 1007.

[0026] **In some embodiments, including one complete embodiment, attach computer controlled multi channel relay board and wire motors to it.** In some embodiments, including one complete embodiment, in reference to FIG. 1, an eight channel computer connected relay board 1400, rated for the amps and volts of the motors, is used to control the motors that operate the sorter, where the relay board is connected through an Ethernet crossover cable, because Ethernet is faster than USB, serial, and firewire, and allow for many units to be connected to a single computer through a router, and accessible at a fixed IP address through a protocol such as telnet or secure shell. In some embodiments, the functional equivalent of a relay board may be used.

[0027] In some embodiments, including one complete embodiment, in reference to FIG. 1, motor wires are stripped as necessary, and where motor 1200 has positive wire connected to relays 1400 and 1403, and negative wire connected to relays 1401 and 1402, while motor 1201 has positive wire connected to relays 1404 and relay 1407, and negative wire connected to relays 1405 and 1406.

[0028] **In some embodiments, including one complete embodiment, write and run the computer code that controls the sorter.** In some embodiments, the functionality of the computer code to control the embodiment may reassemble the functionality explained in the example code in FIG. 3 and FIG. 4, where FIG. 3 provides an image recognition library and FIG. 4 controls the embodiment. The software code will initially be used to take photos of blood components, so that a library can be built of components to sort and not sort.

[0029] In some embodiments, including one complete embodiment, the image recognition software provides functionality comprising:
> import image analysis libraries that provide identification of image pixel colors, extracting and comparing shapes in images, and comparing two images overall similarity.
> define a class for image processing;
>> create variables holding the minimum and maximum percent of image expansion for matching, the image expansion step percent, and the image rotation degree step;
>>> create a function that determines whether or not a larger image contains a smaller image.
>>> initialize an array to hold every possible cut out of the large image of the size of the smaller image.
>>> initialize a variable to hold the current small image degree rotation.
>>> iterate though each image degree rotation while rotation is less than 360 degrees, with all images always being circles since a microscope provides circles both for live images and stored images, which also happens to allow for easy image rotation.
>>>> iterate through each image percent size between the previously specified image size minimum and maximum, utilizing the specified step size.
>>>> generate the image using the determined rotation and sizing.
>>>> append the image to the modified images array.

>>>> increment the image step size.
>>>> increment the image degree rotation.
>>> iterate through each image in the modified images array.
>>>> create variables holding the dimensions of the large image and the small image.
>>>> initialize top left starting point x and y variables to iterate through large image.
>>> loop through each top left x and y point within the large image that and take a cut out the size of the small image.
>>>> if the cut out of the large image matches the small image at or above the previously specified match threshold, return that a match was found.
>>>> increment the y position.
>>> increment the x position.
>>> if no match was found, return false.
>> create a function to check if an image contains a color;
>>> load the individual pixel colors from the image into an array using the corresponding previously imported library;
>>> initialize variables holding the number of pixels processed and found;
>>> iterate through each image pixel, increment the pixel counter, and if the pixel color matches the specified color, increment the pixel found counter;
>>> if the percent of pixels matching the specified color are above the previously set threshold return true, otherwise return false;
>> create a function to determine if the image contains a shape;
>>> create a variable specifying if the shape was found in the image;
>>> initialize a coordinate pairs array;
>>> split the coordinate pairs from the line in the settings file into an array;
>>> iterate through each coordinate pair;
>>> split the coordinate pair into x and y;
>>>> append the x,y coordinates to the coordinates array;
>>>> generate a polygon from the coordinates using the corresponding imported library;
>>>> generate an array of all polygons in the image using the corresponding imported library.
>>> iterate through each polygon found in the image;
>>>> check if the polygon in the image matches the polygon specified in the settings file within the threshold previously specified and if so set the found variable equal to true;
>>> return the variable indicating whether or not the shape was found in the image.

[0030] In some embodiments, including one complete embodiment, in reference to FIG. 4, the software controlling the fluid sorter provides functionality comprising:

> import the previously created image processing library.

> import a library to connect to the relay.

> import a library to pause program execution.

> import a library to receive the system time.

> import a library to connect to access http(s) resources.

> create variables holding relay on off codes.

> create variables each holding arrays of the relay states of each controlled embodiment.

> create variables holding motor millimeters and seconds per step.

> initialize variables to hold the settings and log file.

> create variables holding the fluid chamber cubic millimeters, and running totals for cubic millimeters of fluid processed, and cubic millimeters of fluid sorted.

> establish an array holding embodiments camera IPs.

> create variables holding the cameras' username and password.

> create a variable to hold an array of embodiment relay IPs.

> create variables to hold relays' username and password.

> initialize a variable to hold an array of relay connections.

> create a function that when called, iterates through each relay IP address, established a connection if one has not been initialized or has been dropped, then creates a string for of the relay board states, and sends the relay states to the relay board.

> create a function that will save a snapshot of the image of the fluid under the microscope, by opening a connection to the camera, accessing the current camera image, and writing the image to a file.

> create a function to process the fluid in the compartment using the previously created image recognition library.

>> create an array of sort states of the embodiments.

>> create a variable to hold the current embodiment index.

>> initialize to a variable to hold the saved image integer index for naming saved images of fluid sorted.

>> create an array to hold the retrieved image from each embodiment camera.

>> iterate through each embodiment camera IP address;

>>> retrieve the image from the current embodiment camera, and append it to the images array;

>>> increment the embodiment index;

>> reset the embodiment index to 0.

>> iterate through each embodiment camera image;

>>> default the sort state for each embodiment to False.

>>> iterate through each setting, creating variables to hold the image file, color, or shape, and the match threshold and or and if the fluid should be sorted and if a match was found.
>>>> initialize a variable to hold the recognition mode and default it to image, then determine based on the setting, if it should be color or shape.
>>>> if the recognition mode is color, call the function to determine the match between the current image and the color specified in the setting, and set the sort state for the fluid chamber at the current index based on the result.
>>>> if the recognition mode is image, call the function to determine the match between the current image and the image specified in the setting, and set the sort state for the fluid chamber at the current index based on the result.
>>>> if the recognition mode is shape, call the function to determine the match between the current image and the shape specified in the setting, and set the sort state for the fluid chamber at the current index based on the result.
>>> iterate through each embodiment sort state, open the outflow door to correspond to the sort state of either keeping or discarding the fluid.
>>> for each sorter, open the inflow door.
>>> for each sorter, set the inflow and outflow doors relays to close.
>>> if sorted, add the fluid chamber volume to the total volume of fluid sorted out, save the image, naming it after the saved image index and the embodiment index increment the saved image index, and record to the log file the current system time and the name of the image file.
>>> add the fluid chamber volume to the total volume of fluid processed.
>>> update the relays.
>> display to the operator the total cubic centimeters processed and sorted.
> create a function to be called when the program is executed;
>> retrieve any command line arguments, including the settings file name, the log file name, and any requested embodiment action;
>> if the requested function is to save a snapshot from an embodiment video camera, call the corresponding function;
>> if a snapshot was not requested, read in the settings from the previously specified file, create a log file with the previously specified name, then continuously call the function to process fluid flow.

[0031] **In some embodiments, including one complete embodiment, create images of fluid components.** In some embodiments, including one complete embodiment, in reference to FIG. 1, images are taken of each fluid component, to be used by image recognition software, where the functionality to take the photos was created in the previous step, and can be executed by at the command line, in the same folder as the previously created script, typing "python run.py snapshot", which will save an image file in the same folder of the image of fluid in the sorting device from the digital camera connected to the microscope. After an image has been captured, use standard image editing software such as Photoshop, to cut out the individual components, while maintaining the original size of the image, then saving with the same file type and settings such as resolution as the original image.

[0032] In some embodiments, the image library can include every known blood component, so that it fires when an object is not recognized. The software may be run to recognize either images and or colors and or shapes. Colors are most appropriate when looking to sort out cancer cells that bind with a contrast agent such as Indocyanine Green, after the dye has been administered, where the colors are those that are the shades of green displayed by cancer cells when exposed to the contrast agent. The software accepts at the command line a file that names the image to recognize, the percent match required for the laser to fire, and whether or not the firing is to be live or in test mode. In test mode, a log of the laser fire is created but the laser itself is not fired. Computer controlled motor valve. More than one picture may be taken of the blood in the compartment as cells move around allows greater compartment depth of the height of a fluid tube.

[0033] In some embodiments, including one complete embodiment, matching may be done by shape using pixel coordinates, where the outline of a shape is zoomed in on and manually converted to a series of x,y coordinates. Export image outline from an illustration program after tracing the outline that supports Scalable Vector Graphics format, open the file in a text editor, and take the coordinates of the outline. The scale of the original image should be maintained. May also write out the x,y coordinates in sequence of the points that make up the border. Coordinates should be on the same line as a sequence of x,y values, for example "x,y x,y x,y x,y". If shapes will be identified in the images, because code will identify all continuous polygons in an image, so err on the side of higher contrast images, so that all shapes are correctly identified.

[0034] **In some embodiments, including one complete embodiment, develop a settings file of components to recognize by images and or shape and or color.** In some embodiments, including one complete embodiment, in reference to FIG. 2, a setting file is created, to list fluid components to recognize, by image file names and or colors of fluid and or geometric shape coordinates, along with whether or not to sort the component. If the settings file uses a library of images not taken by the system operator, may adjust camera color balance and light until a photo from the camera matches the cell sizes and color and contrast of library of images, library must be normalized so all images have the same color balance, contrast, and exposure.

[0035] In reference to FIG. 2, 2000 lists four images file names of images of leukemia with each followed by the action to take or sort and the match recognition percent to sort; 2001 lists three image file names of images of HIV with the action to take of sort and the percent match for the sort to occur; 2002 lists four image file names of images of Ebola; 2003 lists two image file names of images of red blood cells that state no action and will therefore simply be counted and specify a percent recognition rate to count; 2004 lists two image file names of images of white blood cells that state no action and will therefore simply be counted and specify a percent recognition rate to count; 2005 lists the coordinates of a shape with the action to sort if the shape is found. Text lines 2006 and above colors are displayed to sort out of the blood, which are shades of green displayed by cells when exposed to an imaging agent such as Indocyanine Green.

[0036] **In some embodiments, including one complete embodiment, optionally build additional units and split fluid flow between them to maximize throughput.** To increase throughput, multiple units may be created with the fluid flow split among them, utilizing tube splitters and joiners, to allow fluid flow may be split between multiple units may be used concurrently to allow fluid to flow at its natural speed. There are 5000 cubic centimeters (5 liters, 5000 milliliters) of blood in a human body, a milliliter of blood can have less than a hundred or over a million copies of a virus, so assuming a maximum of one operations to remove each, means a maximum of 5 billion operations, at .1 second per operation, requires 36 months, which when divided by 360 machines, is 1 day to clean 100% of the blood in a human body. Of course only one machine is required for blood analysis. Unlimited number of IP cameras and relay boards hooked up to a router, with the computer on the same network can get feeds from all cameras at once and operates all boards at once. To build a multi-unit sorting system, use fiber optic camera lines placed over the blood slides, to make them easy to switch out, and could allow a single blood slide unit to be created that has a hundred compartments in the same unit, and wouldn't have space constraints because of the equipment that must sit over them.

[0037] **In some embodiments, including one complete embodiment, run the computer code.** After the code has been written is may be run the computer code on the computer to which the components have been attached. In some embodiments, including one complete embodiment, in reference to FIG. 4, the computer code is written in Python, and is named run.py, and therefore the command executed at the command line "python run.py" will start and run the code continuously.

[0038] **In some embodiments, including one complete embodiment, verify unit effectiveness**. In some embodiments, to verify proper construction before use on a human, the unit may be run with discarded livestock blood.

[0039] **In some embodiments, including one complete embodiment, attach fluid sorting device to tubes, fluid bag(s), and or fluid source.** In some embodiments, including one complete embodiment, in reference to FIG. 1, a standard dialysis needle allows for blood to flow in and out of a patient, and is attached to fluid tubes 1700 and 1701, while fluid tube 1702 connects to sorted fluid bag 1702. In some embodiments, only a sample of fluid may be used, provided by means which may include a test tube or drop of fluid on a slide. In some embodiments, any fluid source may be connected.

[0040] **In some embodiments, including one complete embodiment, sort the fluid.** In some embodiments, including one complete embodiment, the system is run on the blood stream of anyone with detectable foreign objects that the computer can identify, to neutralize them from the person's bloodstream, by attaching the needle to the appropriate point in their body that will allow blood to circulate through the system. In some embodiments, the system is run on any fluid stream.

[0041] In some embodiments, where blood is sorted, in some cases exceptionally high concentrations of the components to sort out may exist in the blood thus reducing efficiency. For example, Ebola has a peak of 10 million copies per milliliter, and HIV has a peak of 1 million copies per millimeter, and Leukemia blood typically contains 20% underdeveloped white blood cells called blasts. To reduce concentrations of components, anti-viral drugs may be used, as for example, HIV gets down to 50 copies per milliliter when controlled with drugs. Additionally, the sorting device blood compartment can be reduced in size, and recirculate the bad blood bag through the machine, to have some of it sent back to the patient as cleared. Blood may also be diluted with traditional medical grade intravenous saline solution to push through residual or dilute blood when too dense in cell number, typically one liter of saline used. Drop of blood 5 million red blood cells. 1 milliliter is equal to 1000 cubic millimeters, 1 liter is about 60 cubic inches.

CLAIMS

What is claimed is:

1. A method for analyzing fluid, and or removing foreign objects in fluid, with the invention comprising:

a magnification device, which as part of the same device or an additional device, provides the magnified images to a computer.

a fluid sorting device that utilizes a outflow controller to sort blood, and may sort to two or more channels, with the sorting controller controlled utilizing computer commands.

a computer program that accesses the magnified images, processes them with image recognition software, which may use a range of combinations of rotated and resized input or output images, and if a specified component is identified in the fluid sample, it may sort the sample to the one blood bag, and sorts the rest back to the patient or another blood bag. Additionally, or alternatively, a record of image processing matches may be produced.

2. A non-transitory computer-readable recording medium holding stored instructions, which when executed by one or more processing devices, cause the one or more processing devices to implement a method comprising, with the invention comprising:

a magnification device, and as part of the same device or a separate device, provides the magnified images to a computer.

a fluid sorting device that utilizes an outflow controller to sort blood, and may sort to two or more channels, with the sorting controller controlled utilizing computer commands.

a computer program that accesses the magnified images, processes them with image recognition software, which may use a range of combinations of rotated and resized input or output images, and if a specified component is identified in the fluid sample, it may sort the sample to the one fluid bag, and sorts the rest back to the patient or another fluid bag. Additionally, or alternatively, a record of image processing matches may be produced.

FIG. 1

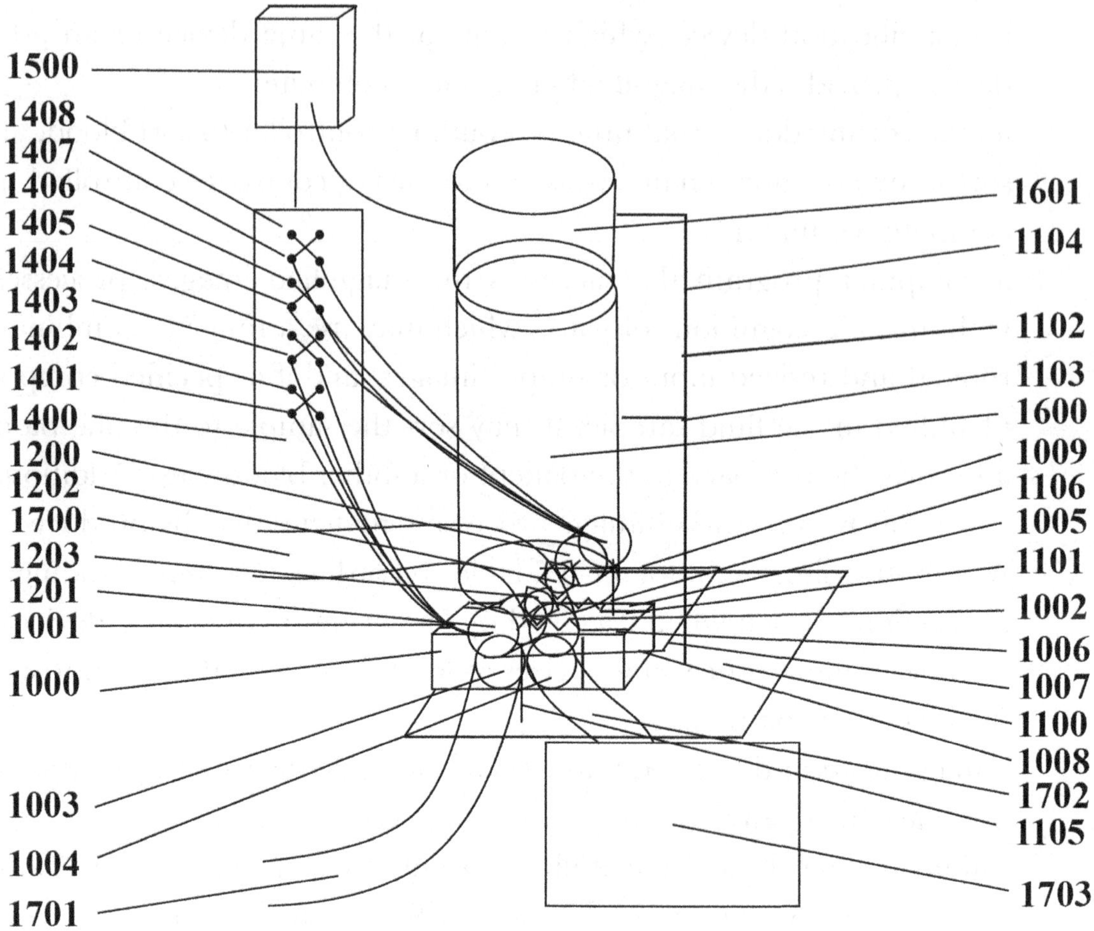

Jonathan Bannon Maher

FIG. 2

2000 – leukemia_1.jpg sort .85
leukemia_2.jpg sort .85
leukemia_3.jpg sort .85
leukemia_4.jpg sort .85
2001 – hiv_1.jpg sort .85
hiv_2.jpg sort .85
hiv_3.jpg sort .85
2002 – ebola_1.jpg sort .80
ebola_2.jpg sort .80
ebola_3.jpg sort .80
ebola_4.jpg sort .80
ebola_5.jpg sort .80
2003 – red_blood_cell_1.jpg .85
red_blood_cell_2.jpg .85
2004 – white_blood_cell_1.jpg .85
white_blood_cell_2.jpg .85
white_blood_cell_3.jpg .85
2005 – 0,0;10,10;10,20;20,20;20,10 sort
2006 – 5A6351 remove
636F57 remove
526F35 remove
3B5323 remove
78AB46 remove
4A7023 remove
458B00 remove
66CD00 remove
76EE00 remove
7CFC00 remove
7FFF00 remove
7F9A65 remove
8BA870 remove
9CBA7F remove
3D5229 remove

FIG. 3

3000 — import image analysis libraries that provide identifying image pixel colors, extracting and comparing shapes in images, and comparing two images overall similarity

```
import cv2
from shapely import geometry
from PIL import Image
import ImageChops
```

3001 — define a class for image processing

```
class image_processing():
```

3002 — create variables holding the minimum and maximum percent of image expansion for matching, the image expansion step percent, and the image rotation degree step

```
    image_size_minimum = 0.8
    image_size_maximum = 1.2
    image_size_step = .02
    image_degree_step = 2
```

3003 — create a function that determines whether or not a larger image contains a smaller image

```
    def image_contains_image(image_large, image_small):
```

3004 — initialize an array to hold every possible cut out of the large image of the size of the smaller image

```
        image_large_array = []
```

3005 – initialize a variable to hold the current small image degree rotation

```
degree = 0
```

3006 – iterate though each image degree rotation while rotation is less than 360 degrees, with all images always being circles since a microscope provides circles both for live images and stored images, which also happens to allow for easy image rotation

```
while degree < 360:
```

3007 – iterate through each image percent size between the previously specified image size minimum and maximum, utilizing the specified step size

```
    size = image_size_minimum
    while size <= image_size_maximum:
        width = image_large.width*size
        height = image_large.height*size
```

3008 – generate the image using the determined rotation and sizing

```
        image_large_adjusted = img.rotate(
            degrees, expand=True)
        image_large_adjusted = img.resize(
            (width,height), PIL.Image.ANTIALIAS)
```

3009 – append the image to the modified images array

```
        image_large_array.append(
            image_large_adjusted)
```

3010 — increment the image step size

```
size += image_size_step
```

3011 — increment the image degree rotation

```
degree += image_degree_step
```

3012 — iterate through each image in the modified images array

```
for large_adjusted in large_image_array:
```

3013 — create variables holding the dimensions of the large image and the small image

```
    large_width = large_adjusted.width
    large_height = large_adjusted.height
    small_width = image_small.width
    small_height = image_small.height
```

3014 — initialize top left starting point x and y variables to iterate through large image

```
    x = 0
    y = 0
```

3015 — loop through each top left x and y point within the large image that and take a cut out the size of the small image

```
while (x + small_width < = large_width)
    while (y + small_height < = large_height):
        large_cutout = large_adjusted.crop(
```

```
        (x, y, small_width, small_height))
```

3016 – if the cut out of the large image matches the small image at or above the previously specified match threshold, return that a match was found

```
      if ImageChops.difference(
        large_cutout, image_small) >=
        match_threshold:
        return True
```

3017 – increment the y position

```
      y = y+1
```

3018 – increment the x position

```
    x = x+1
```

3019 – if no match was found, return false

```
    return False
```

3020 – create a function to check if an image contains a color

```
def image_contains_color(
  image_source, color, image_match_threshold)
```

3021 – load the individual pixel colors from the image into an array using the corresponding previously imported library

```
  pixels = image_source.load()
```

3022 – initialize variables holding the number of pixels processed and found

```
pixel_count = 0
pixel_found_count = 0
```

3023 – iterate through each image pixel, increment the pixel counter, and if the pixel color matches the specified color, increment the pixel found counter

```
for x in range(image_source.width):
    for y in range(image_source.height):
        pixel_count += 1
        pixel_color = pixels[x,y]
        if pixel_color == color:
            pixel_found_count += 1
```

3024 – if the percent of pixels matching the specified color are above the previously set threshold return true, otherwise return false

```
if pixel_count > 0:
    if (pixel_found_count/pixel_count) >
        image_match_threshold:
        return True
return False
```

3025 – create a function to determine if the image contains a shape

```
def image_contains_shape(
    coordinates, image_large):
```

3026 – create a variable specifying if the shape was found in the image

```
found = False
```

3027 – initialize a coordinate pairs array

```
cordinates_array = []
```

3028 – split the coordinate pairs from the line in the settings file into an array

```
coordinates = coordinates.split(";")
```

3029 – iterate through each coordinate pair

```
for coordinate in coordinates:
```

3030 – split the coordinate pair into x and y

```
    coordinate = coordinate.split(",")
    x = coordinate[0]
    y = coordinate[1]
```

3031 – append the x,y coordinates to the coordinates array

```
    coordinates_array.append([x,y]]
```

3032 – generate a polygon from the coordinates using an the corresponding imported library

```
    coordinates_polygon = geometry.Polygon(
        [[p.x, p.y] for p in points])
```

3033 – generate an array of all polygons in the image using the corresponding imported library image_polygons = cv2.findContours(image_large)

3034 – iterate through each polygon found in the image

for image_polygon in image_polygons:

3035 – check if the polygon in the image matches the polygon specified in the settings file within the threshold previously specified and if so set the found variable equal to true

```
  if image_polygon.almost_equals(
    coordinated_polygon, decimal=
    match_threshold]):
    found = True
```

3036 – return the variable indicating whether or not the shape was found in the image

```
return found
```

FIG. 4

4000 – import the previously created image processing library

from image_processing import image_processing

4001 – import a library to connect to the relay

import telnetlib

4002 – import library to pause program execution

import time

4003 – import a library to receive the system time

from sys import time

4004 – import a library to connect to access http(s) resources

import urllib

4005 – create variables holding relay on off codes

off = 0
on = 1

4006 – create variables each holding arrays of the relay states of each controlled embodiment

relay_1_motor_in_forward = [off]
relay_2_motor_in_backward = [off]

relay_3_motor_out_forward = [off]
relay_4_motor_out_backward = [off]

4007 – create variables holding motor millimeters and seconds per step

motor_step_millimeters = 1
motor_step_seconds = 0.2

4008 – initialize variables to hold the settings and log file

settings = None
log_file = None

4009 – create variables holding the fluid chamber cubic millimeters, and running totals for cubic millimeters of fluid processed, and cubic millimeters of fluid sorted

fluid_chamber_cubic_millimeters = 0.2
fluid_processed_cubic_millimeters = 0
fluid_sorted_cubic_millimeters = 0

4010 – establish an array holding embodiments camera IPs

camera_ips = ["127.0.0.1"]

4011 – create variables holding the cameras' username and password

camera_username = "admin"
camera_password = "admin"

4012 – create a variable to hold an array of embodiment relay IPs

relay_ips = ["127.0.0.2"]

4013 – create variables to hold relays' username and password

```
relay_username = "admin"
relay_password = "admin"
```

4014 – initialize a variable to hold an array of relay connections

```
relays = []
for relay_ip in relay_ips:
    relays.append(None)
```

4015 – create a function that when called, iterates through each relay IP address, established a connection if one has not been initialized or has been dropped, then creates a string for of the relay board states, and sends the relay states to the relay board

```
def update_relays():
    index = 0
    for relay_ip in relay_ips:
        command = str(relay_1_motor_in_1_1[index])
        command+= str(relay_2_motor_in_1_2[index])
        command+= str(relay_3_motor_in_2_1[index])
        command+= str(relay_4_motor_in_2_2[index])
        command = str(relay_5_motor_out_1_1[index])
        command+= str(relay_6_motor_out_1_2[index])
        command+= str(relay_7_motor_out_2_1[index])
        command+= str(relay_8_motor_out_2_2[index])
        if not relay[index]:
            relays[index] =
                telnetlib.Telnet(relay_ips[index], 23)
            relays[index]
                .read_until(b"User Name: ")
            relays[index]
                .write(relay_username.encode('ascii') + b"\n")
            relays[index]
```

```
            .read_until(b"Password: ")
        relays[index].write(
        relay_password.encode('ascii') + b"\n")
    relays[index]
        .write(command.encode('ascii') + b"\n")
    index += 1
```

4016 — create a function that will save a snapshot of the image of the fluid under the microscope, by opening a connection to the camera, accessing the current camera image, and writing the image to a file

```
def snapshot():
    camera_image =
        urllib2.urlopen("http://" + camera_ips[0] +
        "/?username=" + camera_username +
        "&password=" + camera_password + ")
        .read()
    image_file = open("snapshot.jpg","wb")
    image_file.write(image)
    image_file.close()
```

4017 — create a function to process the fluid in the compartment using the previously created image recognition library

```
def process():
```

4018 — create an array of sort states of the embodiments

```
    sort_states = []
```

4019 — create a variable to hold the current embodiment index

```
    index = 0
```

4020 –	inititalize to a variable to hold the saved image intiger index for naming saved images of fluid sorted

```
saved_image_index = 0
```

4021 –	create an array to hold the retrieved image from each embodiment camera

```
camera_images = []
```

4022 –	iterate through each embodiment camera IP address

```
for camera_ip in camera_ips:
```

4023 –	retrieve the image from the current embodiment camera, and append it to the images array

```
    camera_images[index].append(
      urllib2.urlopen(
      "http://" + camera_ip +
      "/?username=" +
      camera_username + "&password=" +
      camera_password + ").read())
```

4024 –	increment the embodiment index

```
      index = index +1
```

4025 –	reset the embodiment index to 0

```
    index = 0
```

4026 –	iterate through each embodiment camera image

```
for camera_image in camera_images:
```

4027 — default the sort state for each embodiment to False

```
    sort_states[index] = False
```

4028 — iterate through each setting, creating variables to hold the image file, color, or shape, and the match threshold and or and if the fluid should be sorted and if a match was found

```
    for setting in settings:
        found = False
        setting = setting.split(" ")
        setting_one = ""
        setting_two = ""
        setting_three = ""
        try:
            setting_mode = setting[0]
        except:
            pass
        try:
            setting_image_color_shape = setting[1]
        except:
            pass
        try:
            setting_match_threshold = setting[2]
        except:
            pass
```

4029 — initialize a variable to hold the recognition mode and default it to image, then determine based on the setting, if it should be color or shape

```
        recognition_mode = "image"
```

```
if setting_image_color_shape
    .length()==6
    and
    setting_image_color_shape
    .find(".")==-1:
    recognition_mode = "color"

if setting_file_or_color.find(",")>-1:
    recognition_mode = "shape"
```

4030 — if the recognition mode is color, call the function to determine the match between the current image and the color specified in the setting, and set the sort state for the fluid chamber at the current index based on the result

```
if recognition_mode == "color":
    if image_processing.
        image_contain_color(
        image, color, threshold):
        sort_states[index] = True
```

4031 — if the recognition mode is image, call the function to determine the match between the current image and the image specified in the setting, and set the sort state for the fluid chamber at the current index based on the result

```
if recognition_mode == "image":
    if image_processing.
        image_contains_image(
        current_image, image_file) >=
        image_match_minimum:
        sort_states[index] = True
```

4032 — if the recognition mode is shape, call the function to determine the match between the current image and the shape specified in the setting, and set the sort state for the fluid chamber at the current index based on the result

```
if recognition_mode == "shape":
    if image_processing.
        image_contains_shape(
        current_image, image_file) >
        image_match_minimum:
        sort_states[index] = True
```

4033 — iterate through each embodiment sort state, open the outflow door to correspond to the sort state of either keeping or discarding the fluid

```
index = 0
for sort in sort_states:
    if sort:
        relay_5_motor_out_1_1[index] = on
    if not sort:
        relay_8_motor_out_2_1[index] = on
    index = index + 1
update_relays()
time.sleep(motor_step_seconds)

index = 0
for sort in sort_states:
    if sort:
        relay_5_motor_out_1_1[index] = on
    if not sort:
        relay_8_motor_out_2_1[index] = on
    index = index + 1
update_relays()
time.sleep(motor_step_seconds)
```

4034 — for each sorter, open the inflow door

```
index = 0
for sort in sort_states:
   relay_5_motor_out_1_1[index] = on
   index = index + 1
update_relays()
time.sleep(motor_step_seconds)
```

4035 — for each sorter, set the inflow and outflow doors relays to close

```
index = 0
for sort in sort_states:
   relay_5_motor_out_1_1[index] = on
```

4036 — if sorted, add the fluid chamber volume to the total volume of fluid sorted out, save the image, naming it after the saved image index and the embodiment index increment the saved image index, and record to the log file the current system time and the name of the image file

```
   if sort:
      fluid_sorted_cubic_centimeters +=
         fluid_chamber_cubic_centimeters
      image_file_name = "image" +
         str(saved_image_index) + "-" +
         str(index) + .jpg"
      sorted_image_index += 1
      image_file = open("snapshot.jpg","wb")
      image_file.write(camera_image)
      image_file.close()
            log_file_line =
               str(sys.time()) + " " + image_file
         log_file.write(log_file_line)
```

4037 – add the fluid chamber volume to the total volume of fluid processed

```
fluid_processed_cubic_centimeters +=
    fluid_chamber_cubic_centimeters
index = index + 1
```

4038 – update the relays

```
update_relays()
time.sleep(motor_step_seconds)
```

4039 – display to the operator the total cubic centimeters processed and sorted

```
print "cubic centimeters sorted: " +
    str(fluid_sorted_cubic_centimeters) + " of " +
    str(fluid_processed_cubic_centimeters)
```

4040 – create a function to be called when the program is executed

```
def main():
```

4041 – retrieve any command line arguments, including the settings file name, the log file name, and any requested embodiment action

```
parameters = sys.argv
settings_file_name = ""
log_file_name = ""
try:
    settings_file_name = parameter[1]
    log_file_name = parameter[2]
    action = parameter[3]
except:
```

```
    pass
```

4042 — if the requested function is to save a snapshot from an embodiment video camera, call the corresponding function

```
if settings_file_name == "snapshot":
    snapshot()
```

4043 — if a snapshot was not requested, read in the settings from the previously specified file, create a log file with the previously specified name, then continuously call the function to process fluid flow

```
else:
    settings = open(images_file_name,"r").readlines()
    log_file = open(log_file_name,"w+")
    while True:
        process()
```

HIGH YIELD LOW COST AGRICULTURAL PRODUCTION SYSTEM

INVENTOR JONATHAN BANNON MAHER

TECHNICAL FIELD

[0001] Embodiments of the invention relate generally to the field of agriculture, and more specifically to hydroponics.

ABSTRACT

[0002] Systems, methods, and apparatuses provide for a high yield low cost agricultural production system, by constructing a support structure with inset angled rows, structured for agricultural growth support, allowing the roots extending from growth structures to be exposed to oxygen and a periodic nutrient circulation, coupled with a support structure with indented rows composed of or supporting reflective material that provides for equal distribution of light from a single source among agriculture growth rows, therefore providing unprecedented yields, by combining vertical stacking, eliminating the spacing between growth, and optimized light distribution, and unprecedented cost reductions, by utilizing inexpensive commodity components while eliminating components for individual compartments.

REFERENCE TO RELATED DOCUMENTS

[0003] This application is provided the benefit and priority date of United States Patent and Trademark Office provisional patent application number 62/522,695, filed June 20th 2017 by inventor Jonathan Bannon Maher, which is incorporated herein in its entirety.

BACKGROUND

[0004] This section is intended to introduce the reader to various aspects of the art that may be related to various aspects of the present techniques, which are described and or claimed. This discussion is believed to be helpful in providing the reader with background information to facilitate a better understanding of the various aspects of the

present disclosure. Accordingly, it is understood that these statements are to be read in this light, and not a citation of prior any art.

[0005] There are currently several methods for agricultural production. Traditional outdoor agricultural production provides lower yields per unit of area than in controlled indoor environments such as hydroponics, while having the limitation of being a horizontal system as opposed to vertical, thus not making optimal use of space. Hydroponic systems are often expensive, require individual growth supports, and lack an efficient method for equal distribution of light.

BRIEF DESCRIPTION OF THE ILLUSTRATIONS

[0006] An embodiment of the invention is illustrated by way of example, and not by way of limitation, in the figures of the accompanying drawings, and embodiments may not contain all components, may contain additional components, and may contain functionally similar components.

[0007] FIG. 1 is an embodiment an example of the apparatus that allows for high yield low cost agricultural production including a growth support structure and a light distribution structure.

DETAILED DESCRIPTION

[0008] It is understood that, as in any engineering or design project, the development of any actual implementation will include numerous implementation specific decisions made to achieve the developers' specific goals, such as compliance with business related and system related constraints, which may vary from one implementation to another. It is understood that such a development effort might be complex and time consuming, but is nevertheless a routine undertaking of design, fabrication, and manufacture for those skilled in the art having the benefit of this disclosure. The disclosed steps may be read as prefaced by "In some embodiments, including one complete embodiment, ", may be executed or performed in other orders or sequences, and are not limited to the order and sequence shown and described, which are provided to enable ease in constructing an embodiment, and along with each components of each step, may be removed, modified, combined, or rearranged, and other steps and or step components may be added, without departing from the scope of this disclosure and or invention. Although embodiments of the invention have been described and illustrated in the disclosed implementations, it is understood that the present disclosed subject matter, including apparatuses, methods, specification, and illustrations, has been made only by way of example, not by way of limitation, and the methods and apparatuses may be used in other systems, and that numerous changes and optimizations in the details of implementation of the invention and or embodiment are made without such modifications departing from the spirit and scope of this disclosure and or embodiments of the invention. Although the disclosure has been shown and described with respect to one or more embodiments, features of the disclosed embodiments can be combined and rearranged in various ways, and changes including equivalent alterations, substitutions, modifications, and additional efficiencies will of course occur to someone of ordinary skill in the art without departing from the spirit and scope of this disclosure and or invention. In particular regard to the various functions performed by the described components, the terms used to describe such components are intended to correspond, unless otherwise indicated, to any component which performs the specified function of the described component, or is functionally equivalent to the described component, even though not structurally equivalent to the disclosed structure which performs the function in the implementations described in this disclosure. In addition, while a particular feature of the disclosure may have been provided with respect to only one of several embodiments, such feature may be combined with one or more other features of other embodiments as may be desired and advantageous for any given or particular application. In some instances, well-known circuits, structures and techniques have not been shown in detail in order not to obscure the understanding of this disclosure. Articles in this disclosure such as "a" "an" and "the" may allow for both singular and plural forms. Verbs in this disclosure such as "is" may be read as "may be". Conjunctions in this disclosure such as "or" as used herein may be interpreted as inclusive or meaning any one or any combination, where "A, B or C" means "any of the following: A; B; C; A and B; A and C; B and C; A, B and C". Relational terms in this disclosure, for example first and second, top and bottom, left and right, are to distinguish one entity or action from another, and may not necessarily require or imply a relationship, or order between, such entities or actions. The disclosure includes the best mode contemplated by the inventor, a completely described specific embodiment, along with optional components and alternative embodiments to best suit the implementer, measurements in imperial and metric units to support universal understanding, and dramatically exceeds claims support requirements and enablement requirements by allowing for selection and or construction of the required components to be carried out easily, quickly, and routinely by persons of ordinary skill in the art, who are provided the additional benefit of utilizing readily available commodity components whenever possible. The present disclosure includes material protected by copyrights, and the owner of the copyrights hereby reserves all rights, but with authorization for publication as required by government patent offices.

Various embodiments of the present invention may provide all, some or none of the disclosed technical advantages.

[0009] The inventor retains absolutely no liability for any implementation of this invention, and the invention is implemented exclusively at the risk and liability of the implementer.

[0010] **In some embodiments, including one complete embodiment, construct the structure of the growth platform and attach mesh netting.** In some embodiments, including one complete embodiment, in reference to FIG. 1, the growth support structure consists on one side of steel punched flat bars 1000 1001 1002 1003, bolted or welded together in a rectangular formation, and on the other side steel punched flat bars 1004 1005 1006 1007, bolted or welded together in a rectangular formation, held together by threaded steel bars 1100 1101 1102 1103 1104, each positioned to form an inset slightly upwardly angled growth pocket when wrapped in mesh netting that allow fruits and vegetables to be secure in their horizontal growth, with the bars held in place by nuts on either side of the bar when resting inside the puncture hole of the punctured steel bars, or welded, with threaded steel bard 1005 1006 atop the structure in such a position that they support pipe 1301, with the structure lined with a mesh plastic net 1200 secured by wire twists. In some embodiments, additional nets may be added to provide additional support as required by the growth being supported. In some embodiments, the net may have a root structure medium added, such as small stones that won't overwhelm or fall through the netting. In some embodiments, which may include one complete embodiment, the support structure is a single continuous component created through manufacturing or 3d printing, which may include the functional equivalent of netting.

[0011] **In some embodiments, including one complete embodiment, construct the structure of the light distribution system and attach reflective material.** In some embodiments, including one complete embodiment, the light distribution system allows for light to come in from one source, such as a commodity hydroponic ultraviolet lamp, and be evenly distributed to each pocket of organic growth, where the light distribution support structure consists of the growth support structure consists on one side of steel punched flat bars 2000 2001 2002 2003, in a rectangular formation, and on the other side steel punched flat bars 2004 2005 2006 2007, in a rectangular formation, , held together by threaded steel bars 2100 2101 2102 2103 2104, each positioned to form an inset angled reflection that when wrapped in reflective sheet 2200 provide even light distribution to the growth structure pockets, with the bars held in place by nuts on either side of the bar when resting inside the puncture hole of the punctured steel bars, or welded, with threaded steel bars 2005 2006 atop the structure in such a position that they hydroponic growth light 2300, with a reflective sheet 2200 is run through the bars, with the bars placed in such a way that each level of the reflecting structure, moves further in the further down it is, so that each pocket in the corresponding growth rack will receive approximately the same amount of light. In some embodiments, which may include one complete embodiment, the light distribution structure is a single continuous component created through manufacturing or 3d printing, and may be made of a reflective material or have a reflective coating.

[0012] **In some embodiments, including one complete embodiment, construct the nutrient circulation system.** The nutrient system consists of a half of a sawed in half 1 inch diameter PVC pipe with caps as the nutrient drop pipe 1300, and half of a sawed in half 12 inch PVC pipe with caps as the nutrient receiving pipe 1301, a pump 1600 having input tube 1601 from pipe 1301, and a pump output tube 1304 to pipe 1300, and a repeat cycle timer 1500 connected to pump 1600 and set on an interval that provides regular circulation of the nutrient solution over the roots. In some embodiments, the nutrient circulation system piping may be manufactured or 3d printed as part of the support structure.

[0013] **In some embodiments, including one complete embodiment, install the system in a controlled environment.** In some embodiments, the system is installed in an enclosed temperature controlled area. In some embodiments, when installed outdoors, it may be put into a hole dug in the ground of sufficient depth to contain the constructed system, then covered with a plastic tarp, to utilize the free year round substantial temperature control provided by the ground. In some embodiments, the embodiment may simply be put above ground in a plastic tent created from a steel or plastic pole structure enclosed by plastic tarps. In some embodiments, the system may be out in the open air.

[0014] **In some embodiments, including one complete embodiment, plant seeds, transfer to growth structure, and harvest when ready.** Seedlings should first be placed in a growth medium until they have developed that will allow them to stay in place once transitioned to the growth support structure and stay in place while absorbing the nutrient solution as it passes.

CLAIMS

What is claimed is:

1. An apparatus for high yield low cost agricultural production with the invention comprising:

   a support structure that provides for inset rows to support agricultural growth;

   a pump to circulate nutrients.

   a support structure that provides for inset rows to support agricultural growth.

2. An apparatus for light distribution for agricultural production with the invention comprising:

   a support structure having indented light reflecting rows providing for equal distribution of light from a single source among agriculture growth rows.

FIG. 1

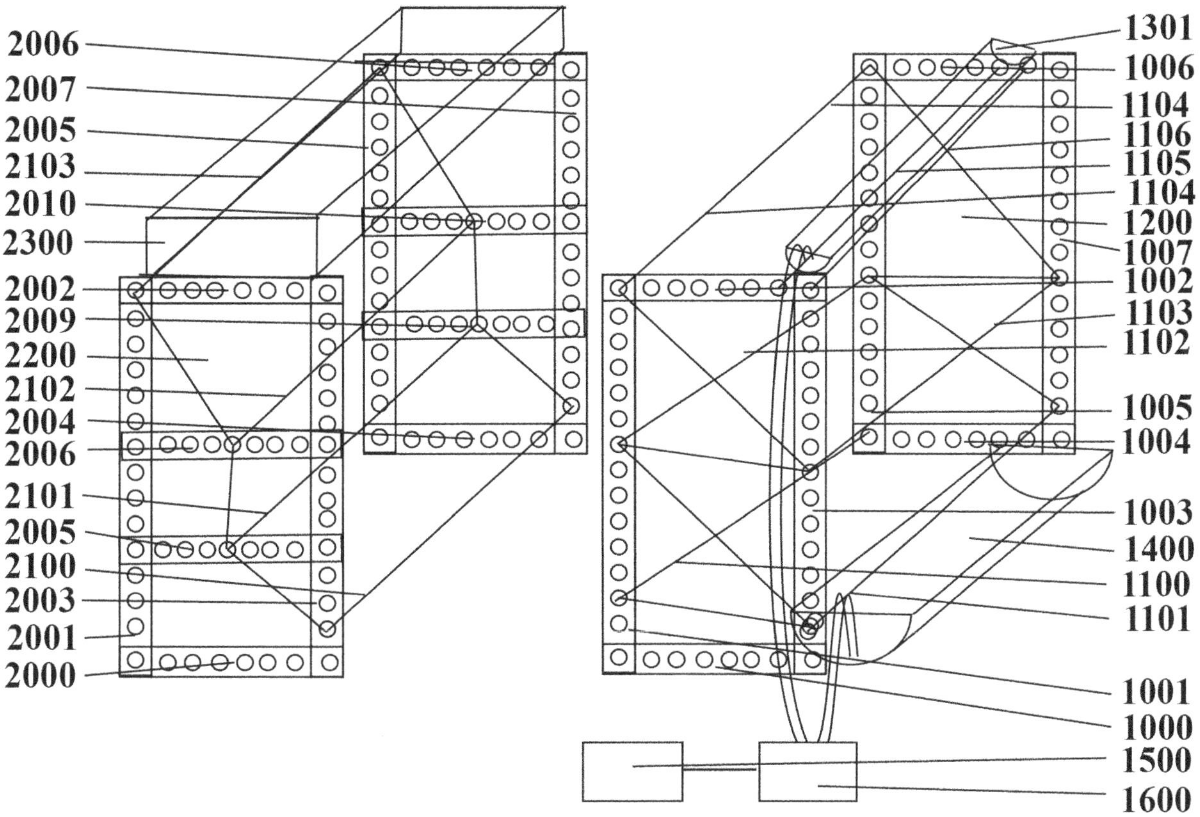

Jonathan Bannon Maher

AUTONOMOUS HIGH SPEED GROUND TRANSPORTATION

INVENTOR JONATHAN BANNON MAHER

TECHNICAL FIELD

[0001] Embodiments of the invention relate to the field of autonomous high speed transportation.

ABSTRACT

[0002] Systems, methods, apparatuses, and computer programs encoded on a computer storage medium, provide for autonomous high speed transport in excess of speeds of other ground based transportation, where a vehicle motor connects to a gear rack track, and the motor may be self-powered, therefore rendering the inefficiency produced by air resistance at high speeds irrelevant, though for additional efficiency, an air evacuated tunnel may be used, while software provides for autonomous control of the vehicle, allowing for the selection of a supported destination anywhere in the world, and then controlling track changes during transport.

REFERENCE TO RELATED DOCUMENTS

[0003] This application is provided the benefit and priority date of United States Patent and Trademark Office provisional patent application number 62/522,699, filed June 20th 2017 by inventor Jonathan Bannon Maher, which is incorporated herein in its entirety.

BACKGROUND

[0004] This section is intended to introduce the reader to various aspects of the art that may be related to various aspects of the present techniques, which are described and or claimed. This discussion is believed to be helpful in providing the reader with background information to facilitate a better understanding of the various aspects of the present disclosure. Accordingly, it is understood that these statements are to be read in this light, and not a citation of prior any art.

[0005] Existing high speed ground based transportation systems use magnetic levitation, and may achieve speeds of up to around 350 miles per hour. A high speed ground based transportation system, as described by Jonathan Bannon Maher in the 2011 book The Destiny of Humanity, utilizes self driving pods propelled by magnetic levitation in an air evacuated high speed tunnel based system over land already government controlled, with a third party proposed variation on the system using fan blades instead of magnetic levitation, except that when air is evacuated from the system the fan blades become ineffective.

BRIEF DESCRIPTION OF THE ILLUSTRATIONS

[0006] An embodiment of the invention is illustrated by way of example, and not by way of limitation, in the figures of the accompanying drawings, and embodiments may not contain all components, may contain additional components, and may contain functionally similar components.

[0007] FIG. 1 is an embodiment of the vehicle and track that allows for optimized high speed ground based transportation.

[0008] FIG. 2 is an embodiment of the transportation station allows for optimized high speed ground based transportation.

[0009] FIG. 3 is an embodiment of the station network that allows for optimized high speed ground based transportation.

[0010] FIG. 4 is an embodiment of text containing network routes and paths for use by the software.

[0011] FIG. 5 is an embodiment of computer code embedded flow diagram, the provides software for the operation of the embodiment.

DETAILED DESCRIPTION

[0012] It is understood that, as in any engineering or design project, the development of any actual implementation will include numerous implementation specific decisions made to achieve the developers' specific goals, such as compliance with business related and system related constraints, which may vary from one implementation to another. It is understood that such a development effort might be complex and time consuming, but

is nevertheless a routine undertaking of design, fabrication, and manufacture for those skilled in the art having the benefit of this disclosure. The disclosed steps may be read as prefaced by "In some embodiments, including one complete embodiment, ", may be executed or performed in other orders or sequences, and are not limited to the order and sequence shown and described, which are provided to enable ease in constructing an embodiment, and along with each components of each step, may be removed, modified, combined, or rearranged, and other steps and or step components may be added, without departing from the scope of this disclosure and or invention. Although embodiments of the invention have been described and illustrated in the disclosed implementations, it is understood that the present disclosed subject matter, including apparatuses, methods, specification, and illustrations, has been made only by way of example, not by way of limitation, and the methods and apparatuses may be used in other systems, and that numerous changes and optimizations in the details of implementation of the invention and or embodiment are made without such modifications departing from the spirit and scope of this disclosure and or embodiments of the invention. Although the disclosure has been shown and described with respect to one or more embodiments, features of the disclosed embodiments can be combined and rearranged in various ways, and changes including equivalent alterations, substitutions, modifications, and additional efficiencies will of course occur to someone of ordinary skill in the art without departing from the spirit and scope of this disclosure and or invention. In particular regard to the various functions performed by the described components, the terms used to describe such components are intended to correspond, unless otherwise indicated, to any component which performs the specified function of the described component, or is functionally equivalent to the described component, even though not structurally equivalent to the disclosed structure which performs the function in the implementations described in this disclosure. In addition, while a particular feature of the disclosure may have been provided with respect to only one of several embodiments, such feature may be combined with one or more other features of other embodiments as may be desired and advantageous for any given or particular application. In some instances, well-known circuits, structures and techniques have not been shown in detail in order not to obscure the understanding of this disclosure. Articles in this disclosure such as "a" "an" and "the" may allow for both singular and plural forms. Verbs in this disclosure such as "is" may be read as "may be". Conjunctions in this disclosure such as "or" as used herein may be interpreted as inclusive or meaning any one or any combination, where "A, B or C" means "any of the following: A; B; C; A and B; A and C; B and C; A, B and C". Relational terms in this disclosure, for example first and second, top and bottom, left and right, are to distinguish one entity or action from another, and may not necessarily require or imply a relationship, or order between, such

entities or actions. The disclosure includes the best mode contemplated by the inventor, a completely described specific embodiment, along with optional components and alternative embodiments to best suit the implementer, measurements in imperial and metric units to support universal understanding, and dramatically exceeds claims support requirements and enablement requirements by allowing for selection and or construction of the required components to be carried out easily, quickly, and routinely by persons of ordinary skill in the art, who are provided the additional benefit of utilizing readily available commodity components whenever possible. The present disclosure includes material protected by copyrights, and the owner of the copyrights hereby reserves all rights, but with authorization for publication as required by government patent offices. Various embodiments of the present invention may provide all, some or none of the disclosed technical advantages.

[0013] The computer code descriptions disclosed, in order to provide comprehensive enabling disclosure, rather than utilizing flow charts, which according to Patent Cooperation Treaty 11.11a are prohibited from containing "text matter, except a single word or words, when absolutely indispensable, such as... a few short catchwords indispensable for understanding", are provided in a text only format where the number of arrows preceding a line indicate logical block level, semicolons indicate a new segment of a logical block, and periods indicate the closure of one or more logical blocks. It is understood that any computer code representations in this disclosure are merely illustrative, rather than restrictive. While code may be written in nearly any computer language, including Java and C++, the illustrative computer code descriptions were derived from code written the Python language, which may be run through the Python interpreter, with appropriate supportive libraries, which at the time of disclosure, may run on nearly any computer, for example one with an Intel or AMD processor, running a current version of Linux, Windows, or Mac OS. All code components may read as if prefaced by "In some embodiments, including one complete embodiment, ". In some embodiments, functionality may be modified, rearranged, excluded, and added. To provide more fundamental computer system details, in some embodiments, the functionality associated with the disclosed computer code descriptions may be referred to as a script, module, software, software application, or code, and can be written in any form of language, including compiled, interpreted, declarative, or procedural, able to be deployed in any form suitable for use in a computing environment, including as an independent or integrated program, module, component, or subroutine, for execution by the computer system, implemented on one or more independent or integrated computers, utilizing a central processing unit in the form of one or more general or special purpose microprocessors, in conjunction with digital electronic circuitry, which may include special purpose logic circuitry such as a field programmable gate array or application specific integrated circuit, with the computer controlled by and operatively coupled to tangibly embodied software and or firmware, which may include code that creates an environment for code execution, including individual or combined use of processor firmware, a protocol stack, a database management system, and an operating system, where such software and or firmware may exist in one or more parts in memory on one or more computers, and is encoded on one or more tangible non transitory software carriers, such as individual or combined use of a random or serial access device or substrate, a semiconductor memory device, transient or persistent random access memory, a magnetic, magnetic optical, or optical disk, or encoded on an artificially generated transmitted signal, for example, optical, electrical, or electromagnetic, transmitted using a sending and a receiving apparatus, where the interaction between the user and the software may be implemented by operatively coupling, to the local implementing computer, or a local computer connected to one or more remote computers through a local or wide area network, a display device which may implement liquid crystals or light emitting diodes, a keyboard, and a pointing device.

[0014] The inventor retains absolutely no liability for any implementation of this invention, and the invention is implemented exclusively at the risk and liability of the implementer.

[0015] In some embodiments, including one complete embodiment, a single rail gear rack track is built, with wheels on both sides of the track, providing stability and to preventing the possibility of derailment, where the vehicle motor connects through a gear to the gear rack track, where the motor may be self-powered, thus eliminating the need for fuel efficiency when dealing with air resistance, and potentially able to travel up to thousands of miles an hour without requiring air evacuated tunnels, thus providing exceptionally low construction costs relative to all current and proposed forms of high speed transportation. Additionally, the lower wheels of the unit are shaped to allow them to run on existing train and subway tracks, and all existing train and subway tracks globally should be interconnected to allow for non-stop travel to anywhere in the world, allowing for the use of existing transportation investments, while allowing vehicles to achieve maximum speed when operating on the specialized track.

[0016] **In some embodiments, including one complete embodiment, construct the passenger car.** In some embodiments, including one compete embodiment, the passenger car has two radars, one on the front and one on the side, where the side radar allows for detection of oncoming traffic, to see if the vehicle can move onto the main track, and a front radar to know if it needs to slow down for traffic ahead, with a video camera whose image feed is read by software to monitor the track curve angle and adjust speed to the predetermined speed for that angle, with wheels both above and below the track hold the passenger compartment in place to make it impossible for a train to ever derail, and that may turn, to extent the track allows, through a steering wheel, to allow the passenger car to enter and exit stations. In some embodiments, the radars systems are attached in pairs to prevent a single failure from potentially causing an accident. The track will always move passenger cars in one direction clockwise direction, so the side radar is only required on one side.

[0017] In some embodiments, the vehicle consists of a steel shell 1000, two primary fixed axles 1001 1002, with fixed axle 1001 having fixed axle wheel holders 1003 1004, axle 1002 having fixed axle wheel holders 1005 1006, eight steel wheels 1100 1101 1102 1103 1104 1105 1106 1107, each of which has a bearing to rotate around the axle, with the vehicle propelled by motor 1200, with a gear on the motor axle 1201 that allows is it to connect the gear rack track, and front mounted radar 1400 and side mounted radar 1401, and a front mounted video camera 1500 to analyze the track curve, a computer 1600, a computer controlled relay board 1601, and computer controlled linear actuator 1602 that allows the front steering axle to be turned by a computer control. In some embodiments, including one complete embodiment, the lower wheels are designed to also be operable on existing train and subway tracks.

[0018] In some embodiments, which may include one complete embodiment, instead of seats, some vehicles are designed to hold shipping containers, and or some pods are designed to allow a car to drive onto them.

[0019] In some embodiments, the vehicle has horizontal and vertical shock absorbing springs that allow it to shift up to 1 foot in either direction, thereby reducing the feeling of horizontal shift when on curved segment of track.

[0020] **In some embodiments, including one complete embodiment, produce the track.** In some embodiments, each track segment consists of a single steel bar, with a center screw mounted commodity gear rack 1301, for the motor to grip, with beam 1300 supported on either end by vertical beams 1302 1303, with steel beam 1300 having a width equal to the width of the gear track plus the width of two wheels, with the rest simply being flat metal allowing the wheels to grip the track from both sides. Given a commodity cargo container can hold a maximum of 30 tons before breaking, the steel used for the beams and gear racks must hold more than that to never deform, where steel high grade steel may be used such as AISI 4130 which has a strength of 951 megapascals (137,895 pounds or about 69 tons of pressure per square inch). Having a detachable gear rack on the wheel track allows the gear rack track to be easily replaces should it become worn after decades of use. The track being a single metal beam makes it extremely inexpensive to manufacture, while also being exceptionally strong.

[0021] In some embodiments, including one complete embodiment, there are local, express, and high speed tracks, where vehicles travel at different speeds, allowing a person to travel within a city on a low speed track, travel from the city to the high speed connection point on an express track, then take the express track across the country, then transition back to an express track to a major city, and a local track to arrive at the final destination within that city. This allows cities to build this system locally now, and later their tracks and passenger cars to the larger system.

[0022] In some embodiments, including one complete embodiment, the cars are initially manually controlled. In some embodiments, the passenger cars are automatically controlled, where a passenger selects the destination station on a computer screen, and the software determines which exit point to take at each station.

[0023] **In some embodiments, including one complete embodiment, install the track.** The track is installed by digging a hole 10 feet deep, putting steel beams, then surrounding the steel beam with concrete, where all steel beams are installed at a consistent height and gradient that allows the passenger compartments to smoothly travel the track at very high speeds without feeling changes in gradient. Because the tracks are bolted to the tops of the poles, steel gradient discs 1304 1305 may be added or removed at any time between the top of the poles and tracks, which are simply slices of the metal pole in varying heights that allow the track to remain at a consistent level should the ground shift over time, where the gradient discs have the identical height of the metal poles and are made in an assortment of heights to suit the need of that particular track point. The track is then screwed to the top of the steel beam.

[0024] In some embodiments, including one complete embodiment, because it's not structurally, financially, or politically reasonable to substantially tunnel under major cities whose bedrock stabilizes billion dollar buildings weighing hundreds of thousands of tons, especially if the area is on a fault line, for example as much of California is, with at least one major building already leaning from the bedrock being inadequate, the track is installed above ground, and in an elevated manner to maintain as straight a line during travel as possible to reduce uncomfortable momentum.

[0025] **In some embodiments, including one complete embodiment, construct a network of stations.** In some embodiments, which may include one complete embodiment, vehicles enter and exit the track from stations utilizing a segment of track run parallel the main track with a connection on either end. In some embodiments, where the passenger car is automatically controlled, when the no approaching cars are detected, the passenger car starts to enter the track. If the station has connects multiple tracks, the parallel track has multiple exit and entry points.

[0026] In some embodiments, which may include one complete embodiment, the structure of a station will be laid out as specified in FIG. 2.

[0027] In some embodiments, which may include one complete embodiment, the structure of a network of stations is laid out as specified in FIG. 3. Because the costal states are too developed to have lines running directly through them, the proper way to picture the routes is a spoked rectangle, where circle in the center of the country where abundant land is available that allows for the lines to be as straight as possible and therefore allow to maximum speed, with entry and exit lines to major cities using the path of least resistance, direct between major cities is the path of greatest resistance, need to connect to backbone, similar to a computer network, getting from home to access point is dramatically slower than data travelling on backbone.

[0028] **In some embodiments, including one complete embodiment where a network of stations is created, create a settings file for the transportation network for the passenger car to use.** The settings file provides for the maximum speed allowed on each track, as well as the route and station exit point that should be taken between any two points.

[0029] **In some embodiments, including one complete embodiment, write the software for automated control of the passenger cars.** In some embodiments, including one complete embodiment, software for passenger car with side radar and front radar to detect approaching cars to get on track and ones in front to stop. The software specified the passenger car maximum speed on straight track as well as the maximum speed for each curve angle to ensure passenger comfort. The software also reads a station destination list, so that the car can know which exit track to take each time it wants to go to a destination. In some embodiments, the settings file containing the available routes and associated paths resembles that in FIG. 4. In some embodiments, the software resembles that in FIG. 5, providing functionality comprising:

> import a library to connect to the radars.
> import a library to connect to the cameras.
> create a variable holding routes file name.
> create a variable holding the current index of the station on the route.
> create a variable holding the front radar detected meters to nearest object.
> create a variable holding the side radar detected meters to nearest object.
> create a variable holding the kilometers traveled this trip.
> create a variable holding the current speed.
> create a value holding whether or not the embodiment is healthy and set it to true.
> create variables to hold the state of the relays.
> create variable holding the connections for the front and side radars and the relay.
> create variables holding the camera IP address, username, and password.
> create a function to operate the steering motor to turn the embodiment left at the next opportunity.
> create a function to operate the steering motor to turn the embodiment right at the next opportunity.
> create a function to set the percent speed engagement.
> define the function that runs on script execution.
>> create a variable holding the name of the route provided at the command line.
>> create a variable stating if the vehicle has arrived at the destination and set it to false.
>> loop while the vehicle has not arrived at the destination.
>> if there is no side radar connection, initialize a connection.
>> if there is no side radar connection, initialize a connection.
>> if there is no relay connection, initialize a connection.
>> retrieve the distance to an obstacle from the front radar, and if there is an exception, set the variable for vehicle health to false.
>> retrieve the distance to an obstacle from the side radar, and if there is an exception, set the variable for vehicle health to false.
>> check if station using video camera.
>> check if the video camera image indicates a station has been found.
>> proceed if a station was found.
>> turn wheel in the direction specified.
>> Slow the speed of the embodiment to decrease to 1%.

[0030] In some embodiments, the software to provide a user interface to the system resembles that in FIG. 6. In some embodiments, additional safeguards may be written into the software to allow it to safely recover without detectable interruption in the event of a computer software or hardware failure. Additionally, if the unit fails any health

checks, a variable is set that identifies it should be decommissioned at the next station. In some embodiments including one complete embodiment, the software to provide the user interface provides functionality comprising:
> import libraries to serve and respond to http requests.
> import library for operating system access.
> define a function to handle http requests.
>> retrieve any route specified in the user http request command.
>> if a command was provided by the user execute it through a command line call.
>> read all available routes into a variable.
>> create a variable to hold the output to display to the user.
>> initialize a form with a drop down from which the rider can select a route.
>> iterate through each route.
>> add the route to the display as a selectable option.
>> close the route selection drop down.
>> add a button to submit the route selection.
>> close the form and output display.
>> display the output to the user.

[0031] In some embodiments, which may include one complete embodiment, the computer to run connect the devices to will require two USB ports and two Ethernet ports, with the radars attached to the USB ports and the camera and motor control relay connected to the Ethernet ports.

[0032] **In some embodiments, including one complete embodiment, enjoy convenient private high speed low cost autonomous travel.** Once the track has been built between at least two points, the software may be activated on the passenger car, with the rider selecting their starting station and ending station from drop down menus and pressing start, where the software to control the car is accessible through a web browser interface, resembling that in FIG. 7.

CLAIMS

What is claimed is:

1. A system for transportation with the invention comprising:
   a vehicle with a motor on a track.
2. An apparatus for transportation with the invention comprising:
   a vehicle with a motor on a track.
3. A method for point to point transportation with the invention comprising:

a vehicle utilizing software to identify departure and arrival points, stations, and routes, thereby allowing for automated point to point travel.

4. A non-transitory computer-readable recording medium holding stored instructions, which when executed by one or more processing devices, cause the one or more processing devices to implement a method comprising, with the invention comprising:

to identify departure and arrival points, stations, and routes, thereby allowing for automated point to point travel.

FIG. 1

Jonathan Bannon Maher

FIG. 2

Jonathan Bannon Maher

FIG. 3

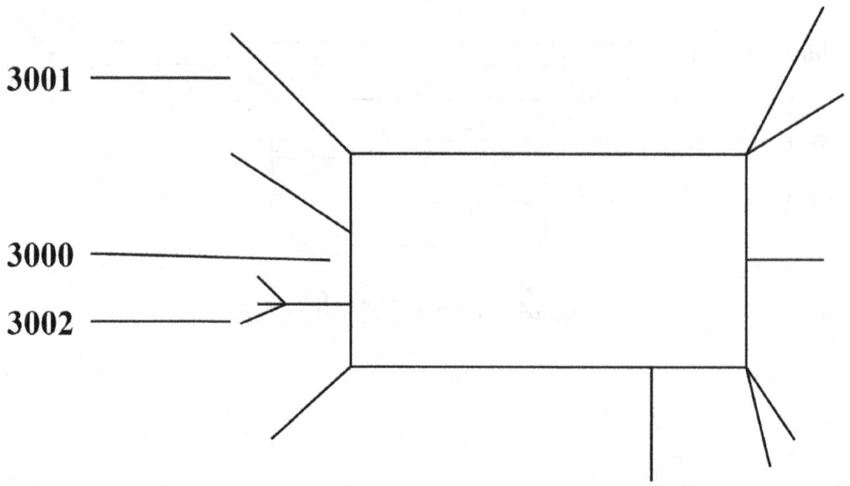

Jonathan Bannon Maher

FIG. 4

Los Angeles > New York; left, left

FIG. 5

5000 – import a library to connect to the radars

```
import serial
```

5001 – import a library to connect to the cameras

```
import urllib
```

5002 – create a variable holding routes file name

```
routes_file = "routes.txt"
```

5003 – create a variable holding the current index of the station on the route

```
station_index = 0
```

5004 – create a variable holding the front radar detected meters to nearest object

```
radar_front_meters = 0
```

5005 – create a variable holding the side radar detected meters to nearest object

```
radar_side_kilometers_per_hour = 0
```

5006 – create a variable holding the kilometers traveled this trip

```
traveled_kilometers = 0
```

5007 – create a variable holding the current speed

current_speed_kilometers = 0

5008	–	create a value holding whether or not the embodiment is healthy and set it to true

unit_healthy = True

5009	–	create vaiables to hold the state of the relays

relay_1_motor_forward_state = 0
relay_2_motor_backward_state = 0

5010	–	create variable holding the connections for the front and side radars and the relay

radar_front = None
radar_side = None
relay = None

5011	–	create variables holding the camera IP address, username, and password

camera_ip = "127.0.0.2"
camera_username = "admin"
camera_password = "admin"

5012	–	create a function to operate the steering motor to turn the embodiment left at the next opportunity

```
def turn_left():
  relay_1_motor_forward_state = on
  update_relay()
```

5013	–	create a function to operate the steering motor to turn the embodiment right at the next opportunity

```
def turn_right():
    relay_1_motor_forward_state = on
    update_relay()
```

5014 – create a function to set the percent speed engagement

```
def speed_engagement(speed_percent):
    relay_3_motor_forward_state = on
    update_relay()
```

5015 – define the function that runs on script execution

```
def main():
```

5016 – create a variable holding the name of the route provided at the command line

```
    route = sys.argv[2]
```

5017 – create a variable stating if the vehicle has arrived at the destination and set it to false

```
    arrived = False
```

5018 – loop while the vehicle has not arrived at the destination

```
    while not arrived:
```

5019 – if there is no side radar connection, initialize a connection

```
        radar_side = serial.Serial()
        radar_side.baudrate = 9600
        radar_side.timeout = 0
        radar_side.port = "/dev/ttyUSB0"
```

```
radar_side.open()
```

5020 — if there is no side radar connection, initialize a connection

```
radar_front = serial.Serial()
radar_front.baudrate = 9600
radar_front.timeout = 0
radar_front.port = "/dev/ttyUSB1"
radar_front.open()
```

5021 — if there is no relay connection, initialize a connection

```
relay_ip = "127.0.0.1"
relay_username = "admin"
relay_password = "admin"

relay = telnetlib.Telnet(relay_ip, 23)
relay.read_until(b"User Name: ")
relay.write(relay_username.encode('ascii') + b"\n")
relay.read_until(b"Password: ")
relay.write(relay_password.encode('ascii') + b"\n")
```

5022 — retrieve the distance to the an obstacle from the front radar, and if there is an exception, set the variable for vehicle health to false

```
try:
  radar_front_distance = radar_front.readline()
except:
  unit_healthy = False
```

5023 — retrieve the distance to the an obstacle from the side radar, and if there is an exception, set the variable for vehicle health to false

```
try:
  radar_side_kilometers_per_hour =
    radar_side.readline()
```

```
except:
    unit_healthy = False
```

5024 — check if station using video camera

```
camera_image =
    urllib2.urlopen("http://" + camera_ip +
    "/?username=" + camera_username +
    "&password=" + camera_password + ").read()
```

5025 — check if the video camera image indicates a station has been found

```
if camera_image.find(green):
    station_found = True
```

5026 — proceed if a station was found

```
if station_found:
```

2027 — turn wheel in the direction specified

```
    if station[station_index] == "left":
        turn_left()
    if station[station_index] == "right":
        turn_right()
```

2028 — Slow the speed of the embodiment to decrease to 1%

```
set_speed (1%)
```

FIG. 6

6000 — import libraries to serve and respond to http requests

```
from flask import Flask
from flask import request
from flask import Response
```

6001 — import library for operating system access

```
import os
```

6002 — define a function to handle http requests

```
@application.route('/*', methods=['GET'])
def display():
```

6003 — retrieve any route specified in the user http request command

```
  route = get_args("route")
```

6004 — if a command was provided by the user execute it through a command line call

```
  if command:
    call = "python run.py " + command
    os.execute(call)
```

6005 — read all available routes into a variable

```
  routes_file = open("routes.txt", "r")
  routes = routes_file.readlines()
  routes_file.close()
```

6006 – create a variable to hold the output to display to the user

```
output = "<html>routes"
```

6009 – initialize a form with a drop down from which the rider can select a route

```
output+= "<form method=get action=/>
  <select name=route>"
```

6010 – iterate through each route

```
for route in routes:
```

6011 – add the route to the display as a selectable option

```
output += "<option>" + route + "</option>"
```

6012 – close the route selection drop down

```
output+="</select>"
```

6013 – add a button to submit the route selection

```
output+="<input type=submit value=Go>"
```

6014 – close the form and output display

```
output+="</form></html>"
```

6015 – display the output to the user

```
return output
```

FIG. 7

7000 – **Route Selection**

7001 – [Los Angeles > New York City]

7002 – [Go]

INTERLOCKING CONSTRUCTION STRUCTURES THAT INCREASE RESILIENCE AND REDUCE COST

INVENTOR JONATHAN BANNON MAHER

TECHNICAL FIELD

[0001] Embodiments of the invention relate to the field of construction and automation.

ABSTRACT

[0002] Systems, methods, and apparatuses provide for interlocking construction structures that reduce construction to assembly – thus providing a stronger connection between structures than can be achieved by any other means, and improving resilience in earthquakes and extreme weather events such as hurricanes and tornadoes – where the structures may be produced in a factory, and may include wiring, pipes, outlets, switches, doors, and windows, as well as panels that integrate sound proofing, weather proofing, interior and exterior designs or design ready surfaces, all of which allow for building construction to be done entirely in an automated manner by a machine, therefore providing for an unprecedented reduction in cost and improvement in quality.

REFERENCE TO RELATED DOCUMENTS

[0003] This application is provided the benefit and priority date of United States Patent and Trademark Office provisional patent application number 62/522,701, filed June 20th 2017 by inventor Jonathan Bannon Maher, which is incorporated herein in its entirety.

BACKGROUND

[0004] This section is intended to introduce the reader to various aspects of the art that may be related to various aspects of the present techniques, which are described and or claimed. This discussion is believed to be helpful in providing the reader with background information to facilitate a better understanding of the various aspects of the present disclosure. Accordingly, it is understood that these statements are to be read in this light, and not a citation of prior any art.

[0005] Constructing structures that are inexpensive and extremely resilient is currently not practical. Additionally, earthquakes, extreme weather, and attacks, can take down buildings, in some cases because they're not constructed properly, and in other cases because welded steel and steel reinforced concrete construction proves inadequate.

BRIEF DESCRIPTION OF THE ILLUSTRATIONS

[0006] Illustrations are provided by way of example, and not by way of limitation, in the figures of the accompanying drawings, and embodiments may not contain all components, may contain additional components, and may contain functionally similar components.

[0007] FIG. 1 illustrates an example of an embodiment of the top view of a construction block.

[0008] FIG. 2 illustrates a side view of an example of an embodiment of the construction blocks being assembled as a floor or roof.

[0009] FIG. 3 illustrates an example of an embodiment of a floor connector block.

[0010] FIG. 4 illustrates an example of an embodiment of a double male coupler.

[0011] FIG. 5 illustrates an example of an embodiment of a double male coupler.

[0012] FIG. 6 is an example of an embodiment of a side view of a construction block.

[0013] FIG. 7 is an example of an embodiment of pattern of multiple interlocked construction blocks.

[0014] FIG. 8 is an example of an embodiment of a panel.

DETAILED DESCRIPTION

[0015] It is understood that, as in any engineering or design project, the development of any actual implementation will include numerous implementation specific decisions made to achieve the developers' specific goals, such as compliance with business related and system related constraints, which may vary from one implementation to another. It is understood that such a development effort might be complex and time consuming, but

is nevertheless a routine undertaking of design, fabrication, and manufacture for those skilled in the art having the benefit of this disclosure. The disclosed steps may be read as prefaced by "In some embodiments, including one complete embodiment, ", may be executed or performed in other orders or sequences, and are not limited to the order and sequence shown and described, which are provided to enable ease in constructing an embodiment, and along with each components of each step, may be removed, modified, combined, or rearranged, and other steps and or step components may be added, without departing from the scope of this disclosure and or invention. Although embodiments of the invention have been described and illustrated in the disclosed implementations, it is understood that the present disclosed subject matter, including apparatuses, methods, specification, and illustrations, has been made only by way of example, not by way of limitation, and the methods and apparatuses may be used in other systems, and that numerous changes and optimizations in the details of implementation of the invention and or embodiment are made without such modifications departing from the spirit and scope of this disclosure and or embodiments of the invention. Although the disclosure has been shown and described with respect to one or more embodiments, features of the disclosed embodiments can be combined and rearranged in various ways, and changes including equivalent alterations, substitutions, modifications, and additional efficiencies will of course occur to someone of ordinary skill in the art without departing from the spirit and scope of this disclosure and or invention. In particular regard to the various functions performed by the described components, the terms used to describe such components are intended to correspond, unless otherwise indicated, to any component which performs the specified function of the described component, or is functionally equivalent to the described component, even though not structurally equivalent to the disclosed structure which performs the function in the implementations described in this disclosure. In addition, while a particular feature of the disclosure may have been provided with respect to only one of several embodiments, such feature may be combined with one or more other features of other embodiments as may be desired and advantageous for any given or particular application. In some instances, well-known circuits, structures and techniques have not been shown in detail in order not to obscure the understanding of this disclosure. Articles in this disclosure such as "a" "an" and "the" may allow for both singular and plural forms. Verbs in this disclosure such as "is" may be read as "may be". Conjunctions in this disclosure such as "or" as used herein may be interpreted as inclusive or meaning any one or any combination, where "A, B or C" means "any of the following: A; B; C; A and B; A and C; B and C; A, B and C". Relational terms in this disclosure, for example first and second, top and bottom, left and right, are to distinguish one entity or action from another, and may not necessarily require or imply a relationship, or order between, such

entities or actions. The disclosure includes the best mode contemplated by the inventor, a completely described specific embodiment, along with optional components and alternative embodiments to best suit the implementer, measurements in imperial and metric units to support universal understanding, and dramatically exceeds claims support requirements and enablement requirements by allowing for selection and or construction of the required components to be carried out easily, quickly, and routinely by persons of ordinary skill in the art, who are provided the additional benefit of utilizing readily available commodity components whenever possible. The present disclosure includes material protected by copyrights, and the owner of the copyrights hereby reserves all rights, but with authorization for publication as required by government patent offices. Various embodiments of the present invention may provide all, some or none of the disclosed technical advantages.

[0016] The inventor retains absolutely no liability for any implementation of this invention, and the invention is implemented exclusively at the risk and liability of the implementer.

[0017] The disclosure resolves the previously cited deficiencies, by providing a design for construction materials that allows them to be interlocked without requiring a connecting material, utilizing a specialized inserter and receiver design, and a method for placing them, utilizing staggering that allows them to form a solid structure. This allows a building to be assembled by an individual without any specialized knowledge, or a machine, similar to the ones that can assemble a concrete block structure, that takes the appropriate structures and slides them into place, thereby providing unprecedented structure strength at an unprecedentedly low cost. Because the entire structure is solid non-welded metal, when the interconnected materials are supported by deep earth support poles, the entire structure may be impervious to earthquakes and extreme weather such as tornadoes and hurricanes, and potentially terrorist attacks. The structures will save countless lives in developing countries because structures won't be able to fail as a result of cut costs in construction workers or materials, since structures come pre-formed. Panels on the structures provide better thermal insulation, soundproofing, and durability, as they are constructed the same way each time by using machines in a controlled factory environment. The structures allow for machine construction of buildings never interrupted by eating, sleeping, or resting, as well as machine construction of space habitats.

[0018] The structures may have an attached panel, which may be made of one or more material that provide thermal and sound insulation, and may include a container holding a flame retarding substance such as water, that breaks open when sufficient heat is applied, in order to diminish or extinguish flames, such as those that corrupted the steel support beams of the World Trade Center on 9/11. The panels may be pre-attached to the previously described support structures. Because of the lightweight nature of the panels, they can be shipped from a factory to a building site anywhere in the world.

[0019] **In some embodiments, including one complete embodiment, design and create the molds for each type of structure.** In some embodiments, multiple structural designs may be used for the construction materials, and may include those functionally equivalent to standard block FIG. 1, with a metal structure 1000 having at opposing ends a male and female coupler, with paneling 1001 1002 1003 1004; a corner block FIG. 2, with a metal structure 2000 having a 90' elbow and paneling 2001; a floor connector inserter block FIG. 3, having a metal structure with on one end male coupler 3000 and the other end female coupler 3003 and on one side female coupler 3002, and panel 3001; and a floor connector block FIG. 4, having double male ends; a base connector receiver block FIG. 5, having a metal structure 5000 with double female ends, and panel 5001; a panel side with a side view in FIG. 6, with interior wall 6002, sound proofing material 6001, metal frame 6000, space for wiring and piping 6003, water container 6004, and exterior wall 6005; a panel front view in FIG. 8 with metal frame 8000 and panel 8001.

[0020] In some embodiments, blocks are to be produced in 1 foot increments in width and height, to allow flexible building designs, and because adjoining blocks are staggered in height to secure blocks both vertically and horizontally, as shown in FIG. 3.

[0021] In some embodiments molds may be etched by hand. In some embodiments, CAD (Computer Aided Design) files may be created with each design, which are then provided to a CNC (Computer Numerical Control) machine, which etches the molds. A mold can be made for each of the blocks, and material poured in. In some embodiments, the structures may be 3d printed from materials which may include steel, titanium, Teflon, and carbon fiber.

[0022] In some embodiments, where metal will be poured into the molds, the molds may be etched into graphite. In some embodiments, where concrete will be poured into the molds, molds should first be etched into graphite, steel poured into the mold, then once the steel is cool, use it to create a mold out of a material such as urethane rubber, into which the concrete can be poured and easily removed after hardening.

[0023] **In some embodiments, including one complete embodiment, pour the material into the molds.** In some embodiments, these blocks may be produced in a traditional material like concrete. In some embodiments, these blocks may be produced as concrete encasing a metal structure. In some embodiments, these may be metal only beams to form the support structure for an otherwise traditionally constructed building.

[0024] In some embodiments, molten metal is to be poured into the molds. In some embodiments, high grade steel may be used such as AISI 4130 which has a strength of 951 megapascals (137,895 pounds or about 69 tons of pressure per square inch). In some embodiments, used for the construction of single story structures, a steel may be used such as ASTM A36, which only has a strength of 250 megapascals (36,250 pounds or about 18 tons of pressure per square inch) but is currently a third of the price of the previously cited high grade steel. In some embodiments, titanium may be used, which has a strength rating of 940 megapascals (136,300 pounds or about 68 tons of pressure per square inch), is substantially lighter than steel, and begins to lose strength around 430 °C as opposed to around 300 °C, but currently trades for around 8 times more than the referenced high grade steel due to difficulties in extracting the element despite its abundance.

[0025] In some embodiments, rather than creating molds, a material such as marble or granite could be cut into the form of the disclosed designs by the CNC machine.

[0026] In some embodiments, earth can be hydraulically compressed into molds of these structures, such as for use by a robot using only readily available materials to build colonies on other bodies in space.

[0027] **In some embodiments, including one complete embodiment, optionally construct and attach panels.** In some embodiments, panels may be attached, that may provide sound and temperature insulation, and wiring and piping. In some embodiments, the panels may have frames and be made of materials such as carbon fiber or fiberglass, that are lightweight, fire resistant, temperature insulating, sound resistant, and don't produce toxic smoke or immediately lose structural integrity when burned. Carbon fiber and fiberglass each weigh less than a pound per square foot at a useable thickness, and are relatively inexpensive. Existing uses of carbon fiber include the production of jet exteriors, and existing uses of fiberglass include exteriors for tornado shelters and boats. In a high-rise, these panels will also dramatically reduce the load on the buildings structure, allowing it to better withstand stress. In tall buildings, in the side panels 1/8th inch high strength steel rods may be used, and in the floor panels 1/4 inch high strength steel rods may be used. The panels should have ding resistant coating such as one that is resin based. These structures may also provide soundproofing between floors. Because the panels can be constructed in a factory and simply bolted onto the structures, while providing temperature and sound insulation, they also dramatically reduce construction costs. Panels may be finished with a solid colored resin that's scratch resistant or of carbon fiber, so the building exterior never needs to be painted or cleaned, with panels having 90' angles at edges so they can be fit together without a visible seam. In some embodiments, the panels can have electrical outlets, wires, and water pipes, installed in the factory, floor and ceiling panels have lighting and air conditioning installed, and simply connected on assembly. In some embodiments, the panels have an internal steel structure that replaced steel bars during construction, using a metal frame around the outside, and are to have metal rods at spaced distances on the inside. The metal frames may be wrapped with fiberglass and carbon fiber, both non-flammable materials. However, the United States Federal Aviation Administration issued a report that found that carbon fiber as used in airplanes, begins to decompose at 300°C (572'F), which is about the same temperature at which steel begins to lose strength, and the standard home fire is standard home fires 300'F to 1600'F, the panels affected by a consistent fire, will deteriorate, thus part of the reason for a steel bar grate like structure being included inside the panels.

[0028] In some embodiments, including one complete embodiment, the inside of the panels are hollow to allow liquid to fill them to provide non-toxic thermal and sound insulation, as well as to provide water to extinguish flames in the event of sufficient fire, while the outside of the panels may be wrapped with carbon fiber, with the interior facing part of the exterior covered in plaster for aesthetics, and the exterior facing part of the panel exterior optionally pressed into mold and colored to appear as stone, brick, or stucco, screwed on from behind and fitting together form an airtight seal. In some embodiments, panels have integrated windows and doors, or provide a mounting space for them. In some embodiments, water is used for sound insulation, where 1 inch of water may provide for 30 decibels of sound reduction, where the water may be added on site to reduce shipping weight.

[0029] In some embodiments, piping and or wiring may be mounted inside the panels, that can be connected to piping and or wiring in other panels, in order to reduce on-site construction skill required and thus cost.

[0030] **In some embodiments, including one complete embodiment, assemble a reduced cost, high quality, highly resilient structure.** In some embodiments, including one complete embodiment, blocks are slipped together in their naturally connecting formation pattern as disclosed in FIG. 7, where every other block in the first row of a structure may be half size blocks, so as to maximize connectivity between blocks, and thus the strength of the structure.

CLAIMS

What is claimed is:

1. A construction structure, with the invention comprised of:
   a connector shape that allows blocks to interlock.
2. A construction structure, with the invention comprised of:
   a container filled with a flame retarding substance that will release in the event of a sufficient fire.

FIG. 1

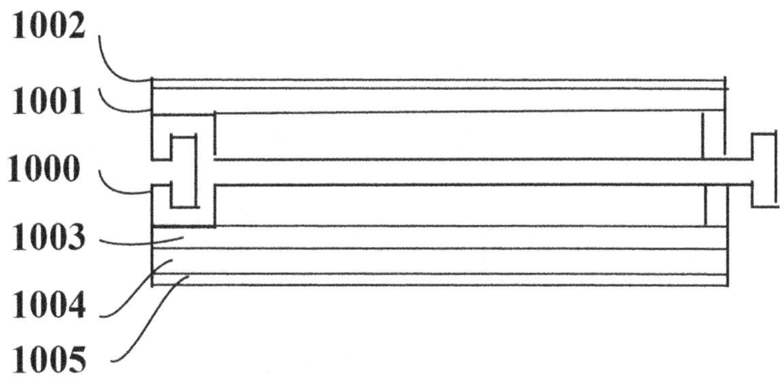

Jonathan Bannon Maher

FIG. 2

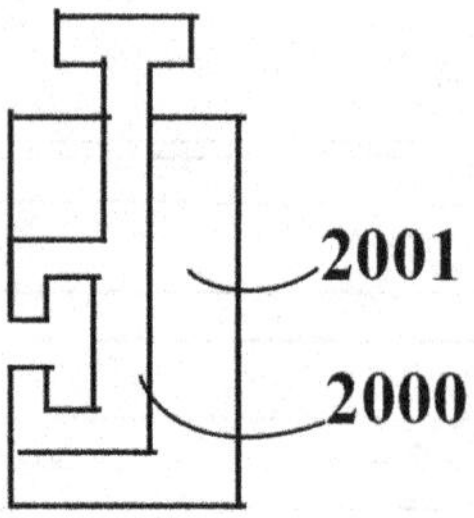

Jonathan Bannon Maher

FIG. 3

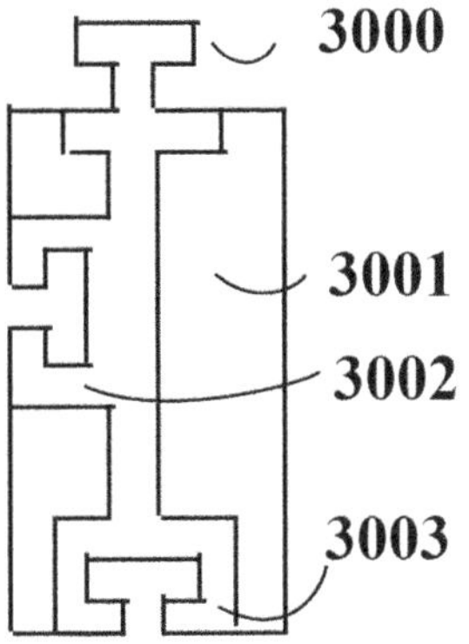

Jonathan Bannon Maher

FIG. 4

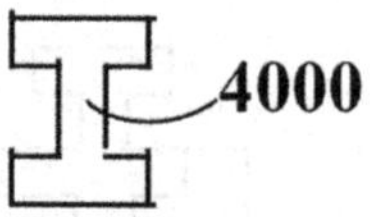

Jonathan Bannon Maher

FIG. 5

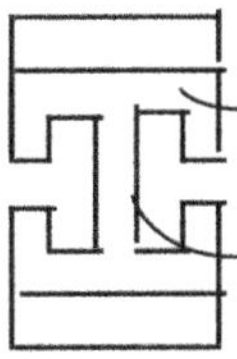

Jonathan Bannon Maher

FIG. 6

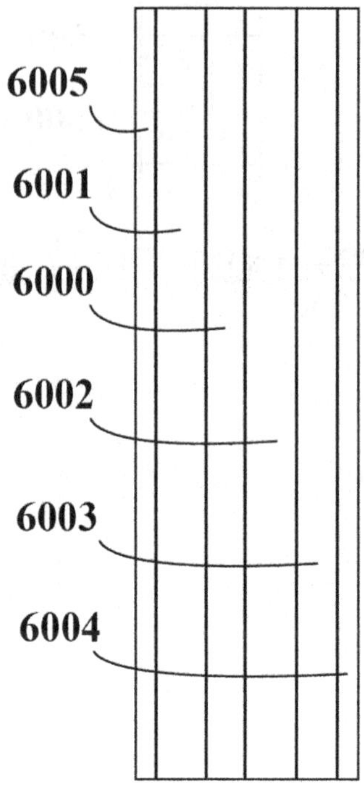

Jonathan Bannon Maher

FIG. 7

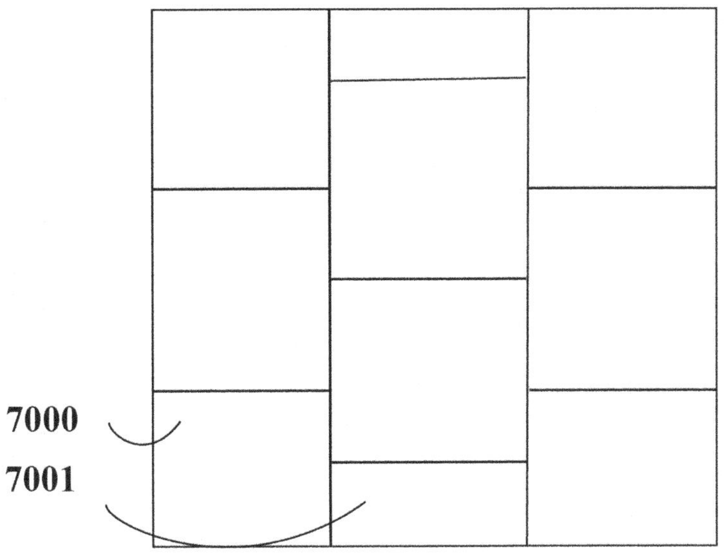

Jonathan Bannon Maher

FIG. 8

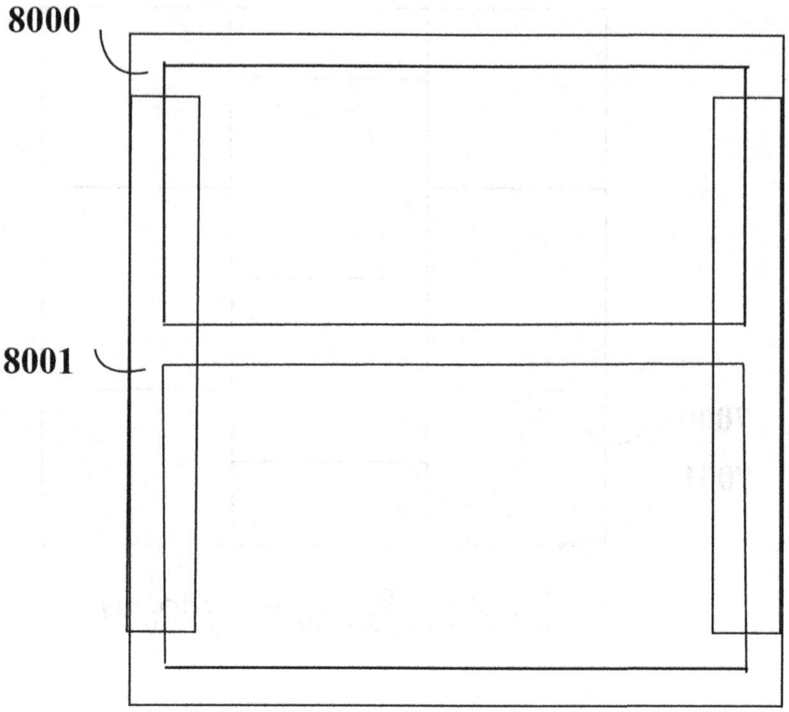

Jonathan Bannon Maher

AUTOMATED DISCOVERY AND CONTROL OF ELECTRONIC DEVICES

INVENTOR JONATHAN BANNON MAHER

TECHNICAL FIELD

[0001] Embodiments of the invention relate generally to the fields of software and electronics, and more specifically to the field of building automation.

ABSTRACT

[0002] Systems, methods, apparatuses, and computer programs encoded on a computer storage medium, provide for the discovery of electronic devices connected to a network, retrieving available commands from the devices, and making the commands available to the end user for execution, through a user interface including a touch screen, mobile device, or voice assistant.

BACKGROUND

[0003] This section is intended to introduce the reader to various aspects of the art that may be related to various aspects of the present techniques, which are described and or claimed. This discussion is believed to be helpful in providing the reader with background information to facilitate a better understanding of the various aspects of the present disclosure. Accordingly, it is understood that these statements are to be read in this light, and not a citation of prior any art.

[0004] Electronic devices such as home appliances and electronics have traditionally not been connected to a network. Devices which currently offer network connectivity, have no agreed upon standard for discovery and control, and are usually only accessible through infrared remotes and unreliable limited range WiFi.

REFERENCE TO RELATED DOCUMENTS

[0005] This application is provided the benefit and priority date of United States Patent and Trademark Office provisional patent application number 62/522,690, filed June 20th 2017 by inventor Jonathan Bannon Maher, which is incorporated herein in its entirety.

BRIEF DESCRIPTION OF THE ILLUSTRATIONS

[0006] Illustrations are provided by way of example, and not by way of limitation, in the figures of the accompanying drawings, and embodiments may not contain all components, may contain additional components, and may contain functionally similar components.

[0007] FIG. 1 is an embodiment of an Ethernet connected computer with an USB connected relay board that controls a light bulb.

[0008] FIG. 2 is an embodiment of a computer code embedded flow diagram, providing computer code to control the computer operating the relay board operating the light bulbs.

[0009] FIG. 3 is an embodiment of a computer code embedded flow diagram, providing computer code to find available devices on the network and record their available commands.

[0010] FIG. 4 is an embodiment of a computer code embedded flow diagram, providing the computer code to display a device remote control to the user.

[0011] FIG. 5 is an embodiment of a computer code embedded flow diagram, providing the computer code to send the command from the users remote control to the device.

DETAILED DESCRIPTION

[0012] It is understood that, as in any engineering or design project, the development of any actual implementation will include numerous implementation specific decisions made to achieve the developers' specific goals, such as compliance with business related and system related constraints, which may vary from one implementation to another. It is understood that such a development effort might be complex and time consuming, but

is nevertheless a routine undertaking of design, fabrication, and manufacture for those skilled in the art having the benefit of this disclosure. The disclosed steps may be read as prefaced by "In some embodiments, including one complete embodiment, ", may be executed or performed in other orders or sequences, and are not limited to the order and sequence shown and described, which are provided to enable ease in constructing an embodiment, and along with each components of each step, may be removed, modified, combined, or rearranged, and other steps and or step components may be added, without departing from the scope of this disclosure and or invention. Although embodiments of the invention have been described and illustrated in the disclosed implementations, it is understood that the present disclosed subject matter, including apparatuses, methods, specification, and illustrations, has been made only by way of example, not by way of limitation, and the methods and apparatuses may be used in other systems, and that numerous changes and optimizations in the details of implementation of the invention and or embodiment are made without such modifications departing from the spirit and scope of this disclosure and or embodiments of the invention. Although the disclosure has been shown and described with respect to one or more embodiments, features of the disclosed embodiments can be combined and rearranged in various ways, and changes including equivalent alterations, substitutions, modifications, and additional efficiencies will of course occur to someone of ordinary skill in the art without departing from the spirit and scope of this disclosure and or invention. In particular regard to the various functions performed by the described components, the terms used to describe such components are intended to correspond, unless otherwise indicated, to any component which performs the specified function of the described component, or is functionally equivalent to the described component, even though not structurally equivalent to the disclosed structure which performs the function in the implementations described in this disclosure. In addition, while a particular feature of the disclosure may have been provided with respect to only one of several embodiments, such feature may be combined with one or more other features of other embodiments as may be desired and advantageous for any given or particular application. In some instances, well-known circuits, structures and techniques have not been shown in detail in order not to obscure the understanding of this disclosure. Articles in this disclosure such as "a" "an" and "the" may allow for both singular and plural forms. Verbs in this disclosure such as "is" may be read as "may be". Conjunctions in this disclosure such as "or" as used herein may be interpreted as inclusive or meaning any one or any combination, where "A, B or C" means "any of the following: A; B; C; A and B; A and C; B and C; A, B and C". Relational terms in this disclosure, for example first and second, top and bottom, left and right, are to distinguish one entity or action from another, and may not necessarily require or imply a relationship, or order between, such

entities or actions. The disclosure includes the best mode contemplated by the inventor, a completely described specific embodiment, along with optional components and alternative embodiments to best suit the implementer, measurements in imperial and metric units to support universal understanding, and dramatically exceeds claims support requirements and enablement requirements by allowing for selection and or construction of the required components to be carried out easily, quickly, and routinely by persons of ordinary skill in the art, who are provided the additional benefit of utilizing readily available commodity components whenever possible. The present disclosure includes material protected by copyrights, and the owner of the copyrights hereby reserves all rights, but with authorization for publication as required by government patent offices. Various embodiments of the present invention may provide all, some or none of the disclosed technical advantages.

[0013] The computer code descriptions disclosed, in order to provide comprehensive enabling disclosure, rather than utilizing flow charts, which according to Patent Cooperation Treaty 11.11a are prohibited from containing "text matter, except a single word or words, when absolutely indispensable, such as... a few short catchwords indispensable for understanding", are provided in a text only format where the number of arrows preceding a line indicate logical block level, semicolons indicate a new segment of a logical block, and periods indicate the closure of one or more logical blocks. It is understood that any computer code representations in this disclosure are merely illustrative, rather than restrictive. While code may be written in nearly any computer language, including Java and C++, the illustrative computer code descriptions were derived from code written the Python language, which may be run through the Python interpreter, with appropriate supportive libraries, which at the time of disclosure, may run on nearly any computer, for example one with an Intel or AMD processor, running a current version of Linux, Windows, or Mac OS. All code components may read as if prefaced by "In some embodiments, including one complete embodiment, ". In some embodiments, functionality may be modified, rearranged, excluded, and added. To provide more fundamental computer system details, in some embodiments, the functionality associated with the disclosed computer code descriptions may be referred to as a script, module, software, software application, or code, and can be written in any form of language, including compiled, interpreted, declarative, or procedural, able to be deployed in any form suitable for use in a computing environment, including as an independent or integrated program, module, component, or subroutine, for execution by the computer system, implemented on one or more independent or integrated computers, utilizing a central processing unit in the form of one or more general or special purpose microprocessors, in conjunction with digital electronic circuitry, which may include special purpose logic circuitry such as a field programmable gate array or application specific integrated circuit, with the computer controlled by and operatively coupled to tangibly embodied software and or firmware, which may include code that creates an environment for code execution, including individual or combined use of processor firmware, a protocol stack, a database management system, and an operating system, where such software and or firmware may exist in one or more parts in memory on one or more computers, and is encoded on one or more tangible non transitory software carriers, such as individual or combined use of a random or serial access device or substrate, a semiconductor memory device, transient or persistent random access memory, a magnetic, magnetic optical, or optical disk, or encoded on an artificially generated transmitted signal, for example, optical, electrical, or electromagnetic, transmitted using a sending and a receiving apparatus, where the interaction between the user and the software may be implemented by operatively coupling, to the local implementing computer, or a local computer connected to one or more remote computers through a local or wide area network, a display device which may implement liquid crystals or light emitting diodes, a keyboard, and a pointing device.

[0014] The inventor retains absolutely no liability for any implementation of this invention, and the invention is implemented exclusively at the risk and liability of the implementer.

[0015] In some embodiments, including one complete embodiment, the discovery and control of electronic devices connected to a network is provided, retrieving available commands for the devices, and making the commands available to the end user for execution. Connected building components may include microwaves, ovens, stove burners, toasters, fireplaces, light bulb sockets, electrical sockets, thermostats, sound systems, televisions, garage doors, and door locks. To enable this functionality, each component has an embedded computer that controls the functionality of the component, through means including a computer controlled relay board which may integrate into the existing circuits of the device, with the computer wired to a local router through Ethernet. This allows for software to be written providing notifications when laundry is done, food to be left in a cooker and turned on so a hot meal is waiting, a recipe to be found online that connects to stove and sets temperature and time, a romantic evening turned on with an active fireplace accompanied by light settings and a music playlist, viewing contents of fridge through built in camera when at a market, a laundry machine using preferred settings, turning off power to all devices if no one is home, setting the conditions in each room in the house from anywhere, and connected devices being easily discovered and controlled through an voice controlled assistant. Electronics manufacturers simply have to allow their devices to be responsive to popular communication protocols and list available commands with a description, and have their devices accept commands, which creates substantial financial returns for electronics manufactures, as it provides all a reason to upgrade, and also allows an ecosystem of developers to provide applications and services for their products. People can then have an electrician or cable installer run Ethernet cables from each device to a central router, which is an effectively identical process of the same cost as running standard television coaxial cable. All new homes should be built in this manner. Light switches should be replaced with simple wall mounted Linux devices offering full home control, potentially using access codes or biometric recognition for certain functions.

[0016] **In some embodiments, including one complete embodiment, enable devices to be responsive to popular communications protocols.** In some embodiments, where manufacturers are adapting existing devices to be discoverable and controllable, a Ethernet enabled miniature computer such as a Raspberry Pi is embedded in the unit along with an attached relay board, where the circuits of the device are wired to the relay board to allow them to be opened and closed in a manner that provides for control of the device, the Ethernet port of the device is connected to the local router through and Ethernet cable, and the miniature computer is provided software to enable its operation of the relay board, which may resemble the software in FIG. 2, providing functionality comprising:
> import a library for connecting to the relay board.
> create variables defining the value of relay on off states, and variables holding the relay states.
> create a variable to hold a connection to the relay board.
> create a function to update the relay board, creating a connection first if one has not been initialized or it has been dropped, where the state of each relay in sequence is posted as a string to the relay board.
> create a function to execute commands on the relay control board that are sent to the script.
> read in commands provided to the script through the command line.
> return a list of available commands.
> set the relay states according to the provided commands and call the function to update the relay board.

[0017] In some embodiments, including one complete embodiment, in reference to FIG. 1, as an example of the method for wiring a component to be controlled through a relay board, a series of light bulb sockets are wired to an Ethernet controlled relay board, where an Ethernet controlled 4 channel relay board 1300 is used where light bulb socket 1000 positive wire is connected to relay 1201 and negative wire to relay 1202, light bulb socket 1001 positive wire is connected to relay 1203, and negative wire is connected to relay 1204, light bulb socket 1002 positive wire is connected to relay 1205 and negative wire to relay 1206, light bulb socket 1003 positive wire is connected to relay 1207, and negative wire is connected to relay 1208, and relay board 1200 is connected to miniature computer 1300, where computer 1300 has software running on startup, where the software resembles that as previously described in FIG. 2.

[0018] In some embodiments, where converting existing electronics to be responsive and controllable through popular protocols, a small computer – some of which are now available for the price of an expensive coffee — may be embedded in unit, switch the existing microcontroller board to one that can be controlled through USB, connect the USB to the computer, where the code to control the microcontroller may resemble that in FIG. 2.

[0019] **In some embodiments, including one complete embodiment, wire devices to a local router.** In some embodiments, including one complete embodiment, all devices that can be controlled over Ethernet are wired to a local computer network router 1400. In some embodiments, including one complete embodiment, computer 1300 is connected to router 1400 through a standard Ethernet cable, and router 1400 is connected through a standard Ethernet cable to computer 1500, which runs the device discovery software in FIG. 3, and the user interface server software in FIG. 4, providing a user interface resembling that in FIG. 5.

[0020] **In some embodiments, including one complete embodiment, create the computer code to search the network for devices and record their available commands.** In some embodiments, including one complete embodiment, in reference to FIG. 3, computer code is written to scan the computer network for responsive electronic devices. In some embodiments, including one complete embodiment, in reference to FIG. 3, the device discovery software provides functionality comprising:

> import the libraries to connect to devices through popular protocols including telnet, ssh, http, and https.

> specify username and password if applicable of devices on the network.

> open the commands file for overwriting.

> create a variable holding the base of the IP address for the local area network.

> create an array to hold all possible IP addresses on the local network.

> generate all possible local network IP address and add them to the IP address array.
> iterate through each local IP address.
> try to retrieve a list of commands for each through each http, https, telnet, ssh, and record which protocol worked.
> if commands were provided by the device, write each command to the commands file.

[0021] In some embodiments, rather than receiving the commands from the device, data is downloaded for each controllable electronic device manufactured, specifying protocol and commands, rather than retrieving the commands.

[0022] **In some embodiments, including one complete embodiment, create the computer code files to display remote control screen that requests commands be sent to devices.** In some embodiments, including one complete embodiment, in reference to FIG. 4, the code to display data to the user is presented, which may be controlled through means including a touch screen, a mobile device, or voice assistant. In some embodiments, including one complete embodiment, the display of the screen to the user resembles that in FIG. 5.

[0023] In some embodiments, including one complete embodiment, in reference to FIG. 4, the user interface software provides functionality comprising:
> import libraries enabling a http server.
> define a function that handles server requests;
>> define the default usernames and passwords for the devices;
>> create variables holding any command and IP address provided by the user;
>> proceed if a command was executed by the user;
>> if the command is to refresh the list of devices and their available commands, execute a command line call to execute the device discovery script;
>> proceed if the command is not to refresh;
>> if the request protocol is http or https, compile a URL to make the request;
>> if protocol is telnet or SSH connect and execute the command line call;
>> read the list of all available commands from all devices from the text file created by the device discovery script;
>> create a variable storing the output display and building the header;
>> iterate through each command;
>> build a button to execute the command and add it to the display;
>> display the output screen to the user.

[0024] **In some embodiments, including one complete embodiment, run the computer server code.** In some embodiments, including one complete embodiment, run the computer code previously written, which for example if written in the Python language, at the command line type "python automation.py" and the user interface will become available at through a web browser at http://localhost/.

CLAIMS

What is claimed is:

1. A system for the discovery of electronic devices, with the invention comprising:
scanning network addresses through various protocols, and for responsive devices, retrieving and recording a list of commands available for device.
2. A system for the control of electronic devices, with the invention comprising:
utilizing a list of electronic devices;
providing a list of available commands;
providing the ability for the user to execute available commands.
3. A system for the control of electronic devices, with the invention comprising:
software that accepts and executes networked device commands, by connecting to the specified device and executing the specified command.
4. A non-transitory computer-readable recording medium holding stored instructions, which when executed by one or more processing devices, cause the one or more processing devices to implement a method comprising, with the invention comprising:
scanning network addresses through various protocols, and for responsive devices, retrieving and recording a list of commands available for device.
5. A non-transitory computer-readable recording medium holding stored instructions, which when executed by one or more processing devices, cause the one or more processing devices to implement a method comprising, with the invention comprising:
providing a list of devices, and any or all devices, providing a list of available commands, with the ability for the user to execute available commands.
6. A non-transitory computer-readable recording medium holding stored instructions, which when executed by one or more processing devices, cause the one or more processing devices to implement a method comprising, with the invention comprising:
software that accepts and executes networked device commands, by connecting to the specified device and executing the specified command.

FIG. 1

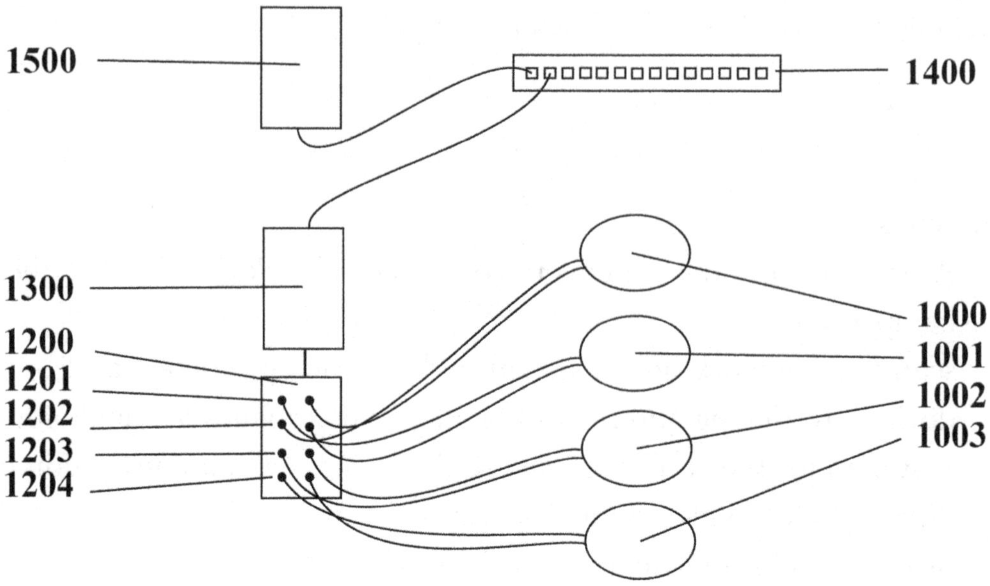

FIG. 2

2000 — Import the library for connecting to the relay board

```
import serial
```

2001 — create variables defining the value of relay on off states, and variables holding the relay states

```
on = 1
off = 0
relay_1_light_1 = off
relay_2_light_2 = off
relay_3_light_3 = off
relay_4_light_4 = off
```

2002 — create a variable to hold a connection to the relay board

```
relay = None
```

2003 — create a function to update the relay board, creating a connection first if one has not been initialized or it has been dropped, where the state of each relay in sequence is posted as a string to the relay board

```
def update_relay():

    if not relay:
        relay = serial.Serial()
        relay.baudrate = 9600
        relay.timeout = 0
        relay.port = "/dev/ttyUSB0"
        relay.open()

    command = str(relay_1_light_1)
```

```
command+= str(relay_2_light_2)
command+= str(relay_3_light_3)
command+= str(relay_4_light_4)
relay.write(command.encode('ascii') + b"\n")
```

2004 – create a function to execute commands sent to the script on the relay control board

```
def main():
```

2005 – read in commands provided to the script through the command line

```
arguments = sys.argv
```

2006 – print a list of available commands

```
print "0 off\n1 on"
```

2007 – Set the relay states according to the provided commands and call the function to update the relay board

```
relay_1_light_1 = off
relay_2_light_2 = off
relay_3_light_3 = off
relay_4_light_4 = off
update_relays()
```

FIG. 3

3000 – import the libraries to connect to devices through popular protocols including telnet, ssh, http, and https

```
import telnetlib
import ssh
import urllib
```

3001 – specify devices username and password

```
username = "admin"
password = "admin"
```

3002 – open the commands file for overwriting

```
commands_file = open("commands","w+)
```

3003 – create a variable holding the base of the IP address for the local area network

```
ip_base = "192.168.0."
```

3004 – create an array to hold all possible IP addresses on the local network

```
ips = []
```

3005 – generate all possible local network IP address and add them to the IP address array

```
index = 1
while index <= 255:
  ips.append(ip_base + str(index))
```

index+=1

3006 – iterate through each local IP address

for ip in ips:

3007 – try to retrieve a list of commands for each through each http, https, telnet, ssh, and record which protocol worked

```
protocol = ""
commands = ""
url = "https:" + "//" + ip + "/?" +
   username=" + username +
   "&password=" + password
commands = urllib.open(url)
if commands:
   protocol = "https"
if not commands:
   commands = urllib.open(url.replace("https","http"))
   commands = urllib.open(url)
if commands:
   protocol = "http"
if not commands:
   connection = telnetlib.Telnet(ip, 23)
if connection:
   protocol = "telnet"
if not connection:
   connection = telnetlib.Telnet(ip, 556)
if connection:
   protocol = "ssh"
if connection:
   connection.read_until(b"User Name: ")
   connection.write(
      username.encode('ascii') + b"\n")
   connection.read_until(b"Password: ")
   connection.write(
      password.encode('ascii') + b"\n")
```

```
command = "automation commands"
commands = connection.write(
   command.encode('ascii') + b"\n")
```

3008 – if commands were provided by the device, write each commands to the commands file

```
commands_entry = ""
commands = commands.split("\n")
for command in commands:
   available_commands = protocol + " " + ip +
   " " + command
```

3009 – write the available commands to the commands file

```
commands_file.write(commands_entry)
commands_file.close()
```

FIG. 4

4000 — import libraries enabling a http server

```
from flask import Flask
from flask import request
from flask import Response
```

4001 — define a function that handles server requests

```
@application.route('/*', methods=['GET'])
def display():
```

4002 — define the default usernames and passwords for the devices

```
  username = "admin"
  password = "admin"
```

4003 — create variables holding any command and IP address provided by the user

```
  command = request.parameters("command")
  ip_address = request.parameters("ip_address")
```

4004 — proceed if a command was executed by the user

```
  if command:
```

4005 — if the command is to refresh the list of devices and their available commands, execute a command line call to execute the device discovery script

```
    if command == "refresh":
```

```
os.execute("python device_scanner.py")
```

4006 — proceed if the command is not to refresh

```
if not command == "refresh":
```

4006 — if the request protocol is http or https, compile a url to make the request

```
    request = ""
    if protocol == "http" or protocol == "https":
        request = "http://" + ip_address +
            "/?command=" + command
        if protocol == "https":
            request = request.replace("http:","https:")
        urllib.open(url)
```

4007 — if protocol is telnet or ssh connect and execute the command line call

```
    if protocol == "telnet" or protocol == "ssh":
        port = 23
        request = "automation " + command
        if protocol = "ssh":
            port = 22
        connection = telnetlib.Telnet(ip, port)
        connection.read_until(b"User Name: ")
        connection.write(
            username.encode('ascii') + b"\n")
        connection.read_until(b"Password: ")
        connection.write(
            password.encode('ascii') + b"\n")
        connection.write(
            request.encode('ascii') + b"\n")
```

4008 – read the list of all available commands from all devices from the text file created by the device discovery script

```
commands = open("commands.txt").split("\n")
```

4009 – create a variable storing the output display and building the header

```
output = "<html><head>
  <title>Device Controller</title>
  </head><body>"
```

4010 – iterate through each command

```
for command in commands:
```

4011 – build a button to execute the command and add it to the display

```
  output + = "<table><tr
    style=\"background:black\"
    onclick=\"javascript(
    "http://localhost/api/?" + \
    ip_address=$ip_address& \
    command=$command_code \
    )\"><td>$command_name</td>
    </tr></table>"
```

4012 – display the output screen to the user

```
return output
```

FIG. 5

Device Control

Living Room Lights
[On]
[Off]

Television
[Volume Up]
[Volume Down]
[Channel Up]
[Channel Down]

Dishwasher
[On]
[Off]

SEPARATION AND PLAYBACK OF AUDIO COMPONENTS

INVENTOR JONATHAN BANNON MAHER

TECHNICAL FIELD

[0001] Embodiments of the invention relate to the field of digital audio.

ABSTRACT

[0002] Systems, methods, apparatuses, and computer programs encoded on a computer storage medium, provide a format for audio, and a system for playback, which in contrast to previous audio storage format and playback systems, allows for the separation and overlapped playback of individual audio components, for example, providing playback where vocals are turned off, or on only vocals and guitar are turned on, to provide audio appropriate to use, setting, and mood.

REFERENCE TO RELATED DOCUMENTS

[0003] This application is provided the benefit and priority date of United States Patent and Trademark Office provisional patent application number 62/522,693, filed June 21st 2017 by inventor Jonathan Bannon Maher, which is incorporated herein in its entirety.

BACKGROUND

[0004] This section is intended to introduce the reader to various aspects of the art that may be related to various aspects of the present techniques, which are described and or claimed. This discussion is believed to be helpful in providing the reader with background information to facilitate a better understanding of the various aspects of the present disclosure. Accordingly, it is understood that these statements are to be read in this light, and not a citation of prior any art.

[0005] Previous audio storage and playback systems, to provide different versions of audio, require each version be produced in fully integrated separate files, for example, a file for each a full, acoustic, and instrumental version. This does not allow a listener to make one purchase and receive multiple versions, for example, an acoustic and an instrumental version along with a primary version, or to listen to the soundtrack in a movie without characters speaking, or a sports broadcast without commentators. Additionally, desirable alternative arrangements of an audio recording are often never produced.

BRIEF DESCRIPTION OF THE ILLUSTRATIONS

[0006] Illustrations are presented by way of example, and not by way of limitation, in the figures of the accompanying drawings, and embodiments may not contain all components, may contain additional components, and may contain functionally similar components.

[0007] FIG. 1 is an example of an embodiment of the storage format for the audio components.

[0008] FIG. 2 is an example of a user interface that can be used for the playback of audio components.

DETAILED DESCRIPTION

[0009] It is understood that, as in any engineering or design project, the development of any actual implementation will include numerous implementation specific decisions made to achieve the developers' specific goals, such as compliance with business related and system related constraints, which may vary from one implementation to another. It is understood that such a development effort might be complex and time consuming, but is nevertheless a routine undertaking of design, fabrication, and manufacture for those skilled in the art having the benefit of this disclosure. The disclosed steps may be read as prefaced by "In some embodiments, including one complete embodiment, ", may be executed or performed in other orders or sequences, and are not limited to the order and sequence shown and described, which are provided to enable ease in constructing an embodiment, and along with each components of each step, may be removed, modified, combined, or rearranged, and other steps and or step components may be added, without departing from the scope of this disclosure and or invention. Although embodiments of the invention have been described and illustrated in the disclosed implementations, it is understood that the present disclosed subject matter, including apparatuses, methods, specification, and illustrations, has been made only by way of example, not by way of limitation, and the methods and apparatuses may be used in other systems, and that numerous changes and optimizations in the details of implementation of the invention and or embodiment are made without such modifications departing from the spirit and scope of this disclosure and or embodiments of the invention. Although the disclosure has been shown and described with respect to one or more embodiments, features of the disclosed embodiments can be combined and rearranged in various ways, and changes including equivalent alterations, substitutions, modifications, and additional efficiencies will of course occur to someone of ordinary skill in the art without departing from the spirit and scope of this disclosure and or invention. In particular regard to the various functions performed by the described components, the terms used to describe such components are intended to correspond, unless otherwise indicated, to any component which performs the specified function of the described component, or is functionally equivalent to the described component, even though not structurally equivalent to the disclosed structure which performs the function in the implementations described in this disclosure. In addition, while a particular feature of the disclosure may have been provided with respect to only one of several embodiments, such feature may be combined with one or more other features of other embodiments as may be desired and advantageous for any given or particular application. In some instances, well-known circuits, structures and techniques have not been shown in detail in order not to obscure the understanding of this disclosure. Articles in this disclosure such as "a" "an" and "the" may allow for both singular and plural forms. Verbs in this disclosure such as "is" may be read as "may be". Conjunctions in this disclosure such as "or" as used herein may be interpreted as inclusive or meaning any one or any combination, where "A, B or C" means "any of the following: A; B; C; A and B; A and C; B and C; A, B and C". Relational terms in this disclosure, for example first and second, top and bottom, left and right, are to distinguish one entity or action from another, and may not necessarily require or imply a relationship, or order between, such entities or actions. The disclosure includes the best mode contemplated by the inventor, a completely described specific embodiment, along with optional components and alternative embodiments to best suit the implementer, measurements in imperial and metric units to support universal understanding, and dramatically exceeds claims support requirements and enablement requirements by allowing for selection and or construction of the required components to be carried out easily, quickly, and routinely by persons of ordinary skill in the art, who are provided the additional benefit of utilizing readily available commodity components whenever possible. The present disclosure includes material protected by copyrights, and the owner of the copyrights hereby reserves all rights, but with authorization for publication as required by government patent offices. Various embodiments of the present invention may provide all, some or none of the disclosed technical advantages.

[0010] The computer code descriptions disclosed, in order to provide comprehensive enabling disclosure, rather than utilizing flow charts, which according to Patent Cooperation Treaty 11.11a are prohibited from containing "text matter, except a single word or words, when absolutely indispensable, such as... a few short catchwords indispensable for understanding", are provided in a text only format where the number of arrows preceding a line indicate logical block level, semicolons indicate a new segment of a logical block, and periods indicate the closure of one or more logical blocks. It is understood that any computer code representations in this disclosure are merely illustrative, rather than restrictive. While code may be written in nearly any computer language, including Java, Objective-C, Swift, and C++, the illustrative computer code descriptions were derived from code written the Python language, which may be run through the Python interpreter, with appropriate supportive libraries, which at the time of disclosure, may run on nearly any computer, for example one with an Intel or AMD processor, running a current version of Linux, Windows, or Mac OS. All code components may read as if prefaced by "In some embodiments, including one complete embodiment, ". In some embodiments, functionality may be modified, rearranged, excluded, and added. To provide more fundamental computer system details, in some embodiments, the functionality associated with the disclosed computer code descriptions may be referred to as a script, module, software, software application, or code, and can be written in any form of language, including compiled, interpreted, declarative, or procedural, able to be deployed in any form suitable for use in a computing environment, including as an independent or integrated program, module, component, or subroutine, for execution by the computer system, implemented on one or more independent or integrated computers, utilizing a central processing unit in the form of one or more general or special purpose microprocessors, in conjunction with digital electronic circuitry, which may include special purpose logic circuitry such as a field programmable gate array or application specific integrated circuit, with the computer controlled by and operatively coupled to tangibly embodied software and or firmware, which may include code that creates an environment for code execution, including individual or combined use of processor firmware, a protocol stack, a database management system, and an operating system, where such software and or firmware may exist in one or more parts in memory on one or more computers, and is encoded on one or more tangible non transitory software carriers, such as individual or combined use of a random or serial access device or substrate, a semiconductor memory device, transient or persistent random access memory, a magnetic, magnetic optical, or optical disk, or encoded on an artificially generated transmitted signal, for example, optical, electrical, or electromagnetic, transmitted using a sending and a receiving apparatus, where the interaction between the user and the software may be implemented by operatively coupling, to the local implementing computer, or a local computer connected to one or more remote computers through a local or wide area network, a display device which may implement liquid crystals or light emitting diodes, a keyboard, and a pointing device.

[0011] The disclosure resolves the previously cited deficiencies, by allowing a single audio file to contain individual components, by storing the audio components in a manner that is sequential or otherwise separated, rather than integrated. Playback software may then allow components to be selected for play back individually, with the selected components overlapped as originally intended for playback. This allows for the most desirable audio to be available based on the situation, for example, when exercising selecting inclusion of drums and bass, or when enjoying a romantic evening selecting only guitar and vocals, or when listening to a sports broadcast turning off commentators. In some cases, television shows and movies require that an instrumental only version of audio be available for when vocals might interfere with characters speaking, which this format provides for by default. In some cases, movies have excellent soundtracks, that could be enjoyed by turning off the audio for other components of the movie audio. In some cases, individuals enjoy singing karaoke — even when other individuals don't enjoy listening – singing along to audio without the original vocals. In some cases, musicians are interested in listening to only a specific instrument in an audio recording, to learn to play that part.

[0012] **In some embodiments, including one complete embodiment, ensure audio components are saved in sequence to a file, or as individual files or streams, with audio file components identified, for example, as part of the audio file metadata, as a separate file or stream, or stored in a database.** In some embodiments, including one complete embodiment, in reference to FIG. 1, an audio format is utilized where the audio file components are attached together into a sequence rather than overlapping, and are saved as a single audio file, where a text specification lists the components of the audio file in the metadata, by name, the start time in the single audio file, and the start time for playback relative to the earliest playback component in the audio when components are overlapped, where the end time of the component may be omitted as it is derived from the start of the next component or the end of the file. In some embodiments, including one complete embodiment, because the embodiment uses a single standard audio file – which may be in any audio format, for example MP3 or WAV – to sequentially store all the components, exporting from audio production tools is simply a matter of manually or automatically copying and pasting audio components in sequence or saving them individually. Other audio players will still recognize the audio file containing sequential components, but will play the components in sequence. Audio players can make an adaptation, to play the components of the audio file on top of each other, as detailed in this disclosure. In some embodiments, the times at which each audio file component starts and ends, is recorded as a text string, which may be in the comments or description metadata field of a standard audio file. Therefore, to produce audio files in this format, existing audio production software can be used, without modification for manual sequencing, or with minimal modification for automatic sequencing, and existing audio file formats can be used, without any modification. Therefore, all existing recordings, where the master is still available, can be saved to this format. In some embodiments, including one complete embodiment, an example of this format represented in FIG. 1, containing a listing of components including component name, time in the single sequential audio file, and the start time for playback relative to the earliest component playback in the audio, prefaced by "components" to indicate the format, may be a string comprising:

components|Guitar;00:00:00.00;00:00:00.00|Vocals;00:04:00.00;00:00:00.00|Bass;8:00:00;00:00:00|Drums;12:00:00;00:00:00|Strings;16:00:00;00:00:00

[0013] **In some embodiments, including one complete embodiment, write software code to read in the audio components and corresponding specification, and play back all or selected audio components.** In some embodiments, including one complete embodiment, when no individual streams are selected by the user, the audio playback software plays the audio components in an overlapping manner using the times provided for in the audio component specification. In some embodiments, including one complete embodiment, where individual audio file components are requested by the user, those streams are played back in an overlapping manner at the times provided for in the component specification.

[0014] In some embodiments, including one complete embodiment, the playback software may provide functionality comprising:
> import a file system access library.
> import an audio library or audio libraries that allow for audio data overlap as well as individual and simultaneous play.
> create a function to play an audio file;
>> retrieve the specification identifying the audio components for the specified audio file, for example by reading the metadata associated with the file, or reading from a local or remote data store;
>> initialize an audio component holding data structure, for example an array;

>> read the specified audio file into memory;
>> split the audio file into audio components and store each in the audio components holding data structure, utilizing the time markers provided for the audio file components;
>> create a new audio file in memory;
>> iterate through each component available in the audio file;
>>> if the current component corresponds to one of the requested components, add the component to the newly created audio file in memory at the appropriate position;
>> stop playback of any existing audio previously provided and play back the audio file created and stored in memory;
> define a function called when the application is run;
>> retrieve parameters passed to the function or any provided command line parameters.
>> send the audio file and any component parameters provided to the previously created play function.

[0015] In some embodiments, including one complete embodiment, in reference to FIG. 2, a user interface is created, providing options for the user to turn on and off the specified audio components, with a default of simply playing the audio with all components on, where provided are a file selection button 2000, a list of audio file components 2001, and a play button 2002. In some embodiments, including one complete embodiment, in reference to FIG. 2, the user interface may be created by computer code implementing functionality comprising:
> generate a user interface screen.
> import or implement the previously created play back functionality.
> display the option for a user to select an audio file 2000.
> provide a play button 2002.
> create a function that is activated when an audio file is selected;
>> read in the specification listing audio file components for the selected file;
>> iterate through the audio file components and display selectors for each component to be enabled or disabled during play back 2001.
> create a function that is activated when the play button is pressed;
>> read the states of the audio file component selectors;
>> send the audio file location or data, along with the audio file components selected, to the play backfunction.

CLAIMS

What is claimed is:

1. A system for the storage of audio, with the invention comprising:
an audio file format that contains audio components in sequence rather than overlapping, or maintains audio components in separate files;
a specification identifying each of said audio components.

2. A non-transitory computer-readable recording medium holding stored instructions, which when executed by one or more processing devices, cause the one or more processing devices to implement a method comprising, with the invention comprising:
audio stored as sequential components in a file, or stored in a group of related files;
a character sequence specifying each of said components of an audio file or of each file in said group of related audio files.

3. A system for playback of audio, with the invention comprising:
an audio file containing components in sequence rather than overlapping, or audio components in separate audio files;
a specification identifying each of said components;
computer software that reads said audio, allows for selection of said audio components, and provides integrated playback of selected said audio components.

4. A non-transitory computer-readable recording medium holding stored instructions, which when executed by one or more processing devices, cause the one or more processing devices to implement a method comprising:
utilizing stored audio components and a stored character string describing said audio components to allow a user to play said audio components individually and together.

FIG. 1

Guitar	Vocals	Bass	Drums	Strings
00:00:00	00:04:00	00:08:00	00:12:00	00:16:00

FIG. 2

2000 – audio.mp3 [Select File]

2001

|

[x] Vocals

[x] Guitar

[   ] Bass

[   ] Drums

[x] Strings

2002 – [Play]

ELECTRONIC HEALTH RECORD STORAGE AND COLLABORATION PLATFORM

INVENTOR JONATHAN BANNON MAHER

TECHNICAL FIELD

[0001] Embodiments of the invention relate to the fields of software and healthcare.

ABSTRACT

[0002] Systems, methods, apparatuses, and computer programs encoded on a computer storage medium, provide a platform for electronic health records storage and collaboration between patient, provider, and other authorized parties – which may include all registered users, or the public, while allowing for removal of display of personally identifiable information – by allowing users to register, create and manage medical records, and submit comments and data on medical records, where access and changes to the records may be made by approved individuals or groups of users, where the platform may conform to import and export standards supporting interoperability with other health records systems, therefore allowing for broad and in some cases life saving collaboration in instances including when the person sharing the health record has a difficult health issue, has been otherwise unable to resolve concerns, or for geographic and or financial reasons has limited access to health care professionals.

REFERENCE TO RELATED DOCUMENTS

[0003] This application is provided the benefit and priority date of United States Patent and Trademark Office provisional patent application number 62/522,703, filed June 20th 2017 by inventor Jonathan Bannon Maher, which is incorporated herein in its entirety.

BACKGROUND

[0004] This section is intended to introduce the reader to various aspects of the art that may be related to various aspects of the present techniques, which are described and or claimed. This discussion is believed to be helpful in providing the reader with background information to facilitate a better understanding of the various aspects of the

present disclosure. Accordingly, it is understood that these statements are to be read in this light, and not a citation of prior any art.

[0005] Health records have been traditionally stored in paper format, creating difficulties that prevent sharing and collaboration between providers. The transition to electronic health records has been slow, largely because of incompatibility between major providers of such systems. Additionally, the control of the records by providers, rather than the patients, prohibits easy transfer of records and collaboration between healthcare providers, as well as limiting patient participation. Therefore, current electronic health record systems have proven inadequate.

BRIEF DESCRIPTION OF THE ILLUSTRATIONS

[0006] An embodiment of the invention is illustrated by way of example, and not by way of limitation, in the figures of the accompanying drawings, and embodiments may not contain all components, may contain additional components, and may contain functionally similar components.

[0007] FIG. 1 is an embodiment of a computer code embedded flow diagram, providing a database schema to support the electronic health records system.

[0008] FIG. 2 is an embodiment of a computer code embedded flow diagram, providing the computer code to display components of the system to the user, and allow for processing of user submissions.

[0009] FIG. 3 is an example of an embodiment of an illustration of the screen to register.

[0010] FIG. 4 is an example of an embodiment of an illustration of the screen for the users account, where the user may update their health profile, manage connections with other users, and manage health records.

[0011] FIG. 5 is an example of an embodiment of an illustration of the screen for a health record.

[0012] FIG. 6 is an example of an embodiment of an illustration of the screen listing groups.

[0013] FIG. 7 is an example of an embodiment of an illustration of the screen of a group, allowing for comments to be posted with attachments and medical records.

DETAILED DESCRIPTION

[0014] It is understood that, as in any engineering or design project, the development of any actual implementation will include numerous implementation specific decisions made to achieve the developers' specific goals, such as compliance with business related and system related constraints, which may vary from one implementation to another. It is understood that such a development effort might be complex and time consuming, but is nevertheless a routine undertaking of design, fabrication, and manufacture for those skilled in the art having the benefit of this disclosure. The disclosed steps may be read as prefaced by "In some embodiments, including one complete embodiment, ", may be executed or performed in other orders or sequences, and are not limited to the order and sequence shown and described, which are provided to enable ease in constructing an embodiment, and along with each components of each step, may be removed, modified, combined, or rearranged, and other steps and or step components may be added, without departing from the scope of this disclosure and or invention. Although embodiments of the invention have been described and illustrated in the disclosed implementations, it is understood that the present disclosed subject matter, including apparatuses, methods, specification, and illustrations, has been made only by way of example, not by way of limitation, and the methods and apparatuses may be used in other systems, and that numerous changes and optimizations in the details of implementation of the invention and or embodiment are made without such modifications departing from the spirit and scope of this disclosure and or embodiments of the invention. Although the disclosure has been shown and described with respect to one or more embodiments, features of the disclosed embodiments can be combined and rearranged in various ways, and changes including equivalent alterations, substitutions, modifications, and additional efficiencies will of course occur to someone of ordinary skill in the art without departing from the spirit and scope of this disclosure and or invention. In particular regard to the various functions performed by the described components, the terms used to describe such components are intended to correspond, unless otherwise indicated, to any component which performs the specified function of the described component, or is functionally equivalent to the described component, even though not structurally equivalent to the disclosed structure which performs the function in the implementations described in this disclosure. In addition, while a particular feature of the disclosure may have been provided with respect to only one of several embodiments, such feature may be combined with one or more other features of other embodiments as may be desired and advantageous for any given or particular application. In some instances, well-known circuits, structures and techniques have not been shown in detail in order not to obscure the understanding of this disclosure. Articles in this disclosure such as "a" "an" and "the" may allow for both singular and plural forms. Verbs in this disclosure such as "is" may be read as "may be". Conjunctions in this disclosure such as "or" as used herein may be interpreted as inclusive or meaning any one or any combination, where "A, B or C" means "any of the following: A; B; C; A and B; A and C; B and C; A, B and C". Relational terms in this disclosure, for example first and second, top and bottom, left and right, are to distinguish one entity or action from another, and may not necessarily require or imply a relationship, or order between, such entities or actions. The disclosure includes the best mode contemplated by the inventor, a completely described specific embodiment, along with optional components and alternative embodiments to best suit the implementer, measurements in imperial and metric units to support universal understanding, and dramatically exceeds claims support requirements and enablement requirements by allowing for selection and or construction of the required components to be carried out easily, quickly, and routinely by persons of ordinary skill in the art, who are provided the additional benefit of utilizing readily available commodity components whenever possible. The present disclosure includes material protected by copyrights, and the owner of the copyrights hereby reserves all rights, but with authorization for publication as required by government patent offices. Various embodiments of the present invention may provide all, some or none of the disclosed technical advantages.

[0015] The inventor retains absolutely no liability for any implementation of this invention, and the invention is implemented exclusively at the risk and liability of the implementer.

[0016] It is understood that any computer code in this disclosure is merely illustrative, is not intended to function as written, and is simply to provide greater disclosure and enablement than provided by a traditional software patent computer code flow diagram, in order to exceed the enablement requirement, and allow those with benefit of this disclosure to easily create an embodiment of the invention. While code may be written in other computer languages, including Java and C++, the illustrative computer code is presented in the Python language, which may be run through the Python interpreter, with appropriate supportive libraries, which at the time of disclosure, may run on nearly any computer, for example one with an Intel or AMD processor, running a current version of Linux, Windows, or Mac OS. The computer code is numbered and described by logical block, to allow for descriptions to be used to easily port the code to other languages, where code comments may be considered the functional equivalent of a flow diagram, with comments indented to demonstrate the logical block of which they are a part, and while code flow diagram text may have been copied to the body of the description for convenience, it is best understood in the original intended format provided in the accompanying illustrations. All code comments may read as if prefaced by "In some embodiments, including one complete embodiment, ". In some embodiments, functionality may be modified, rearranged, excluded, and added.

[0017] In some embodiments, the system allows for collaboration on electronic health records, where a person creates an account, and then may fill out information about their health conditions, similar to the intake form at any physician's office, and may then create and post information to health records, create connections with other users of the system, and share records for collaboration with connected users or public groups in order to solicit broad feedback in difficult to address situations. Embodiments are particularly beneficial for those without convenient or affordable access to top doctors, such as in developing nations, or remote areas of developed countries. Individuals living with specific conditions may be able to offer insight to others with the same condition, that may not be provided by doctors who don't live with the condition.

[0018] **In some embodiments, including one complete embodiment, create a database supporting the health record system from a database schema.** In some embodiments, including one complete embodiment, database schema FIG. 1 is created and executed on a database to support the core functionality of the system, providing functionality comprising:
> create a database table for accounts with fields of an auto incrementing integer primary key, a record creation timestamp, an email address, password, and fields for any health details such as allergies, asthma, diabetes, and cancer.
> create a database table for sessions with fields of a record creation timestamp, a session ID, and an account ID.
> create a database table for medical records with fields of an auto incrementing integer primary key, a record creation timestamp, an associated account ID, and a record name.
> create a database table for medical records posts with fields of an auto incrementing integer primary key, a record creation timestamp, an associated record ID, an associated account ID, comments, and a file attachment.
> create a database table for account connections with fields of an auto incrementing integer primary key, a record creation timestamp, the connection requester account ID, the connection receiver account ID, and the connection approval status.

> create a database table for medical record access permissions with fields of an auto incrementing integer primary key, a record creation timestamp, an associated account ID, and a record name.
> create a database table for groups with fields of an auto incrementing integer primary key, a record creation timestamp, an associated account ID, and a record name.
> create a database table for group posts with fields of an auto incrementing integer primary key, a record creation timestamp, an associated account ID, comments, a file attachment, and a referenced record ID.

[0019] **In some embodiments, including one complete embodiment, create the computer server code files to display system screens and allow users to interact with the system.** In some embodiments, including one complete embodiment, in reference to FIG. 2, the code to display user screens in the system, and process submitted data, providing functionality comprising:
> import the libraries providing a http server.
> import a database library.
> establish a connection to the database.
> define a function to handle get requests to the server;
>> if there was a session ID sent by the user, retrieve the users account ID;
>> create a variable holding any http request parameters;
>> create a variable holding any requested action;
>> create a variable holding any requested record ID;
>> create a variable building the user display;
>> if the action is account, select from the database table Accounts name for the requested ID, and add to display;
>> if the action is records, select from the database table Records the medical records for the given ID;
>> if the action is to request a record, if the record is owned by the logged in user, or shared with the logged in user, or the record is public, return the record;
>> if the request is to list connections, select connection names from the database for the logged in user;
> if the request is to display a list of groups, retrieve the name of each from the database and link from the group name to the specific group;
>> if the request is for a specific group, retrieve for display the group title and all of the posts in that group;
>> if the request is to import a health record, parse the fields of the imported electronic health record, which may be in the international standard Continuity of Care XML format and may be parsed by importing an XML parsing library, and insert the data into

the corresponding fields into a record in the health records table and assign the health record to the specified user;
>> if the request is to export a health record, retrieve the medical records for the specified patient, and return each in a standardized electronic health record format, such as the international standard Continuity of Care XML format;
>> replace template components for title and content with corresponding values;
>> return the output to the user.
> define a function to handle posts to the server;
>> read in the data posted to the form;
>> if the user sent a session ID, retrieve the users account ID;
>> initialize a variable to hold a query;
>> if the action is to start, then check if the provided email address is in the database, and if not create an account, then sign in the user with the provided email address and password by providing them a session identifier;
>> if the requested action is to update the account, create a query to update each account field for the logged in user with data posted by the user;
>> if the requested action is to insert a medical record, generate a query to insert the medical record using the values posted by the user;
>> if the requested action is to update a medical record, generate a query to update the medical record if owned by the logged in user, using the values posted by the user;
>> if the requested action is to delete a medical record, generate a query to delete the medical record using the identifier posted by the user;
>> if the requested action is to insert a medical record post, generate a query to insert the medical record using the values posted by the user;
>> if the requested action is to delete a medical record, generate a query to delete the medical record using the identifier posted by the user;
>> if the requested action is to create a connection between two accounts, generate a query to insert the connection using the values posted by the user;
>> if the requested action is to update a connection, generate a query to update the record using the values posted by the user;
>> if the requested action is to update a connection, generate a query to update the record if the user is the sender or recipient of the connection, using the values posted by the user;
>> if the requested action is to update a connection, generate a query to update the record if the user is the sender or the recipient of the connection, using the values posted by the user;
>> if the requested action is to update a group, generate a query to update the record using the values posted by the user;

>> if the requested action is to insert a group post, generate a query to add the post using the values posted by the user;
>> if the requested action is to delete a group post, generate a query to delete the post if it was created by the user;
>> execute any query generated;
>> redirect the user back the screen from with they posted.

CLAIMS

What is claimed is:

1. A system for electronic health record maintenance and collaboration, with the invention comprising:
   the creation of an account in the system;
   creation of a health record associated with an account in the system, to which others can post comments and files;
   a connection between two accounts that allows access to private health records;
   a forum that allows for an account to post comments, which may include reference to a health record.
2. A non-transitory computer-readable recording medium holding stored instructions, which when executed by one or more processing devices, cause the one or more processing devices to implement a method comprising, with the invention comprising:
   a data structure holding an account in the system;
   a data structure holding a health record associated with an account in the system, to which others can post comments and files;
   a data structure holding a connection between two accounts that allows access to private health records;
   a data structure holding a forum that allows for an account to post comments, which may include reference to a health record.

FIG. 1

1000 — create a database table for accounts with fields of an auto incrementing integer primary key, a record creation timestamp, an email address, password, and fields for any health details such as allergies, asthma, diabetes, and cancer.

```
CREATE TABLE Accounts (
  ID INT AUTO_INCREMENT PRIMARY KEY,
  Created TIMESTAMP,
  Email_Address TEXT,
  Password TEXT,
  Asthma BOOL,
  Allergies BOOL,
  Allergies_Details TEXT,
  Diabetes BOOL,
  Cancer BOOL,
  Cancer_Details TEXT)
```

1001 — create a database table for sessions with fields of a record creation timestamp, a session ID, and an account ID.

```
CREATE TABLE Sessions (
  Created TIMESTAMP,
  Session_ID TEXT PRIMARY,
  Account_ID INT)
```

1002 — create a database table for medical records with fields of an auto incrementing integer primary key, a record creation timestamp, an associated account ID, and a record name.

```
CREATE TABLE Records (
  ID INT AUTO_INCREMENT PRIMARY KEY,
  Created TIMESTAMP,
```

```
  Account_ID INT,
  Name TEXT)
```

1003 – create a database table for medical records posts with fields of an auto incrementing integer primary key, a record creation timestamp, an associated record ID, an associated account ID, comments, and a file attachment.

```
CREATE TABLE Record_Post (
  ID INT AUTO_INCREMENT PRIMARY KEY,
  Created TIMESTAMP,
  Record_ID INT,
  Account_ID INT,
  Comments TEXT,
  Attachment BLOB)
```

1004 – create a database table for account connections with fields of an auto incrementing integer primary key, a record creation timestamp, the connection requester account ID, the connection receiver account ID, and the connection approval status.

```
CREATE TABLE Connections (
  ID INT AUTO_INCREMENT PRIMARY KEY,
  Created TIMESTAMP,
  Sender_Account_ID INT,
  Recipient_Account_ID INT,
  Status INT)
```

1005 – create a database table for medical record access permissions with fields of an auto incrementing integer primary key, a record creation timestamp, an associated account ID, and a record name.

```
CREATE TABLE Record_Permissions (
  ID INT AUTO_INCREMENT PRIMARY KEY,
  Created TIMESTAMP,
  Record_ID INT,
```

```
  Permissioned_Account_ID INT,
  Permissions TEXT)
```

1006 – create a database table for groups with fields of an auto incrementing integer primary key, a record creation timestamp, an associated account ID, and a record name.

```
CREATE TABLE Groups (
  ID INT AUTO_INCREMENT PRIMARY KEY,
  Created DATETIME,
  Name TEXT)
```

1007 – create a database table for group posts with fields of an auto incrementing integer primary key, a record creation timestamp, an associated account ID, comments, a file attachment, and a referenced record ID.

```
CREATE TABLE Group_Post (
  ID INT PRIMARY AUTO_INCREMENT
    PRIMARY KEY,
  Created DATETIME,
  Account_ID INT,
  Comments TEXT,
  Attachment BLOB,
  Record_ID INT)
```

FIG. 2

2000 — import the libraries providing a http server

```
from flask import Flask
from flask import request
from flask import Response
```

2001 — import a database library

```
import mysql
```

2002 — establish a connection to the database

```
db = mysql.connect("127.0.0.1", "admin", "admin", "db")
```

2003 — define a function that is executed when the server received get requests

```
@application.route("/*", methods=["GET"])
def display():
```

2004 — if there was a session ID sent by the user, retrieve the users account ID

```
    account_id = None
    if request.cookies("Session_ID"):
        query = "SELECT Account_ID
            FROM Sessions
            WHERE Session_ID=%s" %
            request.cookies("Session_ID")
        records = db.execute(query)
        for record in records:
            account_id = record["Account_ID"]
```

2005 – create a variable holding any http request parameters

```
parameters = request.argv
```

2006 – create a variable holding any requested action

```
action = parameters.get("action")
```

2007 – create a variable holding any requested record ID

```
id = parameters.get("id")
```

2008 – create a variable building the user display

```
output = "<html><head> <title>
   Electronic Health Records | $title</title>
   </head><body><table><tr>
   <td>$content</td><td valign=top>
   Account<br>Connections<br>
   Records<br>Groups</td>
   </tr></table></body></html>'
```

2009 – if the action is account, select from the database table Accounts name for the requested ID, and add to display

```
if action == "account":
   query = "SELECT Name, Email, Password
      FROM Accounts
      WHERE Account_ID=" + id
   records = db.execute(query)
   for record in records:
      output += record["Name"]
```

2010 – if the action is records, select from the database table Records the medical records for the given ID

```
if action == "records":
   query = "SELECT name
      FROM Records
      WHERE Account_ID=" + id
      records = db.execute(query)
   for record in records:
      output += record["name"]
```

2011 – if the action is to request a record, if the record is owned by the logged in user, or shared with the logged in user, or the record is public, return the record

```
if action == "record":
   query = "SELECT Account_ID,
      Title, Content, Permissions
      FROM Records
      WHERE Record_ID=" + id
   records = db.execute(query)
   for record in records:
      output += record["Title"] + "<P>" +
         record["Content"]
```

2012 – if the the request is to list connections, select connection names from the database for the logged in user

```
   if action == "connections":
      query = "SELECT name
         FROM connections
         WHERE Account_ID=" + account_id
      records = db.execute(query)
         for record in records:
            output += record["name"]
```

2013 – if the request is to display a list of groups, retrieve the name of each from the database and link from the group name to the specific group

```
if action == "groups":
   query = "SELECT ID, Name FROM Groups"
   records = db.execute(query)
   for record in records:
      content+= '<A HREF="/group/' +
      record["ID"] + '">' +
      record["Name"] + "<BR>"
```

2014 — if the request is for a specific group, retrieve for display the group title and all of the posts in that group

```
if action == "group":
   query = "SELECT Name
      FROM Group
      WHERE ID=" + id
   records = db.execute(query)
   for record in records:
      content += "<B>" + records["Name"] +
         "</B><P>"
   query = "SELECT Accounts.Name as Name,
      Group_Posts.Comments as Comments,
      Group_Posts.Attachment as Attachment,
      Group_Posts.Record_ID as Record_ID
      FROM Group_Posts, Accounts
      WHERE Group_ID=" + id + "
      AND Group_Posts.Account_ID=
      Accounts.ID
   records = db.execute(query)
   for record in records:
      output += "Name: ' +
      record["Name"] + "<BR>" +
      "Comments: " +
      record["Comments"] + "<BR>" +
      "<IMG DATA='" +
      record["Attachment"] + "'><BR>"
      if record["Record_ID"]:
```

```
    output += '<A HREF="' +
      record["Record_ID"] + ">
      Attached Record</A>'
```

2015 – replace template components for title and content with corresponding values

```
  output = output.replace("$title", title)
    .replace("$content", content)
```

2016 – return the output to the user

```
  return output
```

2017 – define a function to handle http posts

```
@application.route("/*", methods=["POST"])
def api():
```

2018 – read in the data posted to the form

```
  form = post.form
  email_address = form["email_address"]
  password = form["password"]
  action = form["action"]
```

2019 – If the user sent a session ID, retrieve the users account ID

```
  account_id = None
  if request.cookies("Session_ID"):
    query = "SELECT Account_ID
      FROM Sessions
      WHERE Session_ID=%s" %
      request.cookies("Session_ID")
    records = db.execute(query)
```

```
for record in records:
    account_id = record["Account_ID"]
```

2020 – initialize a variable to hold a query

```
query = ""
```

2021 – if the action is to start, then check if the provided email address is in the database, and if not create an account, then sign in the user with the provided email address and password by providing them a session identifier

```
if action == "start":
    create_session = False
    account_ID = None
    email_address = form["Email_Address"]
    password = form["Password"]
    query = "SELECT ID
        FROM Accounts
        WHERE email_address=\"%s\")
    account = db.execute(query)
    if not account:
        query = "INSERT INTO Accounts
            (Email_Address, Password)
            VALUES(\"%s\", \"%s\")" %
            (email_address, password)
        db.execute(query)
        query = "SELECT ID
            FROM Accounts
            WHERE Email_Address=\"%\"
            AND Password=\"%s\"" %
            (email_address, password)
        accounts = db.execute(query)
        account_id = accounts["ID"]
        create_session = True
        if account:
```

```
        if account["Password"] == password:
            create_session = True
            account_id = account["ID"]
    if create_session:
        session_id = sha.sha256(sys.time())
        query = "INSERT INTO Sessions
            (Session_ID, Account_ID)
            VALUES (%s, %s)" %
            (session_id, account_id)
    db.execute(query)
    response.header
        .setCookie("Session_ID", session_id)
```

2022 — if the requested action is to update the account, create a query to update each account field for the logged in user with data posted by the user

```
if action == "account_update":
    query = "UPDATE Accounts
        SET Email_Address="%s",
        Password="%s",
        Asthma="%s",
        Allergies="%s",
        Allergies_Details="%s",
        Diabetes="%s",
        Cancer="%s",
        Cancer_Details="%s"
        WHERE ID="%s"" % (
        form["Email_Address"],
        form["Password"],
        form["Asthma"],
        form["Allergies"],
        form["Allergies_Details"],
        form["Diabetes"],
        form["Cancer"],
        form["Cancer_Details"]),
        account_id)
```

2023 — if the requested action is to insert a medical record, generate a query to insert the medical record using the values posted by the user

```
if action == "record_insert":
    query = "INSERT INTO Records (
        AccountID, Name)
        VALUES (%s, %s)" %
        (form["Account_ID"], form["Name"])
```

2024 — if the requested action is to update a medical record, generate a query to update the medical record if owned by the logged in user, using the values posted by the user

```
if action == "record_update":
    query = "UPDATE Records
        SET Name=\"%s\"
        WHERE ID=%s AND Account_ID=%s" %
        (form["Name"], form["ID"], account_ID)
```

2025 — if the requested action is to delete a medical record, generate a query to delete the medical record using the identifier posted by the user

```
if action == "record_delete":
    query = "DELETE FROM Records
        WHERE ID=%s" % form["ID"]
```

2026 — if the requested action is to insert a medical record post, generate a query to insert the medical record using the values posted by the user

```
if action == "record_post_insert":
    query = "INSERT INTO Record_Posts
    (Record_ID) VALUES (%s)" %
    (form['Record_ID'])
```

2027 – if the requested action is to delete a medical record, generate a query to delete the medical record using the identifier posted by the user

```
if action == "record_post_delete":
  query = "DELETE FROM Record_Posts
    WHERE ID=%s" % (form["ID"])
```

2028 – if the requested action is to create a connection between two accounts, generate a query to insert the connection using the values posted by the user

```
if action == "connection_insert":
  query = 'INSERT INTO Connections
     (Sender_Account_ID,
     Recipient_Account_ID)
     VALUES (%s, %s)" %
      (form['Sender_Account_ID'],
     form['Sender_Recipient_ID'])
```

2029 – if the requested action is to update a connection, generate a query to update the record using the values posted by the user

```
if action == "connection_update":
  query = "UPDATE Connections
    SET Status=\"%s"
    WHERE ID=%s" %
    (form["Status"], form["ID"])
```

2030 – if the requested action is to update a connection, generate a query to update the record if the user is the sender or recipient of the connection, using the values posted by the user

```
if action == "connection_delete":
  query = "DELETE FROM Connections
    WHERE ID=%s AND
    (Sender_Account_ID=%s OR
```

```
    Recipient_Account_ID=%s)" %
    (form['ID'], account_id, account_id)
```

2031 – if the requested action is to update a connection, generate a query to update the record if the user is the sender or the recipient of the connection, using the values posted by the user

```
if action == "group_insert":
  query = "INSERT INTO Groups (Name)
    VALUES (\"%s\")" %
    (form["Name"])
```

2032 – if the requested action is to update a group, generate a query to update the record using the values posted by the user

```
if action == "group_update":
  query = "UPDATE Groups
    SET Name=\"%s\"
    WHERE id=%s" %
     (form["Name"], form["ID"])
```

2033 – if the requested action is to insert a group post, generate a query to add the post using the values posted by the user

```
if action == "group_post_insert":
  query = "INSERT INTO Group_Posts
    (Group_ID, Content) VALUES
     (%s, \"%s\")" %
     (form["Group_ID"], form["Content"])
```

2034 – if the requested action is to delete a group post, generate a query to delete the post if it was created by the user

```
if action == "group_post_delete":
  query = "DELETE FROM Group_Posts
```

```
WHERE ID=%s AND
Account_ID" %
(form['ID'], account_id)
```

2035 – execute any query generated

```
if query:
    db.execute(query)
```

2036 – redirect the user back the screen from with they posted

```
response.redirect(request.referrer)
```

FIG. 3

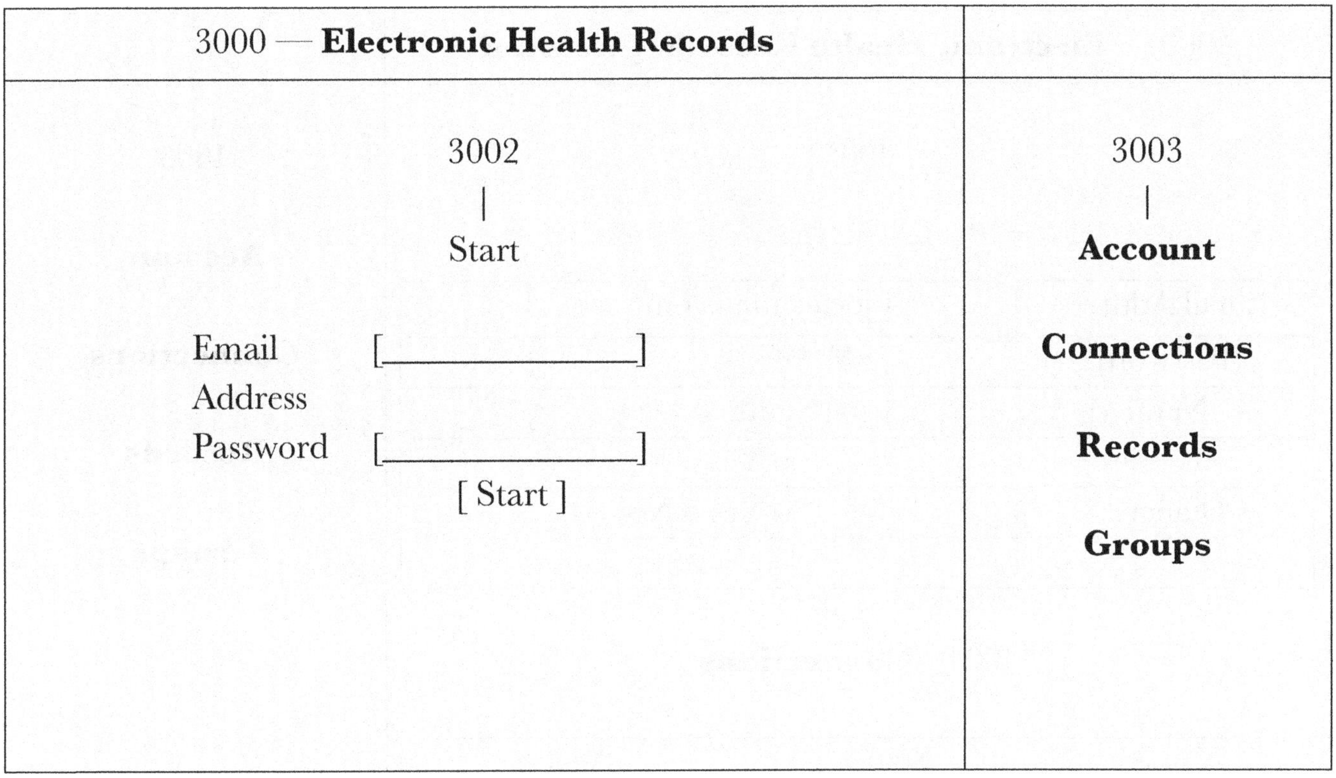
3000 — Electronic Health Records
3002
Start
Email Address
Password
[Start]
3003
Account
Connections
Records
Groups

FIG. 4

4000 — **Electronic Health Records | Account**

4002

Email Address	[joe@email.com]
Password	[**********]
Name	[Joe Smith]
Allergies	o Yes o No
Diabetes	o Yes o No
	[Save]

4003

Account

Connections

Records

Groups

4004 – **Connections**

Name	Status	Action
Dr. Joe Smith	Connected	[remove] [permissions]
Steve Jones	Connected	[remove] [permissions]
	[Email Address]	[Add Connection]

4005 - **Health Records**

Joe's Brain Scan

[Name] [Create Record]

FIG. 5

5000 — **Electronic Health Records | Record**

5002

Joe's Brain Cancer

5004

Name	Post	Action
Dr. Bob Kim	Here's your brain scan. I recommend using the Bannon Maher tumor removal system.	[Remove]
Joe Smith	Thanks Bob.	[Remove]
	[Comment] [Attachment]	[Post]

5003

Account

Connections

Records

Groups

FIG. 6

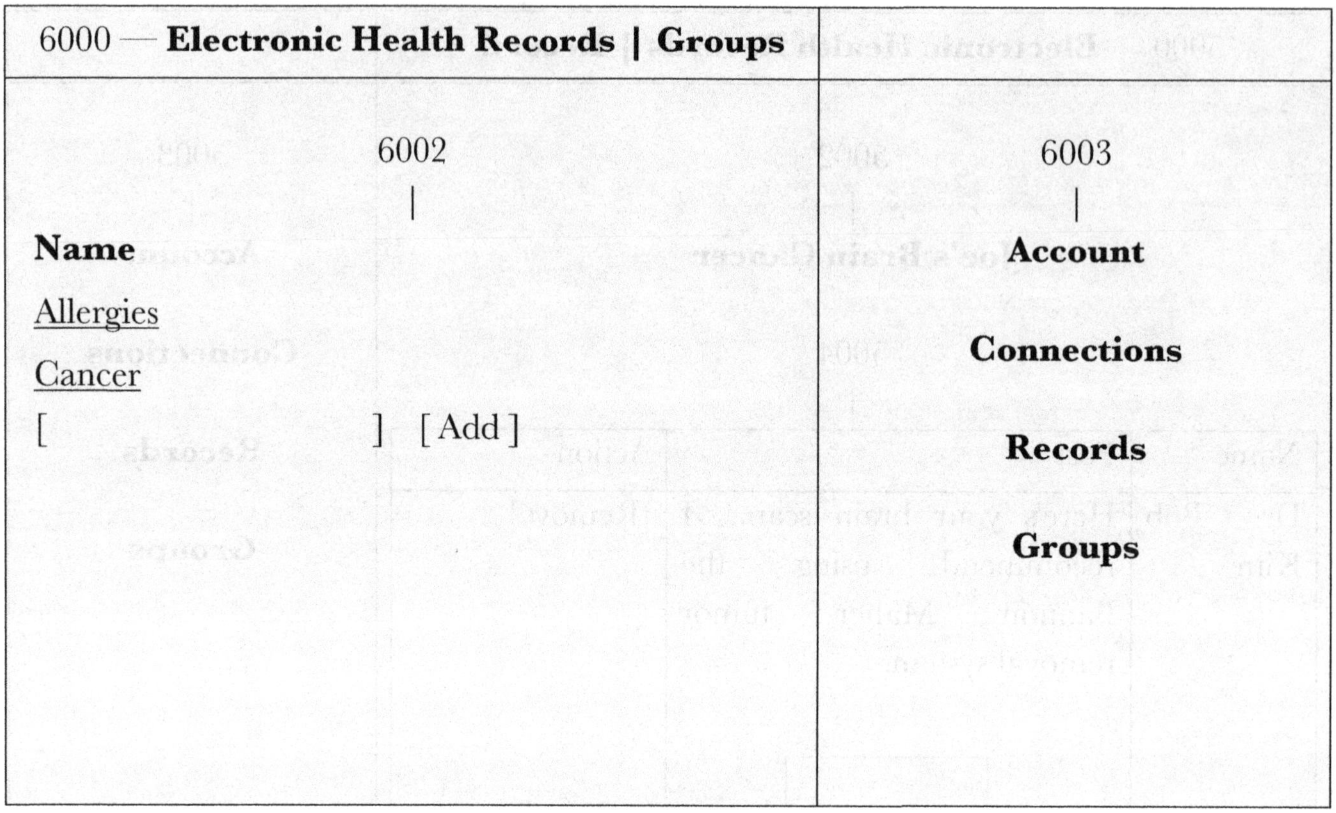
6000 — Electronic Health Records | Groups
6002
Name
Allergies
Cancer
[ ] [Add]
6003
Account
Connections
Records
Groups

FIG. 7

7000 — **Electronic Health Records | Group**

7002

Cancer

Members: Steve Jones, Joe M.D.

[Join]

Date	Poster	Text	Action
Dec 14th	Steve Jones	I had this scan taken of my brain cancer. Bannon Maher device is recommended. Please see the linked record, and share your thoughts. [Attachments]	[Remove]
Dec 17th	Dr. Joe Smith	Others have had great success with the device, as it is leaves healthy tissue intact and comprehensive removal.	[Remove]
		[Comment] [Attachment]	[Post]

7003

Account

Connections

Records

Groups

BED ENVIRONMENT INSULATION

INVENTOR JONATHAN BANNON MAHER

TECHNICAL FIELD

[0001] Embodiments of the invention relate to the fields of sleep and environmental insulation.

ABSTRACT

[0002] Systems, methods, and apparatuses provide for structures and materials that block or reduce the intensity of environmental factors — including sound, vibration, and light – that may otherwise pass through to wake a person on a mattress, while optionally for further environmental control, including speakers able to support a soothing ambience within a header encasement.

REFERENCE TO RELATED DOCUMENTS

[0003] This application is provided the benefit and priority date of United States Patent and Trademark Office provisional patent application number 62/522,696, filed June 20th 2017 by inventor Jonathan Bannon Maher, which is incorporated herein in its entirety.

BACKGROUND

[0004] This section is intended to introduce the reader to various aspects of the art that may be related to various aspects of the present techniques, which are described and or claimed. This discussion is believed to be helpful in providing the reader with background information to facilitate a better understanding of the various aspects of the present disclosure. Accordingly, it is understood that these statements are to be read in this light, and not a citation of prior any art.

[0005] People are often awoken from sleep by noises and vibrations, including from trains, busses, garbage trucks, car horns, airplanes, lawnmowers, leaf blowers, construction, and neighbors. Even people who live luxury housing in quiet areas may not have fully soundproofed homes, and when they do, they may not have soundproofed rooms against noises coming from inside the house, such as a family member who for work or school leaves early or comes home late.

BRIEF DESCRIPTION OF THE ILLUSTRATIONS

[0006] Illustrations are provided by way of example, and not by way of limitation, in the figures of the accompanying drawings, and embodiments may not contain all components, may contain additional components, and may contain functionally similar components.

[0007] FIG. 1 is an illustration of an embodiment of the bed header and footer environmental insulation component, with sound proofing material on all sides of the encasement to diminish ambient noise.

[0008] FIG. 2 is an illustration an embodiment of soundproofing panel to be placed on the under the mattress and or box spring and or on the ground, to reduce sound coming through the floor, which additionally contains springs to absorb vibrations.

[0009] FIG. 3 is an illustration of an embodiment of optional vibration insulating springs, which may be placed under each bed post.

DETAILED DESCRIPTION

[0010] It is understood that, as in any engineering or design project, the development of any actual implementation will include numerous implementation specific decisions made to achieve the developers' specific goals, such as compliance with business related and system related constraints, which may vary from one implementation to another. It is understood that such a development effort might be complex and time consuming, but is nevertheless a routine undertaking of design, fabrication, and manufacture for those skilled in the art having the benefit of this disclosure. The disclosed steps may be read as prefaced by "In some embodiments, including one complete embodiment, ", may be executed or performed in other orders or sequences, and are not limited to the order and sequence shown and described, which are provided to enable ease in constructing an embodiment, and along with each components of each step, may be removed, modified, combined, or rearranged, and other steps and or step components may be added, without departing from the scope of this disclosure and or invention. Although embodiments of the invention have been described and illustrated in the disclosed implementations, it is understood that the present disclosed subject matter, including apparatuses, methods, specification, and illustrations, has been made only by way of example, not by way of limitation, and the methods and apparatuses may be used in other systems, and that numerous changes and optimizations in the details of implementation of the invention and or embodiment are made without such modifications departing from the spirit and scope of this disclosure and or embodiments of the invention. Although the disclosure has been shown and described with respect to one or more embodiments, features of the disclosed embodiments can be combined and rearranged in various ways, and changes including equivalent alterations, substitutions, modifications, and additional efficiencies will of course occur to someone of ordinary skill in the art without departing from the spirit and scope of this disclosure and or invention. In particular regard to the various functions performed by the described components, the terms used to describe such components are intended to correspond, unless otherwise indicated, to any component which performs the specified function of the described component, or is functionally equivalent to the described component, even though not structurally equivalent to the disclosed structure which performs the function in the implementations described in this disclosure. In addition, while a particular feature of the disclosure may have been provided with respect to only one of several embodiments, such feature may be combined with one or more other features of other embodiments as may be desired and advantageous for any given or particular application. In some instances, well-known circuits, structures and techniques have not been shown in detail in order not to obscure the understanding of this disclosure. Articles in this disclosure such as "a" "an" and "the" may allow for both singular and plural forms. Verbs in this disclosure such as "is" may be read as "may be". Conjunctions in this disclosure such as "or" as used herein may be interpreted as inclusive or meaning any one or any combination, where "A, B or C" means "any of the following: A; B; C; A and B; A and C; B and C; A, B and C". Relational terms in this disclosure, for example first and second, top and bottom, left and right, are to distinguish one entity or action from another, and may not necessarily require or imply a relationship, or order between, such entities or actions. The disclosure includes the best mode contemplated by the inventor, a completely described specific embodiment, along with optional components and alternative embodiments to best suit the implementer, measurements in imperial and metric units to support universal understanding, and dramatically exceeds claims support requirements and enablement requirements by allowing for selection and or construction of the required components to be carried out easily, quickly, and routinely by persons of ordinary skill in the art, who are provided the additional benefit of utilizing readily available commodity components whenever possible.

Various embodiments of the present invention may provide all, some or none of the disclosed technical advantages.

[0011] The inventor retains absolutely no liability for any implementation of this invention, and the invention is implemented exclusively at the risk and liability of the implementer.

[0012] The disclosure resolves the afore mentioned deficiencies, by utilizing an insulated encasement encompassing the header, and or footer, and or underside of a mattress, to diminish sound, vibration, and light, where insulation may include a soundproofing material of the type used in earplugs. In some embodiments, the mattress may be supported, directly or indirectly, by vibration dampening springs.

[0013] **In some embodiments, including one complete embodiment, create the molds for the soundproofing material.** In some embodiments, encasements surround each the head and the foot of the bed, encompassing the mattress and box spring, as illustrated in FIG. 1. The header and footer units may be identical, except that in the header unit, audio and light emitting devices may be embedded, and in order for the head unit to effectively block out light and sound, may start at just below the shoulders of a 6 foot tall person to block out sound, to cover around 2 of the approximately 6 feet of the mattress length, while the footer unit provides reduced coverage.

[0014] In some embodiments, the insulting component goes fully around the mattress including underneath the box spring, and may be on both the head and the foot of the bed, to diminish sounds coming from all directions. The material used may diminish and or reflect sound, as well as be lightweight since it will be over peoples' heads while they're sleeping. In some embodiments, the material for the encasement may be durable and light weight, for example carbon fiber. In some embodiments, full encasement may be provided over the mattress allowing for adequate room for comfortable movement of a person in the bed and with one side unenclosed to allow for airflow.

[0015] In some embodiments, materials may be cut and placed by hand to conform to the desired structures. In some embodiments, after having accounted for these details and specifications, graphite molds may be made to cast the components. The molds may be manually etched, or may be setup in a CAD (Computer Aided Design) file, which is then will then processed by a CNC (Computer Numerical Control) machine to etch each mold. While there are many versions of CNC machines, for many the design process provides for a part to be designed in a CAD program, export the design to the vendor neutral Initial Graphics Exchange Specification (IGES) format, developed by the United States Air Force for parts manufacturing, and have the CNC machine's software convert the designs to G-Code instructions, which control the movement of the CNC machine tools. The selected material may then be melted and poured into the graphite molds. In some embodiments, components may be 3D printed, and may include a design that instead of an insulating material, utilizes an insulating structure, which may be in the form of multiple layers of material with air in between layers, and or structures such as honeycomb.

[0016] **In some embodiments, including one compete embodiment, mold and encase the soundproofing material.** Any material surrounding a person's head may be very light, so as to not cause harm if falling on a sleeping person, and not so hard that it would hurt if it hit a person on the head. In some embodiments, the insulating material to be used may be the same as is used to create ear plugs, which are often made of a silicon base foam or rubber. In some embodiments, including one complete embodiment, in reference to FIG. 1, the header encasement and the footer encasement is insulated through coating or filling the interior of the encasement with soundproofing foam until it achieves an adequate insulation thickness. If the soundproofing material is first put into molds, the foam is allowed to solidify, and may then be inserted into the encasement, or wrapped in an encasing material such as carbon fiber. It may be important that the encasement be aesthetic, since some people spend substantial amounts of money on their bed and related items.

[0017] In some embodiments, including one complete embodiment, in reference to FIG. 1, the header encasement contains small speakers 1003 1004 mounted to the inside of the header encasement, with power provided to the speakers by a wiring 1006 connected between the speakers and an external power source, and where the header encasement unit connects the speakers through wiring 1007 to an audio input 1005, which can be connected to a sound machine's headphone jack using a double male headphone wire, so that the speakers in the encasement can provide soothing sound to a resting or sleeping person. In some embodiments, in the header encasement, lights with accompanying on off switches and electrical wiring, which may emit soothing light such as blue light, and or reading light, and may be mounted to the header encasement.

[0018] **In some embodiments, including one complete embodiment, cut a sheet of sound insulation material for the bottom and wrap it with accompanying springs in protective material.** This sheet can be placed between the mattress and box spring or under the box spring, to absorb both vibration and sound. In some embodiments, including one complete embodiment, this is of the same material as the other insulating material, and of an adequate thickness to provide suitable insulation from sound, for example at least 1 inch thick. In some embodiments, including one complete embodiment, insert the springs evenly throughout the foam, then sandwich the foam and springs between thin high strength steel plates on either side, then wrap in carbon fiber.

[0019] In order to insulate from vibration coming from passing busses, or construction, in this embodiment, high load bearing springs sit under each bed post to absorb vibration, as detailed in FIG. 2 is used, though alternative construction could be used including creating flexible plastic containers filled with liquid. These springs are mounted between two sheets of metal. The springs must be made of sufficient quality metal so as to maintain their resilience for a decade when supporting a ton of weight.

[0020] In some embodiments, including one complete embodiment, arrange components to provide a peaceful night's sleep free from unwanted sound, vibration, and light.

CLAIMS

What is claimed is:

1. An environmental insulation apparatus for a sleeping unit, with the invention comprising:
   insulating material surrounding the header and or footer of a mattress in a form so as to reduce the intensity of undesirable sound or light.
2. Vibration insulation apparatus for a sleeping unit, with the invention comprising:
   one or springs supporting a mattress, directly or indirectly, so as to reduce ambient vibrations.
3. An environmental sound control apparatus, with the invention comprising:
   one or more speakers that provide sound from an encasement surrounding the head of a mattress.

FIG. 1

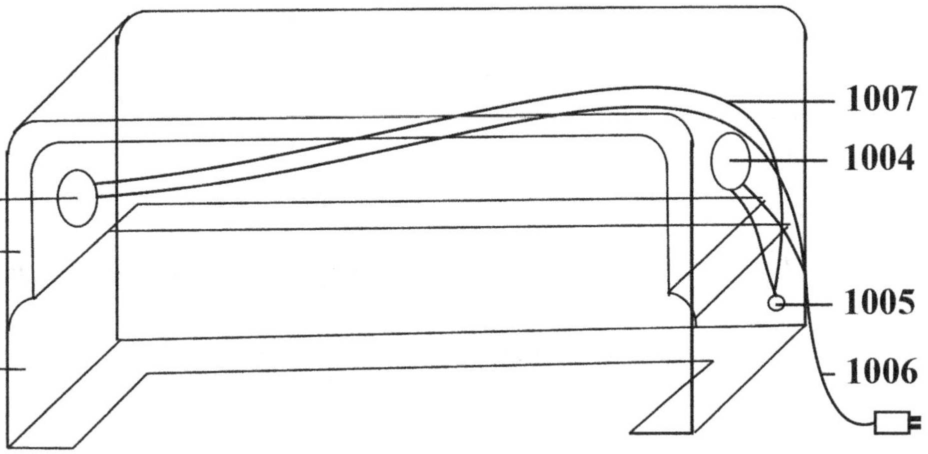

Jonathan Bannon Maher

FIG. 2

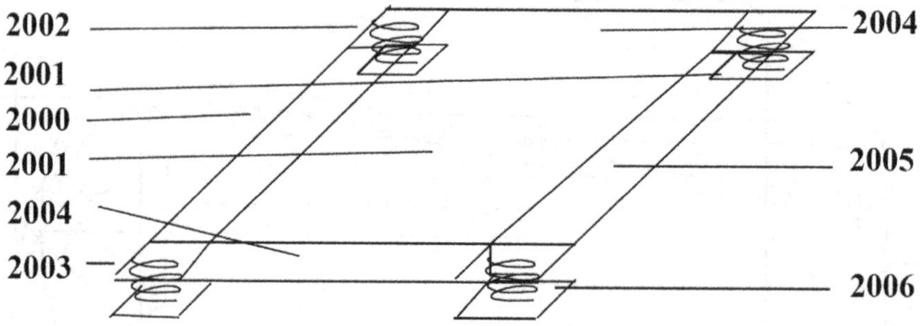

Jonathan Bannon Maher

FIG. 3

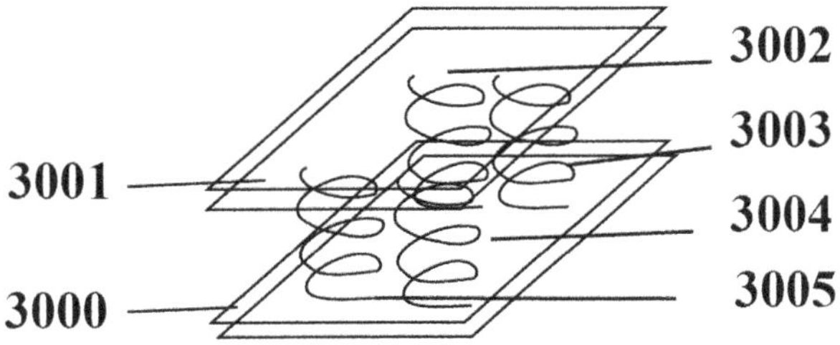

Jonathan Bannon Maher

COMPRESSED DISTRIBUTION OF AIR AND WATER

INVENTOR JONATHAN BANNON MAHER

TECHNICAL FIELD

[0001] Embodiments of the invention relate to the field of air conditioning and water distribution.

ABSTRACT

[0002] Systems, methods, and apparatuses provide for temperature controlled air and or water to be compressed, and passed through during transit, in contrast to traditional systems, conduits of reduced diameter and increased strength, and decompressed upon exit, where the conduits may be flexible and reinforced, thereby reducing construction materials, time, cost and space requirements, and preventing accumulation of condensation, mold, and dust, thereby making the systems self-cleaning.

REFERENCE TO RELATED DOCUMENTS

[0003] This application is provided the benefit and priority date of United States Patent and Trademark Office provisional patent application number 62/522,700, filed June 20th 2017 by inventor Jonathan Bannon Maher, which is incorporated herein in its entirety.

BACKGROUND

[0004] This section is intended to introduce the reader to various aspects of the art that may be related to various aspects of the present techniques, which are described and or claimed. This discussion is believed to be helpful in providing the reader with background information to facilitate a better understanding of the various aspects of the present disclosure. Accordingly, it is understood that these statements are to be read in this light, and not a citation of prior any art.

[0005] Currently temperature controlled air travels through ducts of significant diameter or width and height, which allows for condensation, mold, and dust to accumulate inside the ducts, which is expensive and potentially impossible to fully remove because of the complexity of and limited access to the insides of the ducts.

[0006] Currently hot and cold water travel separately and unpressurized in transit to their final destination.

BRIEF DESCRIPTION OF THE ILLUSTRATIONS

[0007] Illustrations are provided by way of example, and not by way of limitation, in the figures of the accompanying drawings, and embodiments may not contain all components, may contain additional components, and may contain functionally similar components.

[0008] FIG. 1 is an embodiment of an apparatus that allows for compressed tube distribution of temperature controlled air.

[0009] FIG. 2 is an embodiment of an apparatus that allows for compressed water distribution.

DETAILED DESCRIPTION

[0010] It is understood that, as in any engineering or design project, the development of any actual implementation will include numerous implementation specific decisions made to achieve the developers' specific goals, such as compliance with business related and system related constraints, which may vary from one implementation to another. It is understood that such a development effort might be complex and time consuming, but is nevertheless a routine undertaking of design, fabrication, and manufacture for those skilled in the art having the benefit of this disclosure. The disclosed steps may be read as prefaced by "In some embodiments, including one complete embodiment, ", may be executed or performed in other orders or sequences, and are not limited to the order and sequence shown and described, which are provided to enable ease in constructing an embodiment, and along with each components of each step, may be removed, modified, combined, or rearranged, and other steps and or step components may be added, without departing from the scope of this disclosure and or invention. Although embodiments of the invention have been described and illustrated in the disclosed implementations, it is understood that the present disclosed subject matter, including apparatuses, methods, specification, and illustrations, has been made only by way of example, not by way of limitation, and the methods and apparatuses may be used in

other systems, and that numerous changes and optimizations in the details of implementation of the invention and or embodiment are made without such modifications departing from the spirit and scope of this disclosure and or embodiments of the invention. Although the disclosure has been shown and described with respect to one or more embodiments, features of the disclosed embodiments can be combined and rearranged in various ways, and changes including equivalent alterations, substitutions, modifications, and additional efficiencies will of course occur to someone of ordinary skill in the art without departing from the spirit and scope of this disclosure and or invention. In particular regard to the various functions performed by the described components, the terms used to describe such components are intended to correspond, unless otherwise indicated, to any component which performs the specified function of the described component, or is functionally equivalent to the described component, even though not structurally equivalent to the disclosed structure which performs the function in the implementations described in this disclosure. In addition, while a particular feature of the disclosure may have been provided with respect to only one of several embodiments, such feature may be combined with one or more other features of other embodiments as may be desired and advantageous for any given or particular application. In some instances, well-known circuits, structures and techniques have not been shown in detail in order not to obscure the understanding of this disclosure. Articles in this disclosure such as "a" "an" and "the" may allow for both singular and plural forms. Verbs in this disclosure such as "is" may be read as "may be". Conjunctions in this disclosure such as "or" as used herein may be interpreted as inclusive or meaning any one or any combination, where "A, B or C" means "any of the following: A; B; C; A and B; A and C; B and C; A, B and C". Relational terms in this disclosure, for example first and second, top and bottom, left and right, are to distinguish one entity or action from another, and may not necessarily require or imply a relationship, or order between, such entities or actions. The disclosure includes the best mode contemplated by the inventor, a completely described specific embodiment, along with optional components and alternative embodiments to best suit the implementer, measurements in imperial and metric units to support universal understanding, and dramatically exceeds claims support requirements and enablement requirements by allowing for selection and or construction of the required components to be carried out easily, quickly, and routinely by persons of ordinary skill in the art, who are provided the additional benefit of utilizing readily available commodity components whenever possible. The present disclosure includes material protected by copyrights, and the owner of the copyrights hereby reserves all rights, but with authorization for publication as required by government patent offices. Various embodiments of the present invention may provide all, some or none of the disclosed technical advantages.

[0011] The inventor retains absolutely no liability for any implementation of this invention, and the invention is implemented exclusively at the risk and liability of the implementer.

[0012] The disclosure resolves the previously cited deficiencies, by compressing temperature controlled air and water before sending it through reduced diameter conduits, thereby removing the need for installation of large ducts and piping and their associated maintenance.

[0013] **In some embodiments, including one complete embodiment, install transport conduits for air and or water.** In some embodiments, copper may be the principal conduit of the system, which has historically been used because it is toxin free and rust free. In some embodiments, where compressed air is carried, strong flexible copper tubes are used, removing the need for large and unsightly air conditioning ducts, which are comparatively more expensive to install and maintain. In some embodiments, the tubing can be run from the source to destination, without being cut or bridged, as required by traditional air ducts. In some embodiments, in order to maintaining temperature insulation, flexible copper tubing may be fit snugly inside cross linked polyethylene, defined in ASTM F876, and commonly referred to as PEX. The PEX will provide temperature insulation to the air flow, while the copper prevents the possibility of contamination during transport. In some embodiments, flexible copper tubes may be used that are .5" in diameter of .04" thick walls may withstand 700 PSI. In some embodiments, PEX may be used that has a minimum maximum temperature rating of 120' C (~250' F) and a pressure rating around 180 PSI. The PEX can still connect as it always does, and the copper interior does not need to be connected, as this will still allow the air to be exposed only to the copper around 99% of the time. The two materials together extend the life of each other. In some embodiments, copper alone may be used. In some embodiments, PEX may be used alone. In some embodiments, a PEX distribution unit may be used, which connects each tube from source to destination, and resembles a power box, which may have circuits for areas, appliances, and rooms of a house. In some embodiments, red brass crimp fittings are used to connect PEX. In some embodiments, any number of piping or tubing materials may be used.

[0014] In some embodiments, where water is transported, flexible copper tubes may be fit snugly inside PEX tubes, which transport only cold water. In traditional PEX systems, there is no copper lining, and hot and cold water are transported separately. In some embodiments, the flexible copper tubing prevents the water from coming into contact with the PEX tubing 99% of the time, therefore removing concerns about potential contamination. In some embodiments, the cold water tube is connected to a splitter, with one tube then connected to the cold water input on the output valve, and one tube connected to an electric water heating component, which provides for endless hot water and removes the need for a storage tank, before the a tube from the hot water heating component is connected to a hot water input on the output valve, with hot water heater is electric and connected after the tube. Currently both hot and cold water pipes run to each output point in a building. While the compression could be applied to a water system, it is not needed as less water needs to flow through pipes than in the past, as a result of there are already being showers available for purchase that require only a cold water pipe input, and then recirculate, heat, and filter contaminants from water to provide exact continuous temperature and pressure control. Additionally, there is already no need for water heaters, and thus hot water pipes, as there are inexpensive electric devices that heat water on demand at exit points, and furthermore there are electric showerheads that mix water with air to create the perception of full water output while using 70% less water. Additionally, endpoints on the pressurized water lines, would need to be upgraded in material strength, because even if the water is decompressed immediately before the endpoint, endpoints have withstand the additional pressure at rest, or would need to integrate a valve that only allows water flow to the endpoint when the water is exiting the endpoint.

[0015] **In some embodiments, including one complete embodiment, attach compressors for air and or water.** In some embodiments, including one complete embodiment, in reference to FIG. 1, temperature controlled air is compressed by and a commodity air compressor, where the air is brought in from the commodity air conditioning unit through a flow adapter 1001, and air compressor 1000 pushes temperature controlled air through the tubes, with the compressed air pushed through the transport tubes or pipes 1003, to a sprinkler like valve 1005. Full cone axial spray nozzle uses a swirl chamber to create an evenly distributed airflow in the shape of a cone at up to 170', providing far more even air distribution than can be achieved through tradition air conditioning ducts. Air compressors are not expensive and run at 60 decibels which is the same decibel level of a typical air conditioner. A flat metal shower head may be used to disperse air, and can already handle pressure and temperature of hot water can handle the air, air taken from heating and cooling unit that provides hot or cold air depending on what's needed to maintain the set temperature in the room. In some embodiments, the compressors used may be silent, as a result of being surrounded by sound insulating materials in a manner that doesn't overheat the units.

[0016] In some embodiments, which may include one complete embodiment, in reference to FIG. 2, temperature controlled water is distributed, where water compressor 2000 pushes hot water through the tubes, while water compressor 2001 pushes cold water through the tubes. The water pressure is also better maintained in the house, and helps the public water lines move water along when its being forced through by in home motors. Because a single strong thick but flexible plastic tube can be used, air and water can travel from the source to the destination as a single uncut piece, without the problems associated with connecting metal water pipes. This significantly reduces construction costs and space requirements and likelihood and side effects of leaks, while making temperate systems self-cleaning. This also corrects for any low water pressure problems coming from the water utility. This is positive for the environment because it notably reduces the amount of building materials required. the water compressor circuit may be opened and closed by a ball pulled in water flow caused by water usage. In some embodiments, where water is carried, to increase durability of pipes by reducing pressure when not in use, and reduce energy consumption and improve longevity of the water compressor, a ball 2003 pulled in water flow may be attached to open and close the compressor power circuit 2004.

[0017] **In some embodiments, including one complete embodiment, connect the compressors to a power source and adjust run speed.** In some embodiments, the run speed of compressors may be adjusted to provide the desired level of throughput.

[0018] **In some embodiments, including one complete embodiment, connect compressor intake to source of air or water and output conduit.** If attaching a water source to the tube, use the appropriate commodity hardware components to make the connection. If attaching an air source to the motor tube, simply place the tube inside the contained output of a standard air heating and cooling unit.

CLAIMS

What is claimed is:

1. A system for conditioning and distributing air, with the invention comprising:
   an air temperature control device;
   a compressor to push pressurized air through distribution conduits;

conduits which may consist of flexible copper tubes inside of PEX tubes, thereby reducing conduit size while maintaining volume and reducing costs of construction and maintenance.

2. A system for distributing water, with the invention comprising:
a compressor to push pressurized water through distribution conduits;
conduits which may consist of flexible copper tubes inside of PEX tubes, which may be further surrounded by a high strength metal, or materials of functional equivalency.

FIG. 1

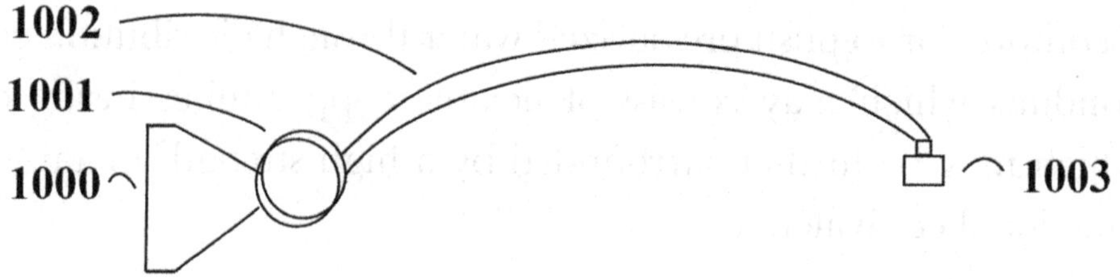

Jonathan Bannon Maher

FIG. 2

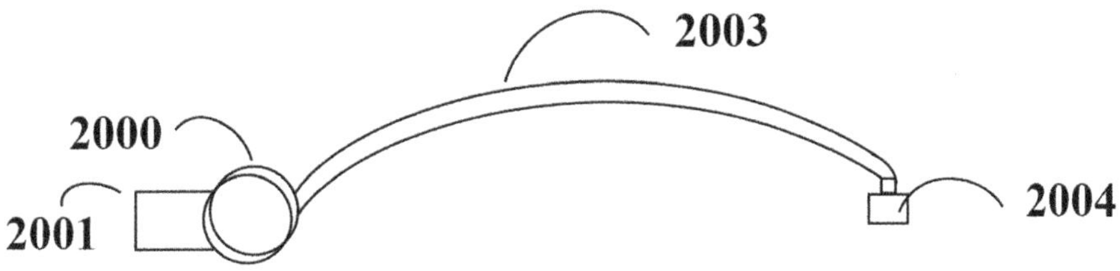

Jonathan Bannon Maher

REUSABLE SPACE VEHICLE LAUNCH SYSTEM

INVENTOR JONATHAN BANNON MAHER

TECHNICAL FIELD

[0001] Embodiments of the invention relate to the fields of space transportation and colonization.

ABSTRACT

[0002] Systems, methods, apparatuses, and in some embodiments computer programs encoded on a computer storage medium, provide reusable space vehicles launchers, for use in an interplanetary space vehicle launch network, by attaching a space vehicle to a launch track – of adequate height and providing vertical release – as well as a chain, with the chain rotated by motors providing adequate force to accelerate the vehicle along the launch track until it is released at high velocity, and optionally in environments with an adequate atmosphere, a platform holding the vehicle and launch track, where the platform is accelerated into the atmosphere by aircraft engines, and at an optimal altitude accelerating the vehicle along the launch track for release, thereby reducing or eliminating the need for traditional launch rockets and their associated costs.

REFERENCE TO RELATED DOCUMENTS

[0003] This application is provided the benefit and priority date of United States Patent and Trademark Office provisional patent application number 62/522,688, filed June 20th 2017 by inventor Jonathan Bannon Maher, which is incorporated herein in its entirety.

BACKGROUND

[0004] This section is intended to introduce the reader to various aspects of the art that may be related to various aspects of the present techniques, which are described and or claimed. This discussion is believed to be helpful in providing the reader with background information to facilitate a better understanding of the various aspects of the present disclosure. Accordingly, it is understood that these statements are to be read in this light, and not a citation of prior any art.

[0005] Current space vehicle launch systems utilize substantial amounts of fuel, which create substantial costs, that make transportation between space based bodies almost prohibitively expensive, where a return trip to Earth from another space based body currently requires fuel for the trip back, which must be brought or somehow produced once on another space based body.

BRIEF DESCRIPTION OF THE ILLUSTRATIONS

[0006] Illustrations are provided by way of example, and not by way of limitation, in the figures of the accompanying drawings, and embodiments may not contain all components, may contain additional components, and may contain functionally similar components.

[0007] FIG. 1 is an embodiment of the apparatus that allows for a reuseable space vehicle launch system.

[0008] FIG. 2 is an embodiment of a link in the chain that accelerates the launch vehicle.

[0009] FIG. 3 is an embodiment of a computer code embedded flow diagram, providing the computer code that activates and controls the launch system.

DETAILED DESCRIPTION

[0010] It is understood that, as in any engineering or design project, the development of any actual implementation will include numerous implementation specific decisions made to achieve the developers' specific goals, such as compliance with business related and system related constraints, which may vary from one implementation to another. It is understood that such a development effort might be complex and time consuming, but is nevertheless a routine undertaking of design, fabrication, and manufacture for those skilled in the art having the benefit of this disclosure. The disclosed steps may be read as prefaced by "In some embodiments, including one complete embodiment, ", may be executed or performed in other orders or sequences, and are not limited to the order and sequence shown and described, which are provided to enable ease in constructing an embodiment, and along with each components of each step, may be removed, modified, combined, or rearranged, and other steps and or step components may be added, without departing from the scope of this disclosure and or invention. Although embodiments of the invention have been described and illustrated in the disclosed implementations, it is understood that the present disclosed subject matter, including apparatuses, methods, specification, and illustrations, has been made only by way of example, not by way of limitation, and the methods and apparatuses may be used in other systems, and that numerous changes and optimizations in the details of implementation of the invention and or embodiment are made without such modifications departing from the spirit and scope of this disclosure and or embodiments of the invention. Although the disclosure has been shown and described with respect to one or more embodiments, features of the disclosed embodiments can be combined and rearranged in various ways, and changes including equivalent alterations, substitutions, modifications, and additional efficiencies will of course occur to someone of ordinary skill in the art without departing from the spirit and scope of this disclosure and or invention. In particular regard to the various functions performed by the described components, the terms used to describe such components are intended to correspond, unless otherwise indicated, to any component which performs the specified function of the described component, or is functionally equivalent to the described component, even though not structurally equivalent to the disclosed structure which performs the function in the implementations described in this disclosure. In addition, while a particular feature of the disclosure may have been provided with respect to only one of several embodiments, such feature may be combined with one or more other features of other embodiments as may be desired and advantageous for any given or particular application. In some instances, well-known circuits, structures and techniques have not been shown in detail in order not to obscure the understanding of this disclosure. Articles in this disclosure such as "a" "an" and "the" may allow for both singular and plural forms. Verbs in this disclosure such as "is" may be read as "may be". Conjunctions in this disclosure such as "or" as used herein may be interpreted as inclusive or meaning any one or any combination, where "A, B or C" means "any of the following: A; B; C; A and B; A and C; B and C; A, B and C". Relational terms in this disclosure, for example first and second, top and bottom, left and right, are to distinguish one entity or action from another, and may not necessarily require or imply a relationship, or order between, such entities or actions. The disclosure includes the best mode contemplated by the inventor, a completely described specific embodiment, along with optional components and alternative embodiments to best suit the implementer, measurements in imperial and metric units to support universal understanding, and dramatically exceeds claims support requirements and enablement requirements by allowing for selection and or construction of the required components to be carried out easily, quickly, and routinely by persons of ordinary skill in the art, who are provided the additional benefit of utilizing readily available commodity components whenever possible. The present disclosure includes material protected by copyrights, and the owner of the copyrights hereby reserves all rights, but with authorization for publication as required by government patent offices. Various embodiments of the present invention may provide all, some or none of the disclosed technical advantages.

[0011] The computer code descriptions disclosed, in order to provide comprehensive enabling disclosure, rather than utilizing flow charts, which according to Patent Cooperation Treaty 11.11a are prohibited from containing "text matter, except a single word or words, when absolutely indispensable, such as... a few short catchwords indispensable for understanding", are provided in a text only format where the number of arrows preceding a line indicate logical block level, semicolons indicate a new segment of a logical block, and periods indicate the closure of one or more logical blocks. It is understood that any computer code representations in this disclosure are merely illustrative, rather than restrictive. While code may be written in nearly any computer language, including Java and C++, the illustrative computer code descriptions were derived from code written the Python language, which may be run through the Python interpreter, with appropriate supportive libraries, which at the time of disclosure, may run on nearly any computer, for example one with an Intel or AMD processor, running a current version of Linux, Windows, or Mac OS. All code components may read as if prefaced by "In some embodiments, including one complete embodiment, ". In some embodiments, functionality may be modified, rearranged, excluded, and added. To provide more fundamental computer system details, in some embodiments, the functionality associated with the disclosed computer code descriptions may be referred to as a script, module, software, software application, or code, and can be written in any form of language, including compiled, interpreted, declarative, or procedural, able to be deployed in any form suitable for use in a computing environment, including as an independent or integrated program, module, component, or subroutine, for execution by the computer system, implemented on one or more independent or integrated computers, utilizing a central processing unit in the form of one or more general or special purpose microprocessors, in conjunction with digital electronic circuitry, which may include special purpose logic circuitry such as a field programmable gate array or application specific integrated circuit, with the computer controlled by and operatively coupled to tangibly embodied software and or firmware, which may include code that creates an environment for code execution, including individual or combined use of processor firmware, a protocol stack, a database management system, and an operating system, where such software and or firmware may exist in one or more parts in memory on one or more computers, and is encoded on one or more tangible non transitory software carriers, such as individual or combined use of a random or serial access device or substrate, a semiconductor memory device, transient or persistent random access memory, a magnetic, magnetic optical, or optical disk, or encoded on an artificially generated transmitted signal, for example, optical, electrical, or electromagnetic, transmitted using a sending and a receiving apparatus, where the interaction between the user and the software may be implemented by operatively coupling, to the local implementing computer, or a local computer connected to one or more remote computers through a local or wide area network, a display device which may implement liquid crystals or light emitting diodes, a keyboard, and a pointing device.

[0012] The inventor retains absolutely no liability for any implementation of this invention, and the invention is implemented exclusively at the risk and liability of the implementer.

[0013] Embodiments provide a reusable system for the launch of space vehicles, by providing a launch track, wrapped by a continuous chain, with the space vehicle having one or more hooks that connect to the chain, with the chain rotated by a fixed motor that propels the space vehicle to the speed necessary to escape gravity and move into space before reaching the end of the track, with the chain and hooks designed to allow naturally disengagement at the end of the track. In environments, where there is an atmosphere, the acceleration structure is first accelerated in the atmosphere by engines to maximum velocity within the atmosphere, prior to acceleration of the vehicle along the launch track. The velocity at which person can accelerate without passing out is 5 to 9 g's, where a g is 32 meters per second per second, so 160 to 288 meters per second per second, or is 576 to 1036 kilometers per hour. The first stage of the platform can accelerate the vehicle to the standard rocket speed at the altitude of 42 miles (68 kilometers) of 6,164 miles per hour (9,920 kilometers per hour). Given the low end of the maximum acceleration speed humans can withstand, an acceleration structure of 5280 feet (1600 meters) can provide the additional following safe speed and distance per second: 1 second: 576 kilometers per hour and 160 meters; 2 seconds 1,152 kilometers per hour, and traveled 480 meters; 3 seconds: 1,728 kilometers per hour and 960 meters; 4 seconds: 2,304 kilometers per hour and 1600 meters. 1600 meters is one mile, and therefore an acceleration structure with a 1 kilometer or 1 mile track can safely accelerate the rocket by an additional 2,304 kilometers per hour. Therefore, allowing a ship to reach escape velocity of Earth without using rocket fuel. The ship would then land on the Moon, which has little gravity and no atmosphere, and use another acceleration track to launch to Mars, which is would unquestionably be able to do without fuel given little gravity and no air resistance. The ship would land on Mars, which has only 38% of the gravity of Earth with almost no atmospheric resistance, and use another accelerator to launch back to the Earth without fuel.

[0014] **In some embodiments, including one complete embodiment, construct the acceleration platform accelerator structure, accelerator chain, accelerator gears, and space vehicle chain and track connector.** In some embodiments, including one complete embodiment, in reference to FIG. 1, horizontal titanium beam 1000 is welded in the center to vertical titanium steel beam 1001, and titanium beam 1000 is welded on either end to custom cast circular titanium structure 1002, with beam 1001 having gears 1200 1201 bolted along the side on to either end, with the gears wrapped by titanium accelerator chain 1202 constructed from chain links provided for in FIG. 2 and later described, with engines 1100 1101 1102 1103 1104 1105 1106 mounted around the exterior of support circular titanium structure 1002, by means which may include welding or bolting, acceleration engines 1200 1201 are mounted to beam 1001, by means which may include welding or bolting, on either side of acceleration structure support 1001, with the acceleration engines having their axles welded to either side of rotating bottom gear 1200. In some embodiment, including one complete embodiment, the accelerator beam 1001 has a length of 1 mile (1.6 kilometers), and may be adjusted to any length required to achieve desired acceleration at speeds safe to humans.

[0015] In some embodiments, including one complete embodiment, in reference to FIG. 2, a titanium chain is constructed to accelerate the launch vehicle, where the chain hooks around gears 1200 1201, where the chain is constructed from a sequence of links, created where chain links 2000 and 2001, have bolts 2004 2005 passing first through bolt support 2002, then bolt 2004 passing through chain link 2000, and bolt 2005 passing through chain link 2001, with bolts 2004 2005 passing through bolt support 2003, then bolt connector 2006 welded to the ends of bolts 2004 2005, thus creating a full chain link, where all chain component are made out of titanium, and any number of chain links can be connected to achieve the desired length of the chain.

[0016] In some embodiments, construct the track as a horizontal to vertical manner, to eliminate downward pressure on the hovering platform as the vehicle accelerates, until it reaches the vertical transition.

[0017] **In some embodiments, including one complete embodiment, attach accelerator mounts to the top and the bottom of the underside of the space vehicle and connect space vehicle to accelerator.** In some embodiments, including one complete embodiment, the vehicle has mounted to it hooks that connect to the accelerator pole 1201, which maintains the vehicle's linear acceleration trajectory, and accelerator chain 1202 that rotates around mounts on the platform to launch the space vehicle.

[0018] **In some embodiments, including one complete embodiment, wire the motors to be computer controlled, and write computer code to allow the space vehicle commander to control the platform and accelerator.** In some embodiments, including one complete embodiment, platform computer 1400 is connected through a crossover Ethernet cable 1401 to relay board 1401, with the relay board 1401 with one set of relays wired to thrust controls of platform engines 1100 1101 1102 1103 1104 1105 1106 and a second set of relays wired to thrust controls of accelerator engines 1200 1201, where platform computer 1400 is connected to computer of space vehicle 1300 through an Ethernet cable, allowing the space vehicle commander to control both the launch platform and the acceleration structure, by connecting to the platform computer through means including SSH, and utilizing command line commands, provided custom software, with the custom software code providing functionality comprising:
> import a library to connect to the relay.
> create variables to hold each relay state.
> create variables holding the relay board connection, IP address, username, and password.
> create a function to update the relay board, first creating a connection to the relay board if one has not been initialized or has been dropped, then create a string holding the state of each relay, and send it to the relay board.
> create a function that is activated on program execution;
>> read in the command line arguments;
>> if the command is to start the launch platform, then adjust the relays to control the platform motor starters.

[0019] **In some embodiments, including one complete embodiment, validate the space vehicle operates correctly.** In some embodiments, including one complete embodiment, hover the launch platform modestly above off the ground, and launch an empty vehicle with a parachute, and ensure that the system performs as expected, with the vehicle safely floating back down to the ground.

[0020] **In some embodiments, including one complete embodiment, launch vehicles into space with no direct cost per launch.** In some embodiments, including one complete embodiment, the launch platform with space vehicle is accelerated to maximum speed and just before it starts to slow down before reaching the edge of the atmosphere, the rocket thruster is started and as soon as it begins to move the rocket forward, the ejection structure accelerates to full speed to launch the space vehicle. It is recommended that anyone on the first test launch of this system have a will in place prior to take off and pray until successfully reaching space. There is no direct cost per launch, only the amortization of the launch platform cost over its useful life.

[0021] **In some embodiments, including one complete embodiment, establish a network of space based launch stations.** In some embodiments, which may include one complete embodiment, multiple acceleration structures are launched into space, in whole, or as parts for assembly, and transported to space based bodies, initially including the Moon and Mars, to allow cost and speed optimized transit within our universe.

CLAIMS

What is claimed is:

1. A system for space vehicle launch with the invention comprising:
   a motor;
   a vehicle;
   a launch track;
   a chain able to be rotated by said motor on one side of said launch track;
   said vehicle attached to said launch track and said chain, and launched from said launch track utilizing force provided by said motor rotating said chain.
2. An apparatus for space vehicle launch with the invention comprising:
   a motor;
   a vehicle;
   a launch track;
   a chain able to be rotated by said motor on one side of said launch track;
   said vehicle attached to said launch track and said chain, and launched from said launch track utilizing force provided by said motor rotating said chain.
3. A method for conducting the launching a space vehicle, with the invention comprising:
   computer software that allows for control of a floating platform, and or for the the control of a structure that accelerates a vehicle for launch into space.
4. A non-transitory computer-readable recording medium holding stored instructions, which when executed by one or more processing devices, cause the one or more processing devices to implement a method comprising, with the invention comprising:

computer software that allows for control of a floating platform, and or for the control of a structure that accelerates a vehicle for launch into space.

FIG. 1

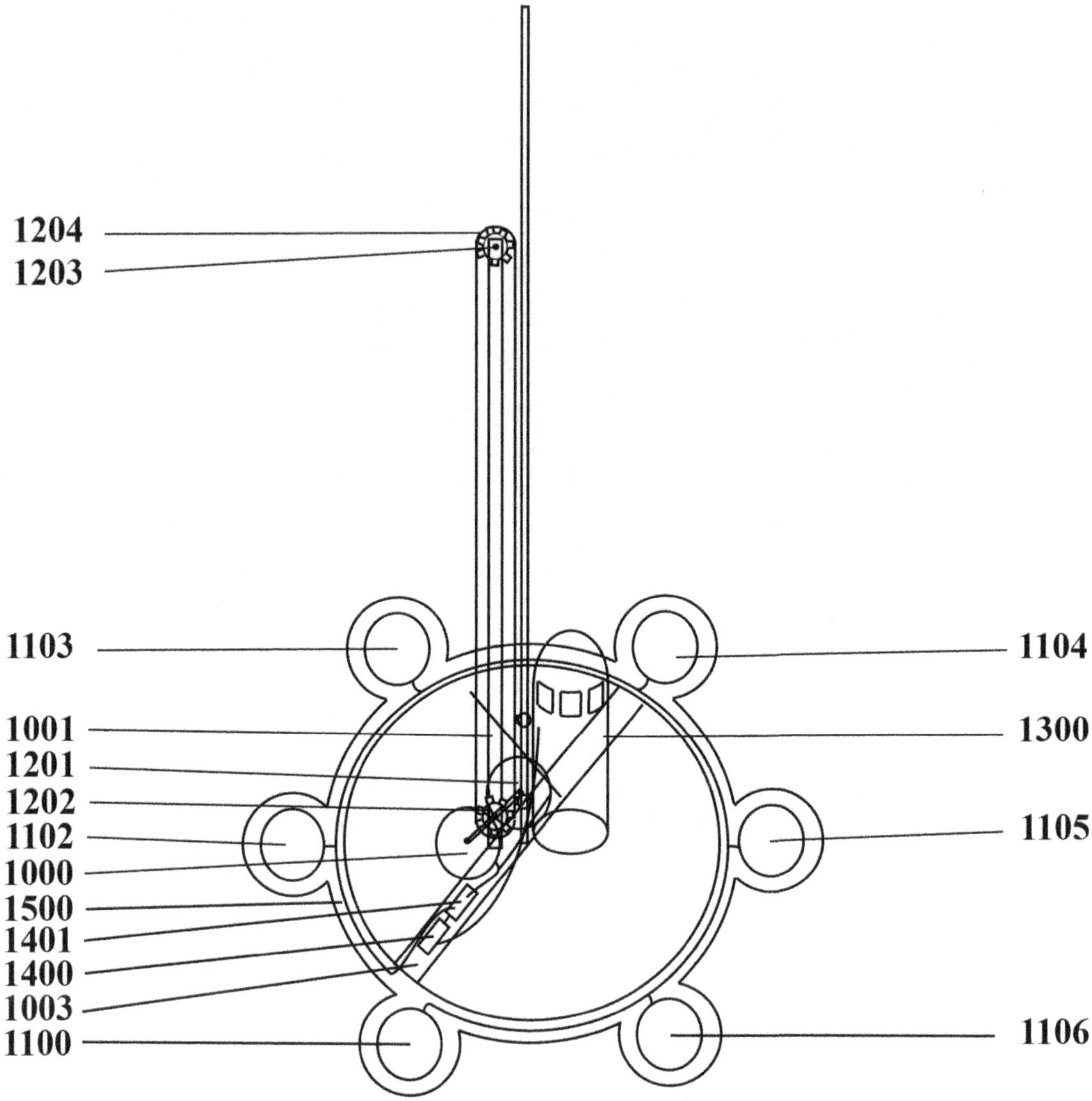

Jonathan Bannon Maher

FIG. 2

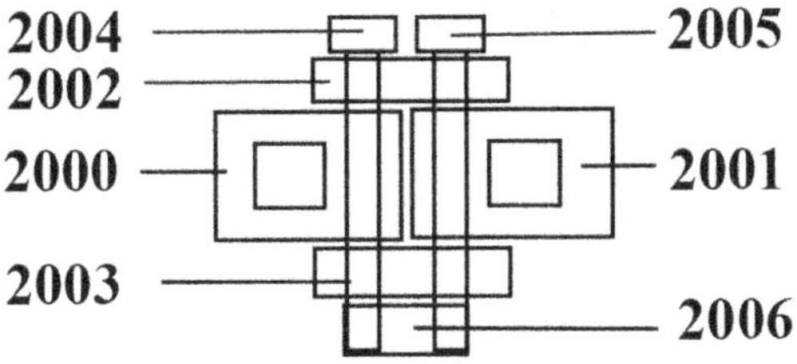

Jonathan Bannon Maher

FIG. 3

3000 – import a library to connect to the relay

```
import telnetlib
```

3001 – create variables to hold each relay state

```
on = 1
off = 0
relay_1_platform_motor_forward = off
relay_2_platform_motor_backward = off
relay_3_launcher_motor_forward = off
relay_4_launcher_motor_backward = off
```

3002 – create variables holding the relay board connection, IP address, username, and password

```
relay = None
relay_ip = "169.254.1.1"
relay_username = "admin"
relay_password = "admin"
```

3003 – create a function to update the relay board, first creating a connection to the relay board if one has not been initialized or has been dropped, then create a string holding the state of each relay, and send it to the relay board

```
def update_relay():

  if not relay:
    relay = telnetlib.Telnet(relay_ip, 23)
    relay.read_until(b"User Name: ")
    relay.write(relay_username.encode('ascii') + b"\n")
```

```
    relay.read_until(b"Password: ")
    relay.write(relay_password.encode('ascii') + b"\n")

  command = str(relay_1_ platform_motor_forward)
  command+= str(relay_2_ platform_motor_backward)
  command+= str(relay_3_ launcher_motor_forward)
  command+= str(relay_4_ launcher_motor_backward)
  relay.write(command.encode('ascii') + b"\n")
```

3004 — create a function that is activated on program execution

```
def __main__():
```

3005 — read in the command line arguments

```
  arguments = sys.argv
  component = arguments[1]
  action = arguments[1]
```

3006 — if the command is to start the launch platform, then adjust the relays to control the platform motor starters.

```
  if component == "platform":
    if action == "up":
      relay_1_platform_motor_forward = on
    if action == "down":
      relay_2_platform_motor_backward = on
    if action == "off":
      relay_1_platform_motor_forward = off
      relay_2_platform_motor_backward = off
```

3007 — if the command is to start the vehicle accelerator, then adjust the relays to control the accelerator motor starters.]

```
  if component == "launcher":
    if action == "up":
```

```
    relay_2_ launcher_motor_forward = on
if action == "down":
    relay_3_ launcher_motor_backward = on
if action == "off":
    relay_2_ launcher_motor_forward = off
    relay_3_ launcher_motor_backward = off
```

Made in the USA
Monee, IL
08 November 2026

67971939R00345